Artificial Intelligence in Neuroscience

Artificial Intelligence in Neuroscience

Edited by

Li Su
University of Cambridge
Cambridge, UK

WILEY

Registered Office
John Wiley & Sons, Inc., 111 River Street, Hoboken, NJ 07030, USA
John Wiley & Sons Ltd, New Era House, 8 Oldlands Way, Bognor Regis, West Sussex, PO22 9NQ, UK

For details of our global editorial offices, customer services, and more information about Wiley products visit us at www.wiley.com.

The manufacturer's authorized representative according to the EU General Product Safety Regulation is Wiley-VCH GmbH, Boschstr. 12, 69469 Weinheim, Germany, e-mail: Product_Safety@wiley.com.

Wiley also publishes its books in a variety of electronic formats and by print-on-demand. Some content that appears in standard print versions of this book may not be available in other formats.

Library of Congress Cataloging-in-Publication Data has been applied for:

Hardback ISBN: 9781394278855
ePDF ISBN: 9781394278879
ePub ISBN: 9781394278862
oBook ISBN: 9781394278886

Cover Design: Wiley
Cover Images: © Ty/stock.adobe.com, © Vooglo.com/stock.adobe.com

Set in 9.5/12.5 STIXTwoText by Lumina Datamatics

Contents

List of Contributors

Juan E Arco
DaSCI Andalusian Institute of Data
Science and Computational Intelligence
University of Granada
Granada
Spain
Mind, Brain and Behavior Research Center
University of Granada
Granada
Spain

Matthew Bracher-Smith
Division of Psychological Medicine &
Clinical Neurosciences
Cardiff University
UK
Dementia Research Institute
Cardiff University
UK

Hao Chen
Department of Clinical Neurosciences
University of Cambridge
Cambridge
UK

Valentina Escott-Price
Division of Psychological Medicine &
Clinical Neurosciences
Cardiff University
UK
Dementia Research Institute
Cardiff University
UK

Juan M Górriz
DaSCI Andalusian Institute of Data
Science and Computational Intelligence
University of Granada
Granada
Spain
Data Science and Computational
Intelligence Institute (DaSCII)
University of Granada
Granada
Spain
Department of Psychiatry
University of Cambridge
Cambridge
UK

Lewis Hotchkiss
Medical School, Swansea University
Swansea
UK

Carmen Jimenez-Mesa
DaSCI Andalusian Institute of Data
Science and Computational Intelligence
University of Granada
Granada
Spain
Department of Communication
Engineering
University of Málaga
Spain

Anna Korhonen
Language Technology Lab
Centre for Human-Inspired AI
University of Cambridge
Cambridge
UK

Chao Li
Department of Applied Mathematics and Theoretical Physics
University of Cambridge
Cambridge
UK
Department of Clinical Neurosciences
University of Cambridge
Cambridge
UK
School of Science and Engineering
School of Medicine
University of Dundee
Dundee
UK

Yong Liu
School of Artificial Intelligence
Beijing University of Posts and Telecommunications
Beijing
China

David López
DaSCI Andalusian Institute of Data Science and Computational Intelligence
University of Granada
Granada
Spain
Mind, Brain and Behavior Research Center
University of Granada
Granada
Spain

Francisco J Martínez-Murcia
DaSCI Andalusian Institute of Data Science and Computational Intelligence
University of Granada
Granada
Spain
Data Science and Computational Intelligence Institute (DaSCII)
University of Granada
Granada
Spain

Rune Nyrup
Department of Mathematics and Science Studies
Aarhus University
Aarhus Centrum
Denmark

Ulla Petti
Language Technology Lab
University of Cambridge
Cambridge
UK

Javier Ramírez
DaSCI Andalusian Institute of Data Science and Computational Intelligence
University of Granada
Granada
Spain
Data Science and Computational Intelligence Institute (DaSCII)
University of Granada
Granada
Spain

Timothy Rittman
Department of Clinical Neurosciences
University of Cambridge
Cambridge
UK
Cambridgeshire University Hospitals NHS Foundation Trust
Cambridge
UK

María Ruz
Mind, Brain and Behavior Research Center
University of Granada
Granada
Spain

Jeffrey M Skopek
Faculty of Law
Centre for Law, Medicine and Life Sciences
University of Cambridge
Cambridge
UK

Li Su
Department of Psychiatry
University of Cambridge
Cambridge
UK
Neuroscience Institute
University of Sheffield
Sheffield
UK

Zhongzhao Teng
School of Life and Health Sciences
Fujian Fuyao University of Science
and Technology
Fujian
China
Nanjing Jingsan Medical Science
and Technology, Ltd.
Jiangsu
China
Tenoke, Ltd.
Cambridge
UK

Vesna Vuksanovic
Health Data Science, Swansea University
Medical School
Swansea University
Sketty
UK

Xiaofei Wang
Department of Clinical Neurosciences
University of Cambridge
Cambridge
UK

Xiong Xiong
Department of Psychiatry
University of Cambridge
Cambridge
UK

Ruodan Yan
Department of Applied Mathematics and
Theoretical Physics
University of Cambridge
Cambridge
UK

Lu Zhang
Computer Science Department
Indiana University Indianapolis
Indianapolis
IN
USA

Kun Zhao
School of Artificial Intelligence
Beijing University of Posts and
Telecommunications
Beijing
China

Dajiang Zhu
Computer Science and Engineering
University of Texas at Arlington
Arlington
TX
USA

About the Editor

Professor Li Su is the head of the Artificial Intelligence and Computational Neuroscience Lab at the University of Cambridge. He is also the chair of Neuroimaging at the Sheffield Institute of Translational Neuroscience, University of Sheffield. He serves as a panel member for UKRI's AI for Health initiative and sits on the National Institute for Health and Care Excellence (NICE) Specialist Committee. He is a regional co-lead for the National Network for the Application of Data Science and AI to Dementia Research (DEMON), a member of the Alzheimer's Society's Research Strategy Council and a reviewer for many research funders. He is also an associate editor for *Frontiers in Computational Neuroscience*.

1

Introduction

Li Su[1,2]* *and Xiong Xiong*[1]

[1]*Department of Psychiatry, University of Cambridge, Cambridge, UK*
[2]*Neuroscience Institute, University of Sheffield, Sheffield, UK*

*Corresponding author: ls514@cam.ac.uk or l.su@sheffield.ac.uk

1.1 Background and Rationale

1.1.1 What Is Neuroscience?

Neuroscience is a scientific discipline studying the nervous system, aiming to unravel the complex structure and function of the brain, spinal cord and peripheral nerves (Binder et al., 2009). It is inherently multidisciplinary, bridging biology, medicine, psychology and philosophy, reflecting the immense complexity of neural systems. Researchers in neuroscience investigate the nervous system across various scales – from molecules and individual neurons, through intricate networks and circuits, all the way to behaviours and cognition observed at the whole-organism level and even at organisational and social levels.

At the molecular and cellular levels, neuroscientists explore how neurons communicate through specialised connections called synapses, utilising biochemical signalling pathways and electrical impulses. On a broader network scale, the interactions among billions of neurons give rise to cognitive functions such as perception, memory, language and decision-making. Ultimately, these integrated neural processes shape complex behaviours and mental states, including emotions and consciousness. Highlighting this, a multiscale approach is crucial, as subsequent chapters will discuss how artificial intelligence (AI) methods are uniquely suited to integrate and interpret these diverse layers of neural data.

Neuroscience employs a wide array of experimental tools to gather insights into neural activity and structure. Electrophysiological techniques, such as single-cell recordings and electroencephalography (EEG), capture electrical signals from individual neurons or large neuronal populations, respectively, providing high temporal resolution data on neural dynamics. Imaging methods, including magnetic resonance imaging (MRI), functional MRI (fMRI) and positron emission tomography (PET), enable visualisation of brain anatomy, connectivity and metabolic activity in living organisms. Genetic and molecular biology approaches, such as optogenetics and gene editing technologies like CRISPR (Clustered Regularly Interspaced Short Palindromic Repeats), a technique for targeted genetic modification (Yang et al., 2016), further allow precise manipulation and observation of neural circuits (Deisseroth, 2015).

The ultimate goal of neuroscience research extends beyond basic scientific curiosity. By deepening our understanding of the nervous system, neuroscientists aim to uncover the underlying causes of neurological and psychiatric disorders, such as Alzheimer's disease, epilepsy, depression and schizophrenia. Identifying these biological mechanisms provides essential insights for developing targeted interventions and personalised treatments, representing one of the most exciting and impactful frontiers of modern biomedical science.

1.1.2 What Is Artificial Intelligence?

AI refers broadly to computer systems designed to perform tasks that usually require human intelligence, such as learning, reasoning, problem-solving and pattern recognition (Hunt, 2014). When we talk about AI in neuroscience, we are often referring specifically to machine learning (ML), a subfield of AI where computers automatically learn from data, improving their performance over time without needing explicit programming (Zhou, 2021).

To give a clearer idea, think about medical imaging. Doctors traditionally spend a significant amount of time examining MRI or CT scans to detect signs of diseases, such as cancer. AI systems using ML algorithms can now efficiently analyse large datasets of medical images, identifying patterns suggestive of potential abnormalities, such as tumours (Choi, 2022). Similarly, speech recognition technologies – like those you use in smartphone voice assistants – can help paralysed individuals communicate by translating spoken commands into actions. These real-world applications highlight AI's powerful potential to support and sometimes even outperform humans in complex tasks. Deep learning (DL) is a more specialised form of ML, which uses multilayered neural networks loosely inspired by the structure of the human brain (Qiyang et al., 2025). The hierarchical relationship among AI, ML and DL is illustrated in Figure 1.1. These deep neural networks (DNNs) can learn complicated patterns directly from raw data, achieving remarkable success in tasks such as image classification, language processing and even game playing.

However, despite these impressive capabilities, AI still has some limitations. One significant challenge is interpretability, often called the 'black box' problem. This means while AI models can provide highly accurate predictions or classifications, they typically offer little explanation for their decisions. For instance, a DL system might correctly predict a patient's risk of developing Alzheimer's disease based on brain scans, but it might not clearly explain which specific features of the scans led to that conclusion. This lack of transparency and interpretability can be problematic in sensitive areas such as healthcare, where understanding the reasoning behind a diagnosis or treatment recommendation is crucial.

In short, AI encompasses powerful technologies capable of learning from vast amounts of data to assist in making complex decisions. While AI already has transformative applications, especially in healthcare and neuroscience, addressing its limitations is critical for further integration into clinical practice.

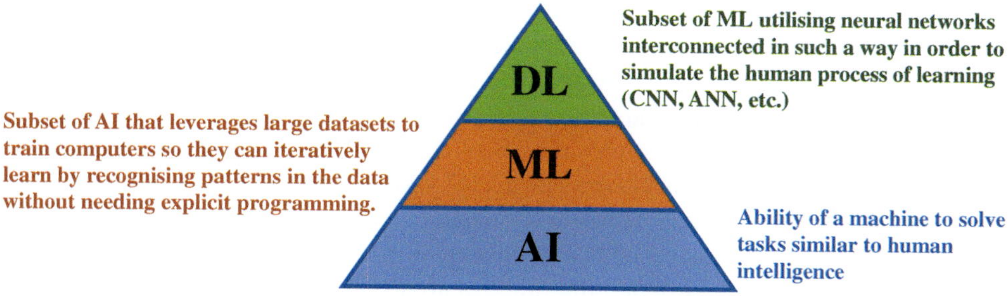

Figure 1.1 A conceptual diagram illustrating the relationship between AI, ML and DL. AI represents the broadest concept – the overall capability of machines to perform tasks that typically require human intelligence. Within this field, ML is a subset that focuses on algorithms allowing computers to learn patterns from large datasets rather than following hard-coded rules. DL is a further subset of ML that uses deeply layered neural networks (e.g., convolutional neural networks [CNNs], RNNs) to simulate human-like learning. This pyramid structure reflects how DL techniques are a specialised part of ML, which in turn is one approach under the umbrella of AI.

1.1.3 Historical Milestones in AI–Neuroscience Intersection

The fields of AI and neuroscience have been intertwined since the very inception of AI. In the 1940s and 1950s, early neuroscientific insights directly inspired AI research. A famous example is Hebbian learning (1949) – the principle 'cells that fire together, wire together' – which provided a biological rule for how synaptic connections strengthen, influencing early neural network models (Hebb, 1950). The first artificial neuron model by McCulloch and Pitts and the creation of the perceptron were grounded in mimicking how real neurons integrate inputs and fire (McCulloch & Pitts, 1943). This era can be seen as an initial collaboration between the disciplines: neuroscience guided the design of AI models. Subsequent decades saw the development of brain-inspired AI architectures such as the neocognitron, a precursor to modern deep convolutional networks inspired by the hierarchical organisation of the visual cortex (Fukushima, 1980). During this period, computational neuroscience also emerged as a field, using mathematical and computer models (often simplified neural networks) to simulate brain processes and test theories about neural function.

By the 1980s and 1990s, AI research experienced a phase of relative 'separation' from direct biological inspiration. The focus shifted towards mathematical and engineering innovations that dramatically improved AI performance but were less constrained by neurobiology. A pivotal breakthrough was the development of the backpropagation algorithm (1986) (Rumelhart et al., 1986), enabling multilayer neural networks to be trained effectively – a technique not directly derived from brain processes. Likewise, the introduction of non-linear activation functions like the rectified linear unit (ReLU) and advanced architectures such as long short-term memory networks (LSTMs; Hochreiter & Schmidhuber, 1997) were driven by computer science goals of improving learning efficiency and stability. These advances led to 'black box' AI models with little transparency, as the internal workings became difficult to interpret in neurobiological terms. During this span, AI's progress (often dubbed the 'deep learning revolution' by 2012) was largely independent of neuroscience, even as neuroscience itself was amassing big data (e.g., high-resolution brain scans, large EEG datasets) that begged for sophisticated analysis methods.

In the last decade, there has been a renewed convergence – effectively a second collaboration between AI and neuroscience. Modern DL models have grown so powerful that neuroscientists are now using them as scientific tools: for example, deep CNNs trained on object recognition have been found to develop intermediate representations with uncanny similarity to visual brain regions, providing computational models of the ventral visual stream (Ajit et al., 2020). Studies have shown that certain AI models can predict neural responses in the visual, auditory, or motor cortices when the brain processes specific stimuli, suggesting that these models offer insight into how real neural circuits might encode information. Conversely, neuroscientific findings are again informing AI – for instance, understanding how the brain balances plasticity and stability has influenced the development of algorithms that avoid catastrophic forgetting. The intersection has thus come full circle: AI began by drawing inspiration from brain structure and function, then advanced through primarily engineering-driven leaps, and is now being used both to explain brain data and to borrow concepts (such as attention mechanisms or sparse coding) from neuroscience to improve AI algorithms, as summarised in the timeline shown in Figure 1.2. This evolving historical interplay underscores a rich feedback loop between the two fields.

1.1.4 Prominent Examples of AI in Neuroscience Applications

To ground the discussion in concrete terms, it is useful to consider a few prominent real-world examples where AI techniques intersect with neuroscience research and healthcare:

- **Brain–Computer Interfaces (BCIs):** BCIs are systems that enable direct communication between the brain and an external device, often to assist individuals with motor or communication impairments. AI algorithms play a crucial role in BCIs by decoding neural signals (recorded via EEG, implanted electrodes) into actionable outputs. For instance, ML models can learn to recognise patterns in a paralysed patient's brain activity corresponding to the intention to move a cursor or a prosthetic limb. In clinical research, AI-driven BCIs have allowed patients with severe paralysis to control robotic arms, type on virtual keyboards by thought, or regain rudimentary speech communication. These successes rely on neural network models capable of interpreting the noisy, complex brain signals in real time, demonstrating AI's power in extracting meaningful information from relevant neural signals. BCIs exemplify how neuroscience knowledge (the brain's electrical signal features) and AI methods (advanced signal processing and classification algorithms) combine to create neuroprosthetic technologies that improve quality of life.
- **Deep Brain Stimulation (DBS) Optimisation:** DBS is a therapeutic technique where electrodes implanted in specific brain regions deliver electrical impulses to modulate neural activity, commonly used for movement disorders like Parkinson's disease and in experimental treatments for depression or obsessive–compulsive disorder. Traditionally, clinicians must manually adjust stimulation parameters to find optimal settings for each patient – a time-consuming process of trial and error. AI is now being employed to enhance DBS precision and outcomes. ML models can analyse patient-specific data (brain signals, symptom profiles, neuroimaging) to predict the most effective stimulation targets and parameter settings. In some cutting-edge

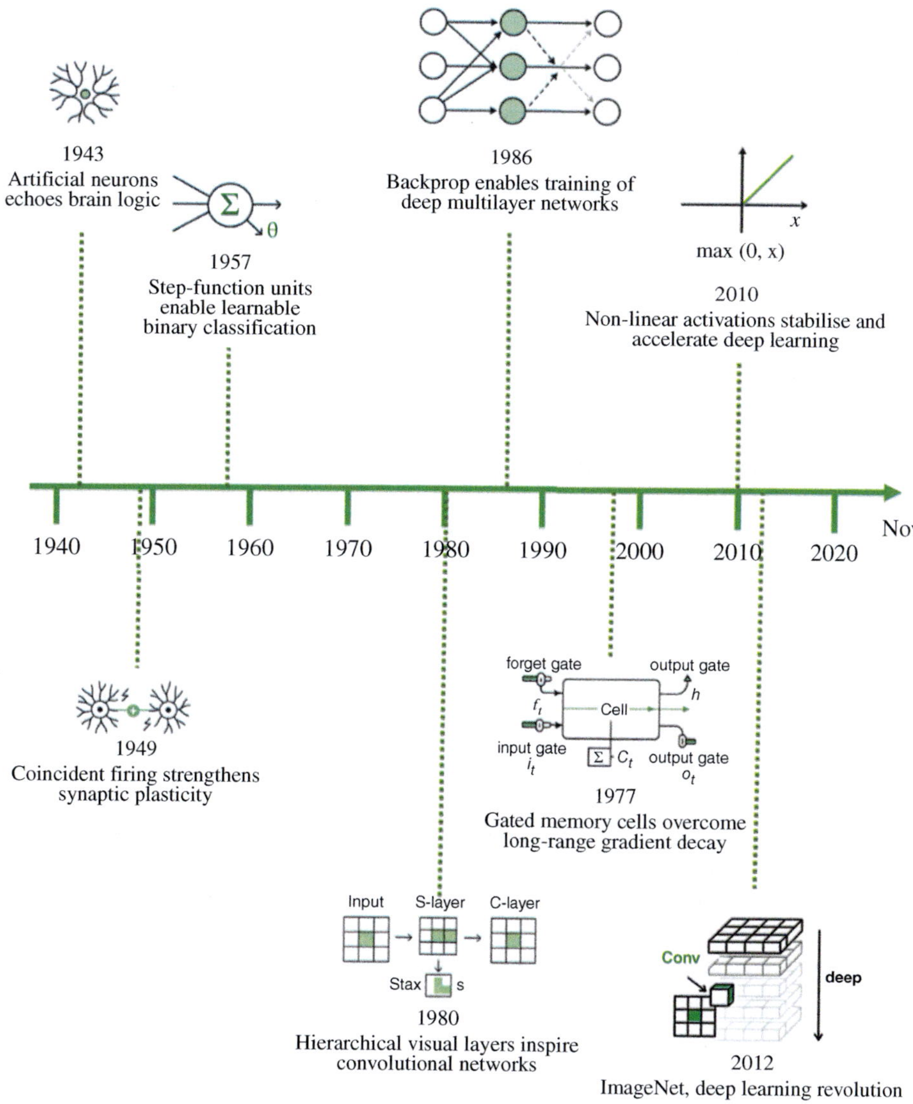

Figure 1.2 Timeline of key milestones at the intersection of neuroscience and AI.

approaches, adaptive DBS systems use AI algorithms to create a closed-loop system: the device monitors the patient's neural signals in real time and automatically adjusts stimulation in response, maintaining therapeutic benefit while minimising side effects. For example, AI-driven signal analysis might detect neural signatures of an impending tremor in a Parkinson's patient and promptly modulate stimulation to prevent it. By intelligently guiding electrode placement and stimulation tuning, AI assists neurosurgeons and neurologists in optimising DBS therapy for each individual, illustrating the synergy of AI with neuromodulation techniques.

These examples underscore the tangible benefits of integrating AI with neuroscience. From enabling direct brain-to-computer communication to refining brain stimulation therapies, AI methodologies serve as powerful tools to decode, interpret and interact with the nervous system. They also highlight a common theme: personalisation and adaptability offered by AI. Whether it's tailoring a BCI to an individual's neural patterns or adjusting DBS in real time to a patient's state, AI helps move interventions from a one-size-fits-all approach towards more precision neuroscience – interventions that are dynamically responsive to each person's unique brain activity.

1.2 Promising Potentials of AI in Neuroscience

1.2.1 AI in Disease Diagnosis and Early Prediction

One of the most promising applications of AI in neuroscience is in the diagnosis of neurological disorders and the early prediction of disease onset or progression. Traditional clinical diagnosis often faces challenges: many brain disorders are heterogeneous (patients may show different symptoms or brain changes), early signs can be subtle or easily missed, and diagnostic criteria may rely on subjective assessments. AI offers tools to overcome these challenges by learning from complex neuroimaging and clinical datasets to detect patterns indicative of disease. For example, in neurodegenerative diseases like Alzheimer's or Parkinson's, DL models have been trained on brain scans (MRI, PET) to automatically identify the hallmark changes (such as patterns of atrophy or reduced glucose metabolism) that even expert radiologists might overlook in early stages. These AI-based diagnostic systems can integrate a breadth of information – imaging, genetic markers and cognitive test scores – to improve accuracy in distinguishing a disease from normal ageing or from other conditions with overlapping features.

The early prediction capability of AI is particularly valuable. ML models can be used to predict which individuals with mild symptoms or risk factors will go on to develop a full-blown disorder. For instance, in mild cognitive impairment (MCI), which sometimes precedes Alzheimer's dementia, AI algorithms have achieved success in predicting conversion to Alzheimer's years before clinical dementia manifests. By analysing subtle combinations of imaging findings, genetic risk profiles (like APOE genotype) and even digital biomarkers (such as slight changes in speech or gait), AI can estimate the likelihood of disease progression. Early diagnosis or risk prediction is crucial because it opens a window for early interventions – even lifestyle changes or experimental treatments – that could slow disease progression. In disorders such as epilepsy, AI models applied to EEG recordings can foretell seizures or identify patients who might not respond to medication, facilitating timely surgical consideration. Similarly, in stroke, AI-driven image analysis can help emergency physicians rapidly distinguish ischemic from haemorrhagic stroke and even gauge the extent of viable brain tissue, thereby speeding up critical treatment decisions. Across these examples, the key advantage is AI's ability to handle high-dimensional data and complex, multifactorial patterns: it can fuse information from MRI voxels, electrophysiology, clinical history, and beyond to deliver diagnostic insights and prognoses that would be exceedingly difficult for humans to derive unaided.

1.2.2 Understanding Disease Mechanisms

Beyond improving diagnostics, AI is also transforming how researchers investigate the mechanistic underpinnings of brain diseases. Many neurological and psychiatric conditions are characterised by intricate changes in brain connectivity, gene expression and other biological networks. AI methods provide a way to make sense of this complexity by identifying patterns and relationships that shed light on disease mechanisms. A prime example is brain connectomes: using graph-theoretical models of the brain's connectivity (or called brain networks), scientists employ AI algorithms to find how these networks differ in disease. Techniques like DNNs and graph convolutional networks can be trained to recognise network configurations associated with disorders – for instance, altered functional connectivity patterns in schizophrenia or characteristic disruptions of brain networks in Alzheimer's disease. By probing these trained models, researchers can gain insight into which connections or network hubs are most implicated in the disease, generating hypotheses about how the disease spreads or which circuits are dysfunctional.

Another avenue is using unsupervised ML to discover hidden structure in biomedical data. Clustering algorithms and dimensionality reduction techniques (such as non-negative matrix factorization or autoencoders) have been applied to high-dimensional datasets (imaging, genomic, proteomic data from patients) to uncover subtypes or end phenotypes of disease that are not obvious from clinical observation alone. For example, in Alzheimer's disease, unsupervised learning on brain scans has revealed distinct patterns of atrophy – some patients have predominant hippocampal degeneration while others show cortical atrophy – suggesting multiple mechanistic pathways leading to the same clinical diagnosis. These data-driven subgroups often correlate with different underlying pathologies (like different protein aggregates) or progression rates, pointing researchers to the existence of mechanistically distinct variants of 'one' disease. Identifying such subtypes improves our understanding of disease heterogeneity and can guide mechanism-specific research (for instance, why one subtype responds to a treatment while another does not). AI is also used to identify biomarkers: by training algorithms on datasets of patients and healthy controls, researchers can let the AI determine which combination of features (imaging pixels, EEG signal characteristics, cytokine levels) best distinguishes disease. The resulting model highlights candidate biomarkers – perhaps a particular brain region's volume or a signature in a brain wave pattern – thus directing scientific attention to those features as potential keys to pathology. In summary, AI serves as a robust analytical framework in neuroscience research, enabling researchers to effectively extract meaningful insights from highly complex brain data. This leads to deeper mechanistic insight, from revealing how diseases perturb brain networks to suggesting novel biological targets for intervention.

1.2.3 Personalised and Precision Medicine

AI's influence in neuroscience is also driving a paradigm shift towards personalised and precision medicine – tailoring healthcare decisions and treatments to the individual characteristics of each patient. The human brain is remarkably variable, and patients with the 'same' diagnosis can differ widely in genetics, brain structure, lifestyle and treatment response. AI systems, with their ability to learn complex, non-linear relationships, are ideally suited to exploit this individuality for better care. In practice, this means AI can assist in creating personalised diagnostic profiles and therapeutic strategies. For instance, consider the management of depression: through ML analysis of a patient's clinical history, brain scan and perhaps blood biomarkers, an AI model might predict which antidepressant medication or therapy is most likely to be effective for that specific person. Such predictions are drawn from patterns learned across many patients, linking personal features to outcomes – a task too multifaceted for clinicians to do alone. Early studies have shown success in using ML to match psychiatric patients to treatments, reducing the trial-and-error cycle of finding an effective therapy.

Personalised prediction extends to prognosis as well. In disorders like multiple sclerosis or traumatic brain injury, AI models can forecast an individual's likely disease course (e.g., rate of progression or risk of relapses) by looking at their unique combination of imaging findings and biomarker levels. This allows clinicians to make proactive, personalised decisions – such as intensifying therapy early for a patient predicted to have aggressive progression, while avoiding over-treatment in someone predicted to remain stable. In neurosurgery and neuro-oncology, AI is aiding personalised planning by predicting surgical outcomes or radiation responses based on patient-specific tumour characteristics extracted via DL from medical images. Even rehabilitation can be personalised: AI-driven adaptive algorithms in neural rehabilitation devices (like smart prosthetics or cognitive training programmes) can continuously adjust their parameters in response to the patient's progress, ensuring the therapy is optimally challenging and effective for that individual. Real-world case studies illustrate the promise of precision neuroscience. For example, in epilepsy, researchers have used AI on EEG data to anticipate which patients will respond to a certain anti-epileptic drug, enabling more informed and individualised medication choices. In stroke recovery, ML models combining brain imaging and clinical data can predict an individual's potential for motor improvement with rehabilitation, thereby helping to personalise therapy intensity and set realistic goals. Overall, by leveraging a wealth of data from each person – genetics, brain scans, symptoms and environment – AI systems facilitate a move away from 'one-size-fits-all' approaches. The outcome is more precise diagnostics, risk assessments and treatments, which ultimately can improve efficacy and reduce unnecessary interventions. This aligns perfectly with the emerging vision of precision medicine in neuroscience: delivering the right intervention to the right patient at the right time, guided by data-driven intelligence.

1.3 Current Advances and Specific Topics Covered in this Book

1.3.1 Network Neuroscience and Deep Learning

One of the cutting-edge themes in modern neuroscience is the network perspective of the brain, often referred to as network neuroscience. This approach treats the brain as an interconnected web of nodes (brain regions or neurons) and edges (functional or structural connections), and it relies on tools from graph theory and network science to understand brain organisation. Chapter 2 delves into how DL, particularly DNNs and graph neural networks (GNNs), are revolutionising brain network analysis. Traditional network analysis in neuroscience often faced limitations in capturing non-linear relationships or handling the very high dimensionality of connectome data. DL addresses these issues by automatically learning features from complex connectivity patterns. For example, DNNs can be trained to predict cognitive traits or disease states directly from whole-brain connectivity matrices, effectively discovering the subtle network signatures that distinguish, say, a healthy brain from a diseased one. GNNs, a class of AI algorithms designed to operate on graph-structured data, are especially well-suited for brain network applications. They can ingest a brain graph (with nodes as regions and weighted edges as connection strengths) and learn to represent each node or the entire network in a way that is informative for a task like classification or community detection. As summarised in this topic, researchers have successfully applied GNNs to identify aberrant network connectivity in disorders (like altered hub connectivity in schizophrenia or disrupted network communities in traumatic brain injury) and to model the brain's structural–functional relationships. A key advantage of these AI methods is their ability to integrate multimodal network data – for instance, combining structural (anatomical connections) and functional (synchronous activity) networks, or layering different scales of networks (from cellular to whole brain). This yields a richer understanding of how network topology underlies cognition and pathology. In short, the current advances show that DNNs are becoming indispensable for extracting knowledge from the brain's connectome, and Chapter 2 provides a comprehensive overview of these techniques, including how DL uncovers new organisational principles of brain networks and aids in detecting network-level biomarkers of disease.

1.3.2 AI Perspectives on Cerebrovascular Drivers of Cognitive Decline

Healthy cerebral vessels are indispensable for sustaining the brain's extraordinary metabolic demand and for clearing metabolic waste. Chapter 3 therefore changes focus from purely neural perspectives to the vascular side of the neuro-AI equation, asking how AI can illuminate the structure, function and pathology of the brain's blood supply network.

Cognitive performance depends on the integrity of the neurovascular unit that delivers oxygen and glucose to active neurons. Disruption of this equilibrium (whether through an acute arterial occlusion or the gradual narrowing of small cerebral vessels) precipitates energy failure and cognitive decline. Clinical series demonstrate that stroke is frequently followed by widespread deficits in memory and executive function, mediated by diffuse hypoperfusion, excitotoxic injury and neuroinflammatory change. Age-related microangiopathy exerts a slower but equally pervasive influence: endothelial dysfunction, capillary rarefaction and secondary white matter damage contribute to reduced processing speed years before a formal diagnosis of dementia. Therefore, Chapter 3 analyses this vascular–cognitive interface from pathophysiological mechanism to bedside observation. It begins with the metabolic consequences of chronic cerebral hypoperfusion, detailing ionic disequilibrium, oxidative stress and synaptic failure. Subsequent sections examine the blood–brain barrier and neurovascular unit, highlighting how endothelial injury, cerebral microbleeds and impaired amyloid-β clearance link vascular lesions to Alzheimer-type neurodegeneration. The chapter then provides a systematic overview of post-stroke cognitive impairment, cerebral small-vessel disease, mixed dementia and rarer hereditary angiopathies, correlating each syndrome with characteristic imaging markers such as white matter hyperintensities, lacunes, cerebral microbleeds and enlarged perivascular spaces.

Chapter 3 also discusses prevention and treatment. Evidence supporting strict control of hypertension, diabetes and dyslipidaemia is evaluated alongside emerging neuroprotective approaches that target endothelial nitric oxide signalling, glymphatic clearance and remyelination. In aggregate, the chapter positions cerebrovascular health as a modifiable determinant of cognitive ageing and provides a rationale for incorporating vascular risk management into dementia-prevention strategies. Subsequent chapters build on this premise by introducing advanced imaging, fluid biomarkers and artificial intelligence-based risk stratification.

1.3.3 Multilayer, Multimodal and Multiscale Brain Network Models

Building on the network theme, Chapter 4 explores multilayer network models in neuroimaging, which represent one of the most advanced ways to capture the brain's complexity. Unlike traditional single-layer network analyses (which might aggregate all data into one network), multilayer approaches maintain separate layers for different types of connections or different conditions and allow interactions across layers. For example, a multilayer brain network model could include one layer for functional connectivity (dynamic correlations between regions during brain activity) and another layer for structural connectivity (physical fibre connections between regions), or multiple layers representing brain connectivity at different frequency bands of EEG signals. The chapter highlights the advantages of multilayer and multimodal integration: by preserving distinctions between modalities or temporal snapshots, researchers avoid losing information that would be averaged out in a single-layer network. This leads to more accurate and biologically meaningful analyses. One study described in Chapter 4 shows that using a multilayer representation of brain networks significantly improved classification accuracy between healthy individuals and patients with a neurological disorder compared to using a traditional collapsed network. Essentially, the multilayer approach can detect subtle cross-layer patterns – such as a certain functional connection becoming abnormally strong only under specific conditions or frequencies, tied to an underlying structural disconnection – which would be missed otherwise.

Moreover, multiscale modelling is discussed, where computational models link different levels of brain organisation. For instance, simulations might integrate neuron-level networks to predict emergent whole-brain dynamics, bridging micro- and macroscales. These complex models are made tractable by AI algorithms that can fit model parameters to data across scales. The state-of-the-art described includes how combining fMRI, EEG and diffusion MRI into unified multilayer frameworks has yielded deeper insight into phenomena like cognitive ageing and brain development. In summary, current advances in this area emphasise integration: the brain is multifaceted, and AI-driven multilayer network models allow neuroscientists to piece together a cohesive picture from disparate data sources. By reading Chapter 4, one will appreciate how such integrated network models are constructed and how they outperform traditional methods in explaining and predicting neural phenomena.

1.3.4 AI-driven Neuroimaging Subtype Analysis

A prominent application of AI in neuroimaging research is the data-driven discovery of disease subtypes – an approach that addresses the heterogeneity within diagnoses. Chapter 5 is dedicated to AI-inspired subtype analysis for brain imaging, illustrating how unsupervised learning methods can unveil distinct groups of patients with similar brain characteristics. The rationale is that many neurodegenerative and psychiatric disorders are not single, uniform entities; instead, they encompass multiple variants with different underlying brain changes and clinical trajectories. In this section, Alzheimer's disease (AD) serves as a primary example. Using clustering algorithms (such as hierarchical clustering, k-means) or matrix factorisation techniques (like non-negative matrix factorization [NMF]) on large MRI datasets of AD patients, researchers have identified several reproducible imaging subtypes of Alzheimer's. One subtype may show predominantly hippocampal and medial temporal lobe atrophy (aligning with classic memory-presentation AD), while another subtype might exhibit more diffuse cortical atrophy with relative sparing of the hippocampus (sometimes corresponding to atypical presentations). These subtypes likely reflect different pathological pathways or co-morbid processes. The chapter explains how discovering such subtypes can lead to more individualised prognoses – for instance, certain AD subtypes progress faster or respond differently to treatments than others.

Beyond Alzheimer's, AI-driven subtyping has been fruitful in other conditions as well. In major depressive disorder (MDD), for example, clustering functional connectivity patterns has revealed subgroups of depression that respond in distinct ways to therapy (one subtype might benefit greatly from repetitive transcranial magnetic stimulation (rTMS), whereas another might respond better to psychotherapy or medication). Similarly, in schizophrenia and autism, subtyping analyses have uncovered variability in brain network alterations that align with differences in symptoms and outcomes. The chapter provides an overview of common algorithms used for these purposes (from Gaussian mixture models to newer approaches like HYDRA – a semi-supervised method for clustering in high dimensions). A key point is that the availability of large neuroimaging datasets and computational power has enabled these sophisticated analyses: AI can disentangle latent patterns only when sufficient data supports learning complex boundaries. Chapter 5 also describes practical aspects, including example code, which shows readers how to replicate subtype analyses – reinforcing that this is an emerging, hands-on area of research. Overall, the current advance highlighted here is how clustering and other unsupervised AI methods are providing new lenses to understand disease heterogeneity. The identification of neuroimaging subtypes promises to refine diagnostic categories and improve personalised medicine, and this book covers both the methodology and specific findings across various brain disorders.

1.3.5 Computer-aided Diagnosis Systems in Neuroscience

AI-powered Computer-aided Diagnosis (CAD) systems have become an integral topic in biomedical imaging, and Chapter 6 focuses on their design and use in the context of neuroimaging for neurological and psychiatric conditions. A CAD system in neuroscience typically involves a pipeline where advanced ML algorithms analyse brain scans (such as MRI, CT or PET images) to detect abnormalities, classify diseases or predict clinical outcomes, acting as an assistive tool for radiologists and clinicians. The chapter outlines how integrating AI with neuroimaging has substantially improved diagnostic accuracy and efficiency. For example, CAD algorithms can automatically segment and measure brain tumours in MRI scans, or detect subtle white matter lesions and microinfarcts that a human might miss, providing a second pair of 'eyes' that never tire. In degenerative diseases like Alzheimer's, CAD tools have been developed that combine features from imaging (like patterns of cortical thinning), along with other biomarkers, to give probability scores that a patient has early Alzheimer's or will convert to it in the near future. These systems often match or exceed expert-level performance in research settings, and they do so in a fraction of the time, enabling quicker diagnoses.

Chapter 6 also emphasises key challenges and how the field is addressing them. One challenge is the limited size of high-quality labelled datasets in certain neurological conditions – AI models require a lot of examples to learn, but rare diseases or subtle pathologies may not have large repositories of annotated images. To tackle this, techniques like data augmentation, transfer learning (using models pretrained on large generic image datasets and fine-tuning them on brain images), or the generation of synthetic training examples (using generative adversarial networks [GANs] to create realistic fake brain images) are being employed. Another challenge is interpretability and trust: clinicians are understandably cautious about relying on a 'black box' AI diagnosis without understanding the rationale. The chapter discusses emerging solutions like explainable AI (XAI) applied to CAD systems – for instance, heat maps that highlight areas of an MRI that most influenced the AI's decision, thereby showing a radiologist why the algorithm flagged a scan as abnormal. Ensuring fairness is also crucial; the text notes the importance of validating CAD performance across diverse patient populations, as models can inadvertently be biased if the training data was not representative (for example, an AI might perform worse on scans from older devices or on patients of demographics underrepresented in the training set). The clinical integration of CAD systems is covered as well: some AI tools for stroke detection on CT or for triaging head CT scans for emergencies have already received regulatory approvals and are in use in hospitals, serving to alert doctors to critical findings faster. By covering these aspects, Chapter 6 provides a state-of-the-art account of how AI-based CAD systems are built for neuroscience applications, their proven benefits in enhancing diagnostic workflows, and the roadmap for making them robust, transparent and reliable in everyday clinical practice. Readers will gain insight into not only success stories (like improved early detection of diseases from scans) but also the collaborative human–AI approach needed to harness these tools safely in healthcare.

1.3.6 Neurodegenerative Diseases and Genetics Research with AI

Chapters 7 and 8 of the book turn towards specific application domains, highlighting how AI techniques are being applied to neurodegenerative disorders and to genetic studies of brain conditions.

Neurodegenerative diseases (such as Alzheimer's, Parkinson's, Huntington's disease, and others like ALS or frontotemporal dementia) are a major focus of current AI research in neuroscience. Chapter 7 surveys the landscape of AI applications in this area, noting that diseases like Alzheimer's and Parkinson's have dominated the research literature.

One reason is the availability of large, well-curated datasets – for instance, the Alzheimer's disease Neuroimaging Initiative (ADNI) and the Parkinson's Progression Markers Initiative (PPMI) – which have enabled researchers to train DL models on thousands of patient scans and clinical records. These models have achieved impressive feats: classifying MRI or PET scans into disease vs. healthy with high accuracy, predicting future cognitive decline in patients with mild symptoms, and even stratifying patients by likely progression rate. An example described might be a DNN that analyses a combination of brain images and cerebrospinal fluid biomarkers to predict whether a patient with MCI will convert to Alzheimer's within three years. By recognising complex patterns across the data (perhaps a subtle cortical thinning pattern combined with an abnormal amyloid level), the AI can provide prognostic information that complements what clinicians can do. Similarly, for Parkinson's disease, ML is used to classify motor signals or dopamine transporter scans, and to predict things like which patients will develop cognitive impairment. Chapter 7 also highlights that while Alzheimer's and Parkinson's are heavily studied, AI is increasingly being applied to less common neurodegenerative conditions as data becomes available (e.g., using ML to detect early Huntington's disease motor signs, or to classify subtypes of frontotemporal dementia). A challenge in neurodegenerative AI research has been generalisation – models trained on one dataset may falter on another due to demographic or protocol differences – and the chapter discusses efforts to ensure models are robust, including using international consortium data and techniques to handle dataset biases.

Chapter 8 complements this by examining the role of AI in the genetics of brain disorders. Brain disorders, whether neurodegenerative or psychiatric, often have a substantial genetic component, but understanding this genetic architecture is complex. Traditionally, polygenic risk scores (PRS) have been the primary tool: they aggregate the effects of many genetic variants into a single risk estimate for a disease. The chapter acknowledges PRS as the current champion in complex trait prediction, but it also discusses why ML approaches have only partially penetrated this field so far. Genetic data for brain disorders can be enormous in scale (millions of variants across thousands of individuals) yet noisy and highly dimensional, which poses challenges that AI is just starting to address. The text outlines how ML methods – ranging from regularised regression to more complex models like gradient boosting trees or neural networks – are being employed to predict disease status from genomic data, potentially capturing non-linear interactions between genes that PRS might miss. It also covers the pitfalls: for example, overfitting is a risk due to limited sample sizes relative to the number of genetic features, and noisy labels are a problem (since diagnoses in neurodegenerative diseases can be imprecise without post-mortem confirmation, some training data might be mislabelled). AI's ability to handle such challenges is improving, with strategies like cross-validation, feature selection and integration of prior biological knowledge to guide learning. Another forward-looking topic in Chapter 8 is the use of federated learning for genetics – recognising that genetic datasets are siloed across institutions and countries due to privacy, federated approaches could allow AI models to be trained on distributed data without pooling it, thus leveraging a much larger collective sample to improve prediction while preserving privacy. The synergy between Chapters 7 and 8 lies in the convergence of clinical and molecular data: as AI helps predict and classify diseases from imaging and clinical markers, it also starts to integrate genetic susceptibility, paving the way for holistic models. Together, these chapters underscore that whether through imaging or DNA, AI is increasingly critical in tackling the puzzle of neurodegenerative and psychiatric disorders – improving early detection, enhancing our understanding of disease pathways, and nudging the field towards more effective, individualised interventions rooted in both neurobiology and genetics.

1.3.7 Neuro-oncology and AI

Brain tumours, and gliomas in particular, represent another area where AI has made significant inroads, as discussed in Chapter 9. Neuro-oncology deals with tumours of the nervous system, and accurate diagnosis and treatment planning for brain tumours are extremely challenging due to the complexity and delicacy of brain tissue. AI technologies are proving invaluable in several respects. Firstly, in glioma diagnosis and grading, ML models (especially deep CNNs) are being used to analyse neuroimaging data – primarily MRI scans – to automatically detect tumours, segment their boundaries and even predict their histological grade (e.g., low-grade vs. high-grade glioma) or molecular subtype. For instance, a DL model can be fed a multi-sequence MRI (T1, T2, FLAIR sequences) and output a detailed tumour mask highlighting the tumour's extent in 3D, saving radiologists many hours of manual contouring and providing consistent results. Moreover, these models can learn subtle imaging cues that correlate with molecular features; researchers have demonstrated 'radiogenomic' AI models that predict genetic mutations like IDH status or 1p/19q deletion in gliomas purely from scans. Such predictions normally require invasive biopsy, so an accurate imaging-based surrogate is a huge boon for non-invasive diagnosis.

Another application highlighted is AI in treatment planning and outcome prediction for neuro-oncology. When planning neurosurgery or radiation therapy for a brain tumour, knowing the precise tumour margins and the involvement of adjacent critical brain regions is crucial. AI algorithms help by combining multimodal data – imaging, intraoperative ultrasound and digital pathology of biopsy samples – to guide surgeons on how much tissue to resect while preserving function. Chapter 9 likely discusses tools that register preoperative MRI with real-time operative data and use AI to update tumour localisation as the brain shifts during surgery, thus acting as a smart assistant in the operating room. Similarly, in radiation oncology, AI can assist in delineating tumour target volumes and protecting healthy tissue by learning from large datasets of expertly contoured cases. The chapter also covers how AI analyses digital pathology slides of brain tumour tissue (an area known as computational pathology). By examining histopathological features with computer vision techniques, AI models can grade tumours or find features (like microvascular proliferation or necrosis) automatically, and even discover novel features that might correlate with prognosis or treatment response, thereby contributing to more informed, personalised therapy choices. Another important facet is tracking tumour progression: AI systems can compare follow-up scans to distinguish true tumour recurrence from treatment-related changes (like radiation necrosis), a notoriously difficult task even for experienced clinicians.

In summary, the current advances described in this section show that AI is becoming an indispensable partner in neuro-oncology, from the moment of diagnosis to treatment decisions. Chapter 9 provides a thorough overview of these technologies, including examples such as how ML improved the classification of glioma subtypes or how integrating MRI with pathology via AI yields a more complete picture of the tumour. The theme emphasises improved precision – AI can parse the heterogeneity of tumours (both in imaging and microscopically) and thus supports clinicians in tailoring interventions to the individual tumour's characteristics. By reading about these developments, one gains insight into a future where brain tumour patients benefit from faster diagnoses, more accurate prognoses, and therapies guided by the collective intelligence of prior cases distilled through AI algorithms.

1.3.8 Ethical Considerations and Privacy Protection

As AI systems become more prevalent in neuroscience research and clinical tools, ethical and privacy considerations have come to the forefront, a topic comprehensively addressed in Chapters 10 and 11. The use of AI in areas like early disease detection raises important ethical questions. Chapter 10, for instance, examines the scenario of using AI to detect early signs of Alzheimer's disease from speech patterns – a novel approach that holds promise for non-invasive screening. While the technology is exciting, it triggers multiple ethical concerns. One concern is autonomy and consent: if people's speech (possibly collected via smart home devices or clinical recordings) is analysed by AI without explicit consent to screen for dementia, it could violate individuals' autonomy and right not to know a potential diagnosis. Even with consent, the issue of how and when to disclose an AI-predicted risk of Alzheimer's is delicate, since early diagnosis in a currently untreatable disease could cause distress. There's also a risk of depersonalisation – reducing a person to a set of risk scores. The chapter discusses these autonomy issues and suggests guidelines for ensuring that individuals maintain control over how their data is used and how results are communicated.

Privacy and data protection are another major focus. Speech and language data are deeply personal; they might inadvertently include sensitive content or identifiable information. An AI analysing speech for cognitive decline needs access to these data, so safeguarding them is critical. Chapter 10 addresses data handling practices – for instance, ensuring that transcripts or audio are stored securely and any personal identifiers are removed. However, as Chapter 11 explores in greater depth, even anonymised neuroscience data can carry re-identification risks. Brain images, for example, can sometimes be reconstructed to reveal a person's face, and combinations of quasi-identifiers (like age, sex and certain health metrics) could uniquely identify individuals in a dataset. There have been cases where supposedly de-identified health records were matched back to individuals using public information. Thus, the book stresses that conventional anonymisation (simply stripping obvious identifiers) is often not sufficient for privacy protection with rich datasets like those in neuroscience.

To mitigate these issues, privacy-preserving technologies are discussed. One approach is federated learning, which Chapter 11 describes as a collaborative model training technique where data never leaves its home institution. Instead, an algorithm is sent to each data site, it learns from local data, and only the model updates (not the raw data) are sent back and aggregated. This way, a hospital's sensitive patient data can be used to train an AI model without exposing the data itself to any outside entity. This technique has been tested in scenarios like distributed training of brain tumour segmentation models across hospitals, achieving performance close to a centrally trained model but with far greater privacy.

Another technology is differential privacy, which involves adding statistical noise to data or model outputs in a way that provides mathematical guarantees that no individual's data can be pinpointed from the released information. For example, if an AI model is querying a database of patient EEG signals, differential privacy methods can ensure that the answer to queries doesn't inadvertently reveal a specific patient's record. The chapter likely describes how applying differential privacy can allow researchers to share aggregate findings or trained models publicly with less risk of privacy leakage. Encryption-based methods (like secure multiparty computation or homomorphic encryption) might also be introduced as cutting-edge ways to perform computations on encrypted data, meaning an AI could be trained or used on sensitive data without ever 'seeing' it in plain form.

Ethical AI deployment also requires fairness and transparency. Bias in AI models is a concern – if a model for, say, speech-based Alzheimer's detection is trained mostly on one demographic group, it might be less accurate (or produce more false-positives/negatives) in others, leading to inequalities in care. The ethical guidance in Chapter 10 emphasises testing AI systems for such biases and ensuring equitable performance. Transparency involves making AI's decision process interpretable to developers, clinicians and patients; for example, clinicians should know the features the AI is weighing, and patients should have the right to an explanation if an AI influences their diagnosis or treatment. The chapter concludes with recommendations for ethical guidelines specific to neuro-AI applications. These likely include points such as involving ethicists early in AI project design, obtaining informed consent that specifically addresses AI usage, continuously monitoring AI outputs for errors or biases and setting up oversight committees to govern AI deployment in clinical settings.

In essence, the advances in AI come paired with a new set of responsibilities. This book doesn't shy away from those, instead providing a thoughtful roadmap on how to harness AI's power in neuroscience responsibly and securely. By covering speech-based Alzheimer's detection as a case study, it grounds the discussion in a real application, then broadens to universal principles of ethics and privacy. The reader gains an understanding that successful AI in neuroscience is not just about high accuracy or efficiency but also about maintaining trust, upholding patient rights and protecting sensitive information. These chapters serve as a vital reminder that as we push the frontiers of neuro-AI, we must do so with careful consideration of the human values at stake.

1.4 Major Challenges and Limitations

1.4.1 Data-related Challenges

The promise of AI in neuroscience is tempered by a number of practical data-related limitations. One fundamental challenge is the small sample size problem that plagues many neuroscience studies. Unlike fields such as computer vision, where millions of images are available to train deep networks, neuroscience datasets (for example, MRI scans or intracranial EEG recordings) are often limited to tens or hundreds of subjects due to the cost and complexity of data collection. This paucity of data can lead to overfitting, where AI models perform brilliantly on training data but fail to generalise to new cases. Efforts like data-sharing initiatives and consortia (e.g., the Human Connectome Project or ENIGMA for neuroimaging) are helping to aggregate larger sample sizes, but issues remain – different sites might use different scanners or protocols, introducing heterogeneity.

Relatedly, data heterogeneity is a serious concern. The human brain varies across individuals in anatomy and function; diseases themselves are heterogeneous; and multicentre data introduces variance in acquisition. AI models, especially complex ones like DNNs, may latch onto site-specific or sample-specific patterns that are not biologically meaningful (for example, slight differences in image contrast between hospitals could confound a model into picking up scanner brand instead of true disease markers). Ensuring models learn the 'right' features require careful dataset curation, normalisation techniques, and sometimes domain adaptation methods that make the model invariant to non-biological differences. Data quality is another facet – neuroscience data can be noisy (think of EEG with lots of artefacts from muscle movement, or fMRI with head motion effects). Poor-quality data can mislead AI algorithms. If, say, some MRI scans have motion blur, a model might erroneously associate that with the patient's condition if not properly handled. Researchers are developing preprocessing pipelines and quality control steps to mitigate this, but the issue cannot be fully eliminated.

Additionally, many neuroscience problems suffer from an unfavourable ratio of features to samples: each brain scan or genomic profile yields thousands of features (voxels, connections, variants), but the number of subjects is much smaller. This high dimensionality can be problematic for ML. Techniques like feature selection, dimensionality reduction (PCA, autoencoders) or regularisation (to penalise overly complex models) are essential to avoid the curse of dimensionality. In summary, the reliability of AI models in neuroscience heavily depends on the quality and quantity of data, and as such, a major limitation is that our datasets often do not meet the ideal criteria. Progress in this area relies not just on better algorithms, but equally on better data practices: larger, cleaner and more standardised datasets, as well as methods to make the most of limited data (like transfer learning from other domains or synthesising data to augment samples).

1.4.2 Model Interpretability and Clinical Translation

The issue of AI model interpretability – often encapsulated as the 'black box problem' – is a critical barrier, especially in clinical neuroscience applications. Many state-of-the-art AI models, such as DNNs with millions of parameters, do not lend themselves to easy understanding by humans. They can achieve high accuracy in predicting or diagnosing, but why they made a particular decision is usually not transparent. In a medical context, this opaqueness poses problems: clinicians are hesitant to trust a diagnosis or treatment recommendation from an algorithm that cannot explain its reasoning. Moreover, regulators often require interpretability for tools that will influence patient care, and without it, the path to approval and adoption is steep.

Another aspect is that interpretability is closely tied to our ability to derive scientific insight from AI models. If an algorithm discovers a complex pattern in brain data that correlates with disease, researchers naturally want to interpret that pattern in neuroscientific terms – which brain regions or connections are implicated, for instance. Black box models don't directly provide that, potentially limiting the knowledge gain from AI studies to just predictive power.

To address these concerns, the field is moving towards Explainable AI (XAI) techniques. XAI methods seek to make the model's decisions more understandable. For neural networks analysing images, a common approach is to use saliency maps or heat maps (like Grad-CAM) that show which parts of the input image were most influential for the model's decision. In an MRI classification task, such a heat map might highlight the hippocampus and temporal lobes if those areas drove an Alzheimer's vs. control distinction, thereby aligning the AI's 'thinking' with known disease markers. Another approach is building inherently interpretable models – for example, using simplified models (like decision trees or generalised additive models) or neural networks that enforce a form of sparsity or alignment with anatomical features, so that the output can be explained in terms of a few key features. Chapter 6's discussion of CAD noted that making AI more interpretable for clinicians (like showing which voxel differences led to a diagnosis) builds trust and aids adoption. The clinical translation of AI – meaning the transition from research proof-of-concept to real-world clinical tool – often hinges on solving interpretability. Clinicians need to validate that the model's outputs make sense and correspond to clinical reality; interpretability helps identify cases where the model might be relying on spurious correlations (e.g., if an AI for tumour detection is keying off a scanner artefact, an attentive review of explanations might catch that).

Furthermore, interpretability ties into accountability and error analysis. When an AI model makes a mistake, understanding its internal logic is essential to fix the issue. For instance, if a model systematically misses a certain subtype of stroke on imaging, examining its decision process may reveal that it wasn't considering a crucial feature, leading developers to adjust the model. In summary, this challenge is about bridging the gap between AI's complexity and human cognition – speaking the language of clinicians and scientists. Without interpretability, even the best models might remain unused in practice. Therefore, a major ongoing effort is to develop methods and standards for AI in neuroscience that are as interpretable and transparent as possible, ensuring that these models can be safely and effectively integrated into clinical workflows.

1.4.3 Multimodal and Cross-scale Data Integration Difficulties

The human brain can be studied at many levels – from genes and molecules to cells, circuits and whole-brain systems – and with various modalities like MRI, EEG, PET, genomics, etc. Integrating these multimodal and cross-scale data to get a unified understanding is a grand challenge where AI could help, but doing so is far from straightforward. One difficulty is the sheer complexity of combining data across different scales. Data at the microscale (e.g., gene expression in neurons or synaptic connectivity from electron microscopy) might be conceptually and quantitatively very different from macroscale data (like fMRI signals or cognitive test scores). Merging them into one model often requires custom approaches to ensure that each type of data is properly represented and that the model can handle their disparate nature (for example, gene data might be high-dimensional and discrete, whereas imaging data is high-dimensional and continuous).

Another issue is that multimodal datasets are often incomplete or unbalanced – not every subject has all types of data. For instance, one study participant might have MRI and EEG data but not genomic data, while another has genomics and MRI but no EEG. This complicates model training, as standard AI models typically expect a complete set of features for each data point. Methods like data imputation, multi-view learning, or sophisticated model architectures that can accept partially missing inputs are needed, and these are active areas of research. Some current approaches in multimodal integration use encoders for each data type (e.g., a convolutional network for MRI, a transformer for genomic sequences) and then a fusion mechanism to combine the learned representations. While powerful, these architectures are complex and require careful tuning to ensure one modality doesn't overwhelm the others or that the model doesn't just latch onto the modality that's easiest to use (a common pitfall where, say, a model given MRI and EEG might learn to ignore EEG if MRI alone can give good accuracy).

Cross-scale integration introduces additional hurdles, particularly when linking micro-level models (like neural simulations) to macro-level observations. Computational neuroscience offers detailed models of small circuits or single neurons, but integrating those into whole-brain models is computationally intensive and conceptually hard (one might need to simplify the micro models significantly, potentially losing fidelity). AI could help by learning surrogate models that approximate fine-grained simulations in a more tractable form, but training such surrogate models requires lots of simulation data and careful validation that they remain biologically realistic.

The current methodological bottlenecks also involve the lack of established standards: each lab might integrate multimodal data differently, making it tough to compare studies or build on each other's work. Moreover, different modalities have different noise characteristics and statistical properties. For example, MRI might have a high signal-to-noise ratio for certain structures but EEG is very noisy; if combined naively, the model might treat EEG noise as equally important as MRI signal, leading to unstable outcomes. Normalising and weighting modalities is thus an important but not fully solved task.

In essence, while the brain's complexity begs for multimodal AI models (since focusing on one modality gives only a partial view), the integration process can feel like assembling a jigsaw puzzle where pieces are not only scattered but of different materials and scales. It's a limitation that many current AI-in-neuroscience studies still handle one modality at a time or do simple pairwise integrations. To move forward, research is directed at creating more unified frameworks and cross-disciplinary methodologies – involving neuroscientists, data scientists and statisticians – that can elegantly combine diverse data. Overcoming this challenge will be key to breakthroughs like connecting molecular pathologies to network dysfunctions to cognitive symptoms within one coherent AI-driven model of a disease.

1.4.4 Ethical and Privacy Concerns in AI

The ethical and privacy issues introduced earlier in the context of specific applications are, in a broader sense, fundamental limitations to what we can or should do with AI in neuroscience. One concern is the risk to patient privacy whenever large datasets of patient information are used. Brain scans, genetic information and even behavioural data are highly personal. If AI models are trained on such data and then shared or deployed, there is a risk that individuals' identities or sensitive conditions could be inferred by malicious actors. There have been demonstrations of re-identification attacks: for instance, researchers reconstructed recognisable facial images from 'anonymised' MRI brain scans, showing that without extra precautions, sharing neuroimages is effectively sharing identifiable information. This risk means that some of the most valuable data (like rare neurological disorder datasets or longitudinal studies) might not be widely accessible to researchers, slowing down AI progress. Federated learning and other privacy-preserving methods (as discussed in Section 1.3.8) are promising, but they come with technical limitations and complexities that are not yet fully resolved (for example, federated models can still be vulnerable to certain attacks or may not perform as well as if data were centralised). So privacy concerns put a limitation on data sharing, which loops back to the data quantity challenge: we need to protect privacy, yet we also need large, diverse datasets for AI – balancing these can be tough.

Ethical guidelines and regulations can also slow down AI research and implementation, which is not to say they are bad (they are absolutely necessary), but it means AI in healthcare cannot just charge forward like in other domains. For example, obtaining consent for AI research is more nuanced; participants might need to consent not just to a study, but specifically to their data being used to train algorithms that could be used in various ways down the line. If a model is developed that indirectly discriminates (say an AI diagnosing autism works less well in girls because the training data had mostly boys, reflecting historical underdiagnosis in females), deploying it would raise fairness issues. Ensuring equitable AI is a challenge: models must be tested across subpopulations, and possibly adjusted (or data re-collected) if disparities are found. These processes can be time-consuming and costly, effectively limiting the speed of bringing AI innovations to practice.

Another ethical concern is managing the impact of AI predictions on patients and clinicians. For instance, if an AI predicts a high risk of Alzheimer's for an asymptomatic individual, delivering that information is ethically sensitive. In some cases, knowing might help prepare or start lifestyle changes; in others, it could cause anxiety or stigma, especially if the prediction isn't 100% certain (and it never is). Thus, an implicit limitation arises: we might have the technical ability to predict something, but we might choose not to use or communicate it straightforwardly until we have guidelines and support systems in place for handling the psychological and social consequences. In clinical use, a known issue is automation bias – clinicians might overrely on AI recommendations, or conversely, distrust them too much. Training and workflow integration are needed so that AI is used as a tool, not an oracle or something to be ignored, which requires cultural and educational shifts in healthcare.

Finally, the development of AI itself needs to adhere to ethical research practices. AI models could inadvertently perpetuate biases present in their training data. For example, if a dataset for brain injury outcomes mostly comes from urban hospitals, an AI might not generalise well to rural populations due to differences in post-care. Not recognising and correcting that would be unethical as it could lead to worse care for underrepresented groups. This necessitates continuous audits of AI systems – another layer of effort that is necessary but can slow deployment.

In summary, responsible AI development introduces constraints that, while absolutely important to follow, act as a reality check on unbridled innovation. Ethical and privacy considerations limit which data can be used, how it can be used, and how results can be disseminated or applied. Far from being negative, these limitations ensure that the pursuit of powerful AI tools does not come at the expense of individual rights and societal trust. The neuroscience community is actively engaging with ethicists, regulators and patients to navigate these issues. Solutions are emerging but must evolve in tandem with technology. In the meantime, the need to respect privacy and ethics can be seen as both a challenge – requiring careful planning, possibly slowing research – and an essential guiding framework that ultimately strengthens the quality and acceptability of AI innovations.

1.5 Future Directions and Prospects

1.5.1 Stronger Interdisciplinary Collaborations

One clear message when looking ahead is that the future of 'AI for Neuroscience' lies in deep interdisciplinary collaboration. The challenges and complexities discussed throughout this chapter underscore that no single field or specialty can tackle them alone. Neuroscientists, with their domain expertise, need to work hand-in-hand with data scientists, AI engineers, clinicians, and even ethicists and policymakers. In a practical sense, this means building teams and research environments where a neurologist might sit next to an ML expert, jointly formulating research questions that are both clinically relevant and algorithmically feasible. Such collaborations ensure that AI models are developed with an understanding of the brain and clinical context, and conversely, that experimental neuroscientists are aware of what AI methods can and cannot do. We're already seeing new models of collaboration emerge: for example, many large research initiatives or consortia now include a computational core alongside traditional experimental cores. Clinicians in hospitals are teaming up with AI startups to co-develop decision support tools, ensuring that the end product actually fits into clinical workflow and addresses real pain points. Going forward, we can expect more cross-training and hybrid roles – neuroscientists who teach coding and ML, data scientists who learn neuroanatomy and physiology. Universities might create joint programmes (indeed, fields like Neural Informatics and Computational Neuroscience are precursors), and industries might create R&D units that specifically bridge brain science and AI technology. Successful interdisciplinary integration also requires aligning goals and expectations: a data scientist might be used to quick iteration and benchmarks, whereas a clinical trial takes time and careful planning; mutual understanding is key. By fostering stronger collaborations, we not only accelerate innovation but also improve the chances of translation – because a solution co-created by clinicians and engineers is more likely to address regulatory, usability, and efficacy requirements needed for real-world impact. In essence, the convergence of diverse expertise will be the engine that drives the next wave of breakthroughs, whether it's cracking the neural code with the help of AI or deploying an FDA-approved AI device in neurology practice. The vision of the future is a truly multidisciplinary neuroscience, where AI is seamlessly integrated through constant dialogue between those who understand brains and those who understand algorithms.

1.5.2 Towards Precision and Personalised Neuroscience

The concept of precision medicine – delivering the right treatment to the right patient at the right time – is poised to take centre stage in neuroscience with the help of AI. In the future, we foresee a precision neuroscience approach where interventions for brain disorders are tailored to the individual's unique biological profile, much like how oncology now stratifies patients by genetic mutations for targeted therapies. AI will be a crucial enabler of this shift. One future prospect is that before a treatment decision, a patient might undergo a comprehensive evaluation: multigenomic sequencing, advanced brain imaging, cognitive and behavioural profiling, perhaps even wearable sensor monitoring of their daily function. The myriad data from these sources would be processed by AI systems trained to parse such complexity. The output could be, for example, a prediction of which medication or non-invasive brain stimulation protocol would yield the best outcome for that patient, or even suggesting a novel combination of therapies optimised to their profile. This could transform how we manage conditions like depression or epilepsy, where currently finding an effective treatment is often trial and error.

Emerging technologies will bolster this personalised approach. Digital twins for the brain are one promising idea: creating a detailed computational model of an individual's brain (based on their specific data) on which virtual experiments can be run. AI can help build and refine these personalised brain models and simulate how that 'twin' responds to various drugs or stimulations, thus guiding real-world choices. On the therapeutic front, brain–computer interfaces and neuroprosthetics will become more finely tuned to individuals. We expect AI-driven adaptability – for instance, a prosthetic limb's control algorithm adapting in real-time to a user's changing neural signals, or a closed-loop DBS system adjusting stimulation paradigms as a patient's disease progresses or daily activities change. Precision neuroscience also entails early intervention. In the future, AI might detect a person's risk for Alzheimer's a decade in advance via subtle cognitive and imaging markers, and that individual could then embark on a prevention plan (be it lifestyle changes or prophylactic therapies) personalised to their risk factors. Similarly, children with developmental differences could be identified and given customised educational or therapeutic support well before standard diagnoses would allow.

Emerging technologies in genomics (like CRISPR and gene therapies) and stem cells might also feed into precision neuroscience, with AI helping to match patients to these advanced treatments safely. For example, if one day gene therapy can cure certain monogenic neurological disorders, AI will assist in pinpointing the right candidates and predicting any off-target effects by analysing genetic networks. Another aspect is omics integration – combining genomics, proteomics and metabolomics for each patient; this is data-heavy, and AI will be indispensable to interpret it and find actionable insights (like a specific metabolic deficiency contributing to a psychiatric condition, treatable by a dietary supplement).

In summary, the future is moving away from a 'one-size-fits-all' paradigm. Neuroscience will likely emulate oncology's success in stratified medicine, and AI is the tool that can make sense of the high-dimensional data needed for such stratification. It will enable clinicians to move from population averages to personal specifics. While the full realisation of this vision faces hurdles (data collection, validation of AI recommendations in clinical trials), the trajectory is clear. Each year we see steps in that direction: AI models that predict individual outcomes or recommend patient-specific interventions are increasingly being published, and some are entering clinical evaluation. As those efforts mature, the next decade could witness genuinely personalised neuroscience care – improving efficacy and reducing unnecessary treatments by targeting what works best for each unique brain.

1.5.3 Data and Model Sharing Frameworks

For AI to truly flourish in neuroscience, the community must embrace frameworks that promote data sharing and collaboration, while respecting privacy and security. The future will likely see the establishment of more trusted research environments (TREs) and secure data enclaves dedicated to neuroscience data. These are controlled platforms where sensitive data (like brain scans linked to health records) can be stored and accessed by researchers under strict governance. In a TRE, one might perform analysis on data from hundreds of hospitals without the data ever leaving the secure environment – only the aggregated results or trained models are exported, and even those exports are checked for privacy compliance. By developing such infrastructures, researchers globally could access rich datasets that were previously siloed, accelerating discoveries and improving AI model robustness (since models trained on diverse, large data are generally better). Governments and institutions are recognising this; for example, national brain initiatives might set up central TREs for brain data, akin to how genetics has platforms like UK Biobank (which is not exactly a TRE but has controlled access).

In tandem with data sharing, model sharing and openness will grow. There is a push towards open-source in the AI field, and neuroscience AI should be no exception. Researchers might publish not just their paper but also the trained model (if it's not too large) or at least the model architecture and code. We could envision repositories of pretrained neural networks for various neuroscience tasks – analogous to how computer vision has models like ImageNet pretrained networks. For instance, a model pretrained on a huge EEG dataset could be shared for others to fine-tune on their smaller EEG study. This kind of model sharing saves resources and allows wider use of state-of-the-art techniques even by teams that don't have the means to train from scratch. Initiatives to promote reproducibility and transparency (like the Journal of Open Source Software, or requirements by funding agencies to have a data sharing plan) are likely to become more standard in neuroscience as well.

Standards and data format harmonisation will also be pivotal. The community is likely to converge on certain standard formats for data (we already see things like Brain Imaging Data Structure [BIDS] for organising MRI/EEG data). When every lab structures and labels its data similarly, it becomes much easier to merge datasets or apply the same AI pipeline across them. Furthermore, collaborative networks may emerge where multiple hospitals or labs jointly train AI models via federated learning (as touched upon in privacy discussions). We might see formal consortia where each member agrees to run a bit of code on their site and share model updates. This will effectively create giant virtual datasets without centralising data. To facilitate this, frameworks and tools for federated ML will become more user-friendly and robust.

Another aspect is encouraging data openness for older or completed studies. There is a lot of valuable neuroscience data sitting in archives from past clinical trials or studies. Incentives or mandates to make those datasets available (in de-identified form) for secondary analysis by AI could unlock new findings and prevent redundant data collection. In fact, some discoveries might only emerge when multiple datasets are combined to give enough statistical power – something AI can assist with by handling the integration and analysis.

In summary, the future likely holds a more collaborative and open culture in neuro-AI research. By sharing data and models within secure, standardised frameworks, the field can avoid the bottleneck of small, isolated studies and move towards large-scale endeavours that yield generalisable knowledge. Progress is already visible: multicentre neuroimaging challenges, open databases like OpenNeuro, and cloud-based collaboration hubs are gaining traction. In the next decade, these will mature into even more seamless ecosystems that make doing global, data-driven neuroscience as easy as doing local research – a development that will undoubtedly accelerate innovation and application.

1.5.4 From Explainability to Reliability

While recent efforts have focused on making AI models more explainable (as discussed in the challenges section), the ultimate goal in a clinical or high-stakes context is reliability. We want AI systems that not only can explain themselves, but more importantly, consistently make correct and safe decisions in the real world. Future research will likely emphasise strategies to enhance the robustness and trustworthiness of AI models in neuroscience. One direction is developing rigorous validation frameworks, analogous to clinical trial phases, for AI tools. Before an AI algorithm – say, seizure prediction is widely adopted – it might undergo multisite prospective validation where it's tested on new patients in real time, and its performance and impact on outcomes are measured. These studies will inform regulatory bodies and healthcare providers about the model's reliability under practical conditions (which may include distribution shifts like using a different EEG machine brand than it was trained on, etc.).

We will also see advances in robustness techniques. AI models can be surprisingly brittle – small perturbations in input can sometimes lead to large changes in output (as shown in adversarial example research). In a medical context, this could be dangerous. There is growing work on making models more robust to noise or adversarial manipulation, which might include training with adversarial examples, using Bayesian neural networks that quantify uncertainty or hybrid models that combine data-driven learning with rule-based components to ensure critical safety rules are never violated. For example, an AI that suggests brain stimulation settings might be constrained by hard-coded safety limits learned from decades of neurosurgical experience, so it will never propose a parameter outside the safe range, regardless of what the learned model part says. These kinds of guardrails will bolster reliability. Another facet of reliability is model monitoring and updating. Once an AI system is deployed, it shouldn't be static. The future will bring AI that can detect if the input data distribution is drifting (perhaps the hospital installed a new MRI machine with slightly different image characteristics) and either alert users or adjust itself (maybe via periodic retraining on new data). This continuous learning must be done carefully to avoid degradation, but it is important for maintaining performance over time. Moreover, a reliable AI should know when it's not confident. Expect to see more work on predictive uncertainty – models that don't just output a result, but also a confidence measure, and perhaps an automatic refrain when confidence is low ('the system is not sure in this case, human expert please review'). This type of introspection in models can prevent overtrust in AI and foster appropriate use: like a triage tool that flags easy cases automatically but flags unusual ones for expert diagnosis.

Finally, reliability encompasses generalisability – the idea that an AI model should work on new patients, in different hospitals, among diverse populations, not just the training cohort. This will be addressed both by the data/model sharing efforts (ensuring training data is broad) and by explicitly testing generalisation during development. Techniques like cross-validation across multiple cohorts or training adversarial networks to minimise site or demographic biases will become standard. The community might establish benchmark datasets that include intentionally challenging and diverse cases to test whether new algorithms truly generalise.

In summary, the shift from explainability to reliability means that beyond understanding AI decisions, we want to guarantee (as much as possible) the correctness and safety of those decisions. In the medical field, this is paramount. The future likely holds AI systems that come with performance guarantees, are resilient to perturbations, adapt to new conditions, and integrate fail-safes. As these aspects improve, clinicians and patients will grow more comfortable with AI, leading to its increased adoption. Reliability is the bridge from experimental AI to a routine clinical tool. By focusing research on reliability, the field acknowledges that an AI model is only as good as its worst failures; minimising those failures and ensuring graceful handling of uncertainty will make AI a trusted partner in neuroscience endeavours.

1.6 Overview of the Chapters

- **Chapter 2:** Introduces how advanced DL models, including GNNs, are applied to analyse complex brain connectivity data. This chapter highlights network neuroscience approaches enabled by AI, such as modelling the brain as a graph and discovering connectivity patterns related to cognition and disease.
- **Chapter 3:** Adopts a data-driven view of cerebral haemodynamics, showing how DL surrogates of fluid dynamics and GNNs convert multimodal imaging into real-time maps of pressure, flow, and collateral routes. Case studies – from thrombectomy planning to aneurysm rupture risk – highlight the emerging role of vascular intelligence in precision neurotherapeutics.
- **Chapter 4:** Discusses multilayer network frameworks for integrating multimodal neuroimaging data (e.g., combining fMRI, EEG, and diffusion MRI) and multiscale information. The chapter demonstrates how multilayer models provide richer insights into brain structure–function relationships than traditional single-layer analyses.
- **Chapter 5:** Focuses on techniques for uncovering disease subtypes using AI (especially unsupervised learning). Centred on examples like Alzheimer's disease, it shows how clustering and matrix factorization methods can reveal distinct neuroimaging

phenotypes within a disorder, improving understanding of heterogeneity and personalised medicine approaches.

- **Chapter 6:** Explores the development of ML systems that assist in diagnosing neurological conditions from neuroimaging. It covers practical CAD design, from feature extraction and multivariate analysis of brain scans to challenges like small sample sizes and the need for interpretability. Applications span Alzheimer's, Parkinson's, schizophrenia, and more.
- **Chapter 7:** Reviews the use of AI in studying and managing neurodegenerative diseases such as Alzheimer's and Parkinson's. This includes AI-driven early diagnosis from biomarkers, progression prediction, and the analysis of large-scale patient data to identify patterns (with an emphasis on how public datasets and DL have advanced the field).
- **Chapter 8:** Examines how ML techniques are applied to genomic and genetic data in neurology and psychiatry. Topics include polygenic risk score comparison to ML models, using AI to predict disease risk from genetic profiles, handling challenges like noisy labels and high-dimensional data, and emerging methods like federated learning for collaborative genetic analysis.
- **Chapter 9:** Details how AI aids in brain tumour diagnosis and treatment. It covers DL for tumour detection and segmentation on MRI, 'radiogenomic' models predicting tumour genetics from images, AI in guiding surgical and radiotherapy planning, and the integration of pathology and imaging data for comprehensive tumour characterisation.
- **Chapter 10:** Discusses the ethical issues arising from an AI approach that uses speech data to detect early cognitive decline. Key themes include informed consent, privacy of patient data, and potential for bias or discrimination, the impact of predictive information on patients' welfare, and ensuring transparency and fairness in AI-driven diagnostic tools.
- **Chapter 11:** Focuses on methods to safeguard patient privacy when developing and deploying AI models on sensitive neuroscience data. It introduces anonymisation challenges, risks of reidentification, and solutions like federated learning, differential privacy, and secure data environments that allow research and clinical AI use while upholding strict privacy standards.

References

Ajit, A., Acharya, K. and Samanta, A. (2020). A review of convolutional neural networks. 2020 International Conference on Emerging Trends in Information Technology and Engineering (ic-ETITE), (1–5 Feb 2020). https://doi.org/10.1109/ic-ETITE47903.2020.049.

Binder, M.D., Hirokawa, N. and Windhorst, U. (ed.) (2009). Encyclopaedia of Neuroscience. Berlin Heidelberg: Springer. https://doi.org/10.1007/978-3-540-29678-2.

Choi, K.S. (2022). Artificial intelligence in neuroimaging: clinical applications. *Investigative Magnetic Resonance Imaging Korea Science* (accessed 25 Jun 2025). [Online]. https://koreascience.kr/article/JAKO202210450822578.page

Deisseroth, K. (2015). Optogenetics: 10 years of microbial opsins in neuroscience. *Nature Neuroscience* 18 (9): 1213–1225. https://doi.org/10.1038/nn.4091.

Fukushima, K. (1980). Neocognitron: A self-organizing neural network model for a mechanism of pattern recognition unaffected by shift in position. *Biological Cybernetics,* 36(4), 193–202.

Hebb, D.O. (1950). The organization of behaviour: a neuropsychological theory. New York: John Wiley and Sons, Inc., 1949. 335 p. $4.00. *Science Education* 34 (5):336–337. https://doi.org/10.1002/sce.37303405110.

Hochreiter, S. and Schmidhuber, J. (1997). Long short-term memory. *Neural Computation* 9 (8): 1735–1780. https://doi.org/10.1162/neco.1997.9.8.1735.

Hunt, E.B. (2014). Artificial Intelligence. Academic Press.

McCulloch W.S. and Pitts, W. (1943). A logical calculus of the ideas immanent in nervous activity. *Bulletin of Mathematical Biophysics* 5 (4): 115–133. https://doi.org/10.1007/BF02478259.

Qiyang, S., Alican, A., Björn, W.S. et al. (2025). Explainable Artificial Intelligence for Medical Applications: A Review ACM Transactions on Computing for Healthcare. (accessed 25 Jun 2025). [Online]. https://dl.acm.org/doi/full/10.1145/3709367?casa_token=BVlCZjz5v3EAAAAA%3AU5GEr94NF7jAT0LlRlAEf1u7WfZTK6tqZfksJQkvSnBdnHwV71puKL4iV-wQNyrX6mYwd31YDqSA

Rumelhart, D.E., Hinton, G.E., and Williams, R.J. (1986). Learning representations by back-propagating errors. *Nature* 323 (6088): 533–536. https://doi.org/10.1038/323533a0.

Yang, W., Tu, Z., Sun, Q. et al. (2016). CRISPR/Cas9: implications for modeling and therapy of neurodegenerative diseases. *Frontiers Molecular Neuroscience* 9: 30. https://doi.org/10.3389/fnmol.2016.00030.

Zhou, Z.-H. (2021). Machine Learning. Springer Nature.

2

Deep Neural Networks in Brain Networks Study

Lu Zhang[1] and Dajiang Zhu[2]

[1]Computer Science Department, Indiana University Indianapolis, Indianapolis, IN, USA
[2]Computer Science and Engineering, University of Texas at Arlington, Arlington, TX, USA

2.1 Introduction

Our brain operates as a complex communication network in which interconnected neural elements form a structural foundation for information exchange (Park & Frinton, 2013). Building upon this structural substrate, signalling and information transmission occur across various levels and spatial scales of brain activity, governing the function, cognition and behaviour of a healthy brain (Petersen & Sporns, 2015). Moreover, research on brain disorders has revealed that pathological perturbations often spread through axonal pathways – i.e., the network connections, impacting distant regions rather than remaining confined to a single locus (Fornito et al. 2015). These findings suggest that disease propagation patterns are constrained by brain's intricate and highly organised network topology. Consequently, adopting a network perspective to study the brain has proven to have significant potential for uncovering organisational principles of the brain and their connections to cognitive functions and neurological disorders.

The human brain network is widely regarded as the most complex network known to science. Comprising approximately 100 billion neurons interconnected by nearly 100 trillion synapses, it operates across multiple spatial scales and exhibits dynamic interactions over diverse temporal domains. Additionally, the brain's intricate structural and functional architectures display hallmark properties such as small-world topology, modular organisation, and highly connected hub regions. Capturing these complex, non-linear relationships poses substantial challenges for traditional analytical approaches, which often fall short in modelling the brain's multiscale and dynamic connectivity patterns.

Recent advances in deep neural networks (DNNs) have transformed the landscape of machine learning (Hinton & Salakhutdinov, 2006; LeCun et al., 2015; Sun et al., 2019; Wang et al., 2015; Sirinukunwattana et al., 2016; Roth et al., 2015), introducing powerful and efficient models for brain network analysis that offer several key advantages over traditional statistical methods (Zhang et al., 2024). First, DNNs can autonomously extract meaningful features directly from raw data, enabling the automatic detection of crucial patterns and complex relationships within high-dimensional connectivity data. This capability greatly reduces the need for manual feature engineering, which is often laborious and subjective. Second, brain networks inherently exhibit intricate, non-linear relationships between nodes and edges that are challenging to capture with linear models. DNNs, with their deep hierarchical architectures and non-linear activation functions, are well suited to model such complexities. Third, DNNs are highly scalable, allowing them to efficiently handle the massive and intricate datasets that are increasingly common in modern neuroscience. This scalability is essential as the data volumes continues to expand across multimodal and longitudinal studies. Finally, brain network research frequently requires integrating diverse data types, such as structural and functional modalities, into a unified analytical framework. DNNs facilitate such multimodal fusion, providing a comprehensive representation of brain connectivity. These significant advantages have driven the rapid adoption of DNNs-based approaches in brain network research, establishing this as a fast-evolving and influential field. Among various DNN architectures, deep graph neural networks (GNNs) (Kipf & Welling, 2016; Wu et al., 2020; Zhang et al., 2020) have emerged as particularly powerful tools due to their ability to process complex graph data, making them especially suitable for modelling the topology and dynamics of brain networks (Ktena et al., 2018; Kazi et al., 2019; Parisot et al., 2018; Zhang et al., 2020, 2021, 2023).

In this chapter, we will provide a comprehensive overview of brain networks across different scales and modalities, the architecture and computational processes of GNNs, and their applications in studying the relationships between brain structure and function as well as advancing brain disease research.

2.2 Brain Networks

A brain network generally consists of two fundamental components that shape its architecture: neural elements and neural connections. The definitions of these components can vary across different scales and levels of structural description, which typically focus on three distinct organisational levels (Zhang et al., 2024; Park & Friston, 2013). At the **microscale**, the focus is on individual neurons and synapses. The **macroscale**, on the other hand, examines anatomically distinct brain regions of interest (ROIs) and the pathways between them. The **mesoscale** lies between these extremes, concentrating on neuronal populations. Single neurons are relatively straightforward to define, but their immense number, substantial variability and rapid dynamics make them impractical as the fundamental units of the connectome. Conversely, while identifying and delineating brain regions and neuronal populations is more challenging – due in part to the lack of a universally accepted parcellation scheme for human brain regions – anatomically distinct ROIs and interregional pathways represent the most practical level of organisation for the brain networks.

In the ROI-based macroscale brain networks derived from in vivo neuroimaging, connections between ROIs can be categorised into three types (Zhang et al., 2024; Park & Friston, 2013) (Figure 2.1): **structural connectivity** (SC), representing anatomical links; **functional connectivity** (FC), capturing directed/undirected statistical dependencies;

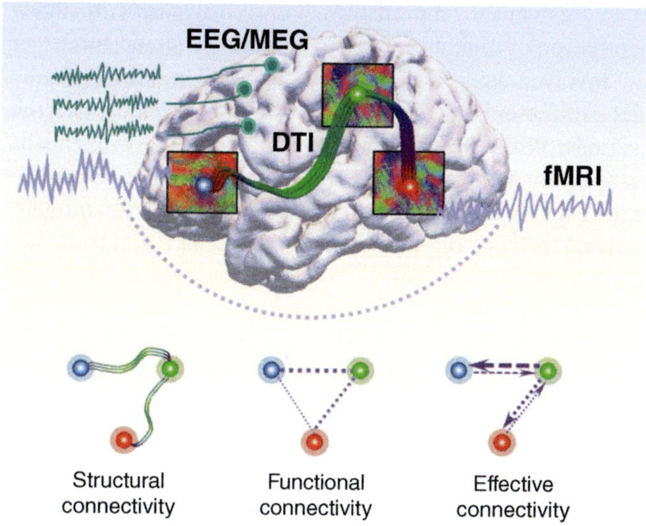

Figure 2.1 Three types of connectivity are commonly analysed in brain networks: SC, which refers to the anatomical links and is typically estimated using fibre bundles derived from diffusion magnetic resonance imaging (MRI); FC and EC, both of which are generally inferred through the correlation of nodal activities based on BOLD-fMRI or EEG/MEG. *Source:* Zhang et al. (2024)/Oxford University Press/CC BY NC 4.0.

and **effective connectivity** (EC), describing directed causal relationships among distributed responses.

Structural connectivity (SC) at the macroscale primarily reflects long-range fibre bundles, which are typically derived from diffusion MRI (Basser et al., 1994). A structural brain network can be constructed based on these fibre bundles by associating them with the specific brain regions they connect (Park et al., 2004; Hagmann et al., 2007). However, it is important to note that SC derived from diffusion MRI is usually undirected and does not distinguish between excitatory and inhibitory connections. This contrasts with SC obtained from tracing studies, which can vary in strength or density in either direction and, in some cases, can be linked to excitatory or inhibitory postsynaptic effects.

Functional connectivity (FC), on the other hand, is generally inferred from the correlation of nodal activities, typically using blood oxygenation level-dependent (BOLD) functional MRI (fMRI) or coherence analysis of electroencephalogram (EEG) or magnetoencephalogram (MEG) signals, recorded during task performance or at rest. In their study (Zhang et al., 2022), the authors provide a comprehensive summary of common measurements used to represent pairwise relationships between two fMRI signals. These measurements include correlation, partial correlation, covariance and both weighted and binary edges. Resting-state fMRI, in particular, has emerged as a cornerstone for analysing functional networks, driven by the discovery of spatially organised endogenous low-frequency fluctuations within BOLD signals (Biswal et al., 1995).

Effective connectivity (EC) explores causal relationships within a network, focusing on how one node (neuronal population) influences another within a specific model of network dynamics. Inferring EC involves using a neuronal integration model, typically requiring the estimation of model parameters (effective connectivity) that best explain the observed BOLD or EEG/MEG signals. While SC can constrain the neuronal model, it is crucial to recognise that SC does not fully determine EC. EC is dynamic, responsive to the system's state and dependent on the experimental context. Furthermore, in certain models, EC can include polysynaptic pathways, not just direct axonal connections.

2.3 Graph Neural Networks

Within the expansive domain of deep neural networks, graph convolutional networks (GCNs) extend traditional convolutional neural networks (CNNs) by performing convolutional operations on graph-structured rather than Euclidean-structured neighbours. Owing to their remarkable capability to model complex graph data, GCNs have become a powerful and widely adopted framework for brain network analysis (Kipf & Welling, 2016; Wu et al., 2020; Zhang et al., 2020, 2021, 2023).

A GCN begins by representing input data as a graph, denoted by $G = (N, E, A)$, where N is the set of nodes. Each node n_i is associated with an attribute vector $x_i \in R^{1 \times D}$, where D is the dimension of attributes (features). The feature matrix of the graph is $X = [x_1, \ldots, x_N] \in R^{N \times D}$. E represents the set of edges, and $A \in R^{N \times N}$ denotes the adjacency matrix. In this graph, $n_i \in N$ refers to the i-th node, $e_{ij} = (n_i, n_j) \in E$ denotes an edge connecting nodes n_i and n_j, and $a_{ij} \in A$ represents the connection strength between n_i and n_j. Based on the graph $G = (N, E, A)$ and the feature matrix X, spectral convolutions on graphs are defined in the Fourier domain as $g_\theta * x_i = U g_\theta U^T x_i$. Here $g_\theta = \text{diag}(\theta)$ is a spectral filter parameterised by a vector of Fourier coefficients $\theta \in R^N$, and U is a set of eigenvectors of normalised graph Laplacian $L = I_N - \mathcal{D}^{-\frac{1}{2}} A \mathcal{D}^{-\frac{1}{2}} = U \Lambda U^T$, where $\mathcal{D}$ is the degree matrix of A and Λ is a diagonal matrix of the corresponding eigenvalues. In this formula, g_θ can be viewed as a function of the eigenvalues, i.e., $g_\theta(\Lambda)$ and $U^T x_i$ represents the graph Fourier transform of x_i. However, directly computing $U g_\theta U^T x_i$ has a computational complexity of $O(N^2)$, and computing the eigendecomposition of graph Laplacian L is prohibitively expensive for large graphs. To address this challenge, Defferrard et al. (2016) approximated g_θ (Λ) using a truncated polynomial expansion: $g_\theta(\Lambda) = \sum_{k=0}^{K-1} \theta_k \Lambda^k$. This formula defines a K-localised kernel, as the filter is represented by the Kth-order polynomial of the graph Laplacian, depending only on nodes within K steps (K^{th}-order neighbourhood) from the central node. The resulting complexity of calculating $\sum_{k=0}^{K-1} \theta_k \Lambda^k$ becomes $O(|E|)$, which is linear with the number of edges. A GCN can thus be constructed by stacking multiple convolutional layers of this polynomial form: $g_\theta(\Lambda) = \sum_{k=0}^{K-1} \theta_k \Lambda^k$. For example, in Kipf and Welling (2016), a two-layer GCN is formulated by $Z = f(X, A) = f(\widehat{A}\, ReLU(\widehat{A} X W^{(0)})\, W^{(1)})$, where $f(\bullet)$ is the activation function and $\widehat{A} = \mathcal{D}^{-\frac{1}{2}} A \mathcal{D}^{-\frac{1}{2}}$ is the Laplacian transformation of A. In a subsequent study, Zhang et al. (2019) compared several Laplacian transformations, including $\widehat{A} = A$, $\widehat{A} = \mathscr{D} - A$, $\widehat{A} = \mathscr{D}^{-\frac{1}{2}} A \mathscr{D}^{-\frac{1}{2}}$ and $\widehat{A} = \mathscr{D}^{-1}A$. Their experiments demonstrated that $\widehat{A} = A$ and $\widehat{A} = \mathscr{D}^{-\frac{1}{2}} A \mathscr{D}^{-\frac{1}{2}}$ give comparable classification performances.

In brain network studies, brain connectivity data are represented as a graph G and input into a GCN, enabling the model to learn the intricate topological and functional relationships within the brain networks. This facilitates a deeper understanding of normal brain organization and supports the investigation of various brain disorders.

2.4 GCNs in Understanding Brain Structure–Function Relationship

Uncovering the intricate relationship between brain structure and function remains one of the central challenges in neuroscience. This relationship exhibits a distributed and heterogeneous pattern: at the network level, many functional relationships arise from non-overlapping sets of anatomical connections, indicating that functional networks do not necessarily correspond to the underlying structural substrate with a simple node-to-node mapping. Furthermore, this complex relationship is further complicated by individual variations in both structural and functional organization, making it an especially challenging research topic. Several pioneering studies have employed graph convolutional networks (GCNs) to investigate this structure–function coupling by leveraging their powerful capacity to model graph-structured data. Zhang et al. (2020) introduced multiple GCNs into a generative adversarial network (GAN), developing a novel graph-based GAN model to predict individual structural networks from the corresponding functional networks. Each GCN component acts as a graph filter to capture a specific type of relationship between the structural and functional networks – similar to how different filters in a CNN capture various visual features such as edges and shapes. A structure-preserving loss function was further designed to maximise the similarity between the ground truth and the graphs predicted by the generator. Building on this idea, Gao et al. (2023) combined a GCN with a multi-context discriminator-based GAN to predict diverse functional networks from structural networks. This model introduces a novel multi-context discriminator that provides three levels of supervision – edge-level, node-level and graph-level – to enhance the generator's performance. Similarly, Bessadok et al. (2021) proposed a dual-discriminator GCN-GAN framework to align the source and target brain networks during training and to adversarially regularise the embeddings of the source graphs for both training and testing. Collectively, these works have achieved remarkable success in predicting and mapping brain networks, highlighting the strong potential and effectiveness of GCNs in brain network research.

2.5 GCNs in Investigating Brain Disease

Another major application of GCNs lies in the study of brain diseases, particularly Alzheimer's disease (AD). AD is a progressive neurodegenerative disorder characterised by the accumulation of pathological proteins in specific brain regions, which subsequently propagate through the large-scale brain networks. This progression leads to disruptions in both regional integrity and inter-regional connectivity. Consequently, analysing brain networks is crucial for understanding the development and progression of AD, and GCNs have become a natural methodological choice for such network-based investigations.

Existing studies have introduced innovative approaches for organising brain network data and designing model architectures tailored to diverse analytical goals. From a data representation perspective, two primary paradigms have emerged: population-level graphs and individual-level graphs. At the population level, each node represents an individual subject, and edges denote correlations between subjects. For example, Parisot et al. (2018) utilised a sparse graph representation in which nodes correspond to imaging-based feature vectors and edges incorporate phenotypic information as weights. Huang and Chung (2022) advanced this method by proposing a framework that automatically learns population graphs with variational edges, while Zhu et al. (2022) adopted a dynamic graph learning strategy that adaptively adjusts the neighbourhood relationships to generate robust node embeddings and improve classification performance. Additionally, Li et al. (2022) enhanced interpretability by incorporating feature and sample interpretation modules into the population graph learning process.

At the individual level, each node represents a region of interest ROI and edges signify connectivity or correlation between ROIs. Zhang et al. (2021) developed a deep brain connectome to model both individual structural and functional networks for brain disease analysis. In this model, the individual structural network serves as the initial graph topology, which was progressively updated using functional information to maximise classification performance. The resulting deep connectome integrated multiple brain networks to form an 'individual connectome signature'. Similarly, Zhang et al. (2019, 2020) explored complementary designs using distinct neural architectures to analyse fMRI time-series data: Zhang et al. (2019) employed recurrent neural networks (RNNs) to capture temporal dependencies within fMRI data, while Zhang et al. (2020) used an attention layer to extract disease-related features, enhancing the interpretability of the deep model.

Other studies have adopted hierarchical or hybrid frameworks that integrate both individual- and population-level representations. For example, Jiang et al. (2020) used two GCNs to model individual brain functional networks and the whole population network, respectively. The framework learned compact *representations* of individual brain networks through graph-level embedding and updated the population network by aggregating node representations. Li et al. (2021) applied multiple thresholds to generate connectivity networks capturing varying degrees of topological sparsity, ensuring consistent network structure across subjects.

In summary, a wide range of graph representations – individual, population or multi-scale – are used to capture different aspects of brain network organisation. To enhance the representational power of GCNs for complex, multi-scale and multi-modal brain networks, advanced models and techniques, such as RNNs (Zhang et al., 2019, 2020), attention mechanisms (Zhang et al., 2020) and transfer learning (Li et al., 2021), have been integrated. Moreover, innovations in dynamic GCN designs (Zhang et al., 2021; Zhu et al., 2022), which adaptively update graph topology or learn task-specific node/edge features during training, offer significant advantages over static models.

2.6 Future Work

Applying GCNs to brain network analysis has yielded valuable insights into graph data organisation and representation, analytical methodologies and model design, establishing a solid foundation for future investigations. Nevertheless, several critical challenges remain to be addressed in order to deepen our understanding of the human brain. First, the inherent complexity of GCNs, like other DNNs, makes them opaque and difficult to interpret. Understanding the mechanisms behind their decision-making processes is crucial in neuroscience applications, where interpretability is not merely desirable but essential. Interpretable models can shed light on the neural mechanisms that underpin healthy brain function and, at the same time, foster the trust of clinicians and patients in AI-assisted medical systems. Although progress has been made with interpretation techniques for graph-based deep models, such as GNNExplainer (Ying et al., 2019) and GraphLIME (Huang et al., 2022), assessing the validity and reliability of these explanations remains a significant challenge. This limitation arises from our incomplete understanding of both brain networks and DNNs. Continued efforts to develop innovative, domain-aware interpretability frameworks are therefore essential for advancing the use of GCNs in neuroscience and medicine. Second, most existing GCN architectures are designed for undirected graphs and require uniform node feature dimensions, which restricts their flexibility in handling the diverse and heterogeneous graph data encountered in brain studies. Future research should focus on developing more adaptable models capable of processing multi-scale, multi-modal, and dynamically structured graph data, thereby enhancing the generalisability and applicability of GCNs to the complex and variable nature of brain connectivity.

References

Basser, P.J., Mattiello, J., and LeBihan, D. (1994). MR diffusion tensor spectroscopy and imaging. *Biophysical Journal* 66 (1): 259–267.

Bessadok, A., Mahjoub, M.A., and Rekik, I. (2021). Brain graph synthesis by dual adversarial domain alignment and target graph prediction from a source graph. *Medical Image Analysis* 68: 101902.

Biswal, B., Zerrin Yetkin, F., Haughton, V.M. et al. (1995). Functional connectivity in the motor cortex of resting human brain using echo-planar MRI. *Magnetic Resonance in Medicine* 34 (4): 537–541.

Defferrard, M., Bresson, X., and Vandergheynst, P. (2016). Convolutional neural networks on graphs with fast localized spectral filtering. *Advances in Neural Information Processing Systems* 29. https://doi.org/10.48550/arXiv.1606.09375.

Fornito, A., Zalesky, A., and Breakspear, M. (2015). The connectomics of brain disorders. *Nature Reviews Neuroscience* 16 (3): 159–172.

Gao, X., Zhang, X., Zhang, L. et al. (2023). Predicting diverse functional connectivity from structural connectivity based on multi-contexts discriminator GAN. International Conference on Medical Image Computing and Computer-Assisted Intervention (348–357). Cham: Springer Nature Switzerland.

Hagmann, P., Kurant, M., Gigandet, X. et al. (2007). Mapping human whole-brain structural networks with diffusion MRI. *PloS One* 2 (7): e597.

Hinton, G.E. and Salakhutdinov, R.R. (2006). Reducing the dimensionality of data with neural networks. *Science* 313 (5786): 504–507.

Huang, Q., Yamada, M., Tian, Y. et al. (2022). Graphlime: local interpretable model explanations for graph neural networks. *IEEE Transactions on Knowledge and Data Engineering* 35 (7): 6968–6972.

Huang, Y. and Chung, A.C. (2022). Disease prediction with edge-variational graph convolutional networks. *Medical Image Analysis* 77: 102375.

Jiang, H., Cao, P., Xu, M. et al. (2020). Hi-GCN: a hierarchical graph convolution network for graph embedding learning of brain network and brain disorders prediction. *Computers in Biology and Medicine* 127: 104096.

Kazi, A., Shekarforoush, S., Arvind Krishna, S. et al. (2019). Inception GCN: receptive field aware graph convolutional network for disease prediction. Information Processing in Medical Imaging: 26th International Conference, IPMI 2019, Hong Kong, China (2–7 June 2019). Proceedings 26 (73–85). Springer International Publishing.

Kipf, T.N. and Welling, M. (2016). Semi-supervised classification with graph convolutional networks. *arXiv preprint arXiv*:1609. 02907.

Ktena, S.I., Parisot, S., Ferrante, E. et al. (2018). Metric learning with spectral graph convolutions on brain connectivity networks. *Neuro Image* 169: 431–442.

LeCun, Y., Bengio, Y., and Hinton, G. (2015). Deep learning. *Nature* 521 (7553): 436–444.

Li, H., Shi, X., Zhu, X. et al. (2022). FSNet: dual interpretable graph convolutional network for Alzheimer's disease analysis. *IEEE Transactions on Emerging Topics in Computational Intelligence* 7 (1): 15–25.

Li, L., Jiang, H., Wen, G. et al. (2022). TE-HI-GCN: an ensemble of transfer hierarchical graph convolutional networks for disorder diagnosis. *Neuroinformatics* 20 (2): 353–375.

Parisot, S., Ktena, S.I., Ferrante, E. et al. (2018). Disease prediction using graph convolutional networks: application to autism spectrum disorder and Alzheimer's disease. *Medical Image Analysis* 48: 117–130.

Park, H.J. and Friston, K. (2013). Structural and functional brain networks: from connections to cognition. *Science* 342 (6158): 1238411.

Park, H.J., Kubicki, M., Westin, C.F. et al. (2004). Method for combining information from white matter fiber tracking and gray matter parcellation. *American Journal of Neuroradiology* 25 (8): 1318–1324.

Petersen, S.E. and Sporns, O. (2015). Brain networks and cognitive architectures. *Neuron* 88 (1): 207–219.

Roth, H.R., Lu, L., Liu, J. et al. (2015). Improving computer-aided detection using convolutional neural networks and random view aggregation. *IEEE Transactions on Medical Imaging* 35 (5): 1170–1181.

Sirinukunwattana, K., Raza, S.E.A., Tsang, Y.W. et al. (2016). Locality sensitive deep learning for detection and classification of nuclei in routine colon cancer histology images. *IEEE Transactions on Medical Imaging* 35 (5): 1196–1206.

Sun, L., Fan, Z., Ding, X. et al. (2019). Region-of-interest undersampled MRI reconstruction: a deep convolutional neural network approach. *Magnetic Resonance Imaging* 63: 185–192.

Wang, L., Gao, Y., Shi, F. et al. (2015). LINKS: Learning-based multi-source integration framework for segmentation of infant brain images. *Neuro Image* 108: 160–172.

Wu, Z., Pan, S., Chen, F. et al. (2020). A comprehensive survey on graph neural networks. *IEEE Transactions on Neural Networks and Learning Systems* 32 (1): 4–24.

Ying, Z., Bourgeois, D., You, J. et al. (2019). GNNExplainer: generating explanations for graph neural networks. *Advances in Neural Information Processing Systems* 32. https://doi.org/10.48550/arXiv.1903.03894.

Zhang, L., Na, S., Liu, T. et al. (2023). Multimodal deep fusion in hyperbolic space for mild cognitive impairment study. International Conference on Medical Image Computing and Computer-Assisted Intervention (674–684). Cham: Springer Nature Switzerland.

Zhang, L., Qu, J., Ma, H. et al. (2024). Exploring Alzheimer's disease: a comprehensive brain connectome-based survey. *Psychoradiology* 4: kkad033.

Zhang, L., Wang, L., and Zhu, D. (2022). Alzheimer's disease neuroimaging initiative predicting brain structural network using functional connectivity. *Medical Image Analysis* 79: 102463.

Zhang, L., Wang, L., and Zhu, D. (2020). Jointly analyzing Alzheimer's disease related structure-function using deep cross-model attention network. *2020 IEEE 17th International Symposium on Biomedical Imaging (ISBI)* (pp. 563–567). IEEE.

Zhang, L., Wang, L., and Zhu, D. (2020). Recovering brain structural connectivity from functional connectivity via multi-GCN based generative adversarial network. Medical Image Computing and Computer Assisted Intervention–MICCAI 2020: 23rd International Conference, Lima, Peru (4–8 October 2020). Proceedings, Part VII 23 (53–61). Springer International Publishing.

Zhang, L., Wang, L., Gao, J. et al. (2021). Deep fusion of brain structure-function in mild cognitive impairment. *Medical Image Analysis* 72: 102082.

Zhang, L., Zaman, A., Wang, L. et al. (2019). A cascaded multi-modality analysis in mild cognitive impairment. Machine Learning in Medical Imaging: 10th International Workshop, MLMI 2019, Held in Conjunction with MICCAI 2019, Shenzhen, China (13 October 2019). Proceedings 10 (557–565). Springer International Publishing.

Zhang, Z., Cui, P., and Zhu, W. (2020). Deep learning on graphs: a survey. *IEEE Transactions on Knowledge and Data Engineering* 34 (1): 249–270.

Zhu, Y., Ma, J., Yuan, C. et al. (2022). Interpretable learning based dynamic graph convolutional networks for Alzheimer's disease analysis. *Information Fusion* 77: 53–61.

3

Cerebrovascular Disease and Cognitive Function

Zhongzhao Teng[1,2,3*]

[1]*School of Life and Health Sciences, Fujian Fuyao University of Science and Technology, Fujian, China*
[2]*Nanjing Jingsan Medical Science and Technology, Ltd., Jiangsu, China*
[3]*Tenoke, Ltd., Cambridge, UK*

[*]Corresponding author: zt@fyust.edu.cn; zt@tenoke.com

3.1 The Close Relationship Between the Cerebrovascular System and Cognitive Function

3.1.1 An Overview of Cognitive Function

Cognitive function refers to an individual's ability to acquire, process, store and utilise external information, covering a multitude of core domains. Memory serves as a pivotal component of cognitive function, encompassing the processes of encoding, storing and retrieving information, such as committing events to memory and acquiring new knowledge (Squire & Wixted, 2011). In these processes, brain structures like the hippocampus play a critical role. Executive function pertains to the organisation, planning, decision-making and problem-solving of higher-order cognitive activities. For instance, when confronted with complex tasks, it enables individuals to systematically arrange steps and adapt flexibly to unforeseen circumstances, and it is closely associated with the prefrontal cortex. Attention forms the bedrock of cognitive activities, allowing individuals to focus their mental efforts on specific stimuli or tasks while filtering out irrelevant distractions. For example, it enables one to concentrate on reading or listening in a noisy environment. It primarily relies on the coordinated efforts of the brainstem reticular formation and the prefrontal cortex, among other regions. Language ability involves the comprehension, expression and communication of language, and it engages multiple brain areas, including the auditory and motor language centres. Damage to areas such as Broca's area and Wernicke's area can lead to different types of language disorders (Friederici, 2011). Visuospatial function is concerned with an individual's perception, localisation and discrimination of spatial relationships among objects in the surrounding visual space. It is crucial for daily activities such as walking and identifying object shapes. The parietal and occipital lobes, among other brain regions, work together in this process.

These cognitive domains collaborate with each other to jointly support an individual's daily mental and behavioural activities. Their normal functioning is highly dependent on the brain's good condition, including a stable energy supply and a suitable neural microenvironment, and the cerebrovascular system plays an irreplaceable role in this process.

3.1.2 Cerebrovascular System: Structure, Function and the Pathway for Cerebral Energy Supply

The cerebrovascular system has a unique and delicate structure to meet the brain's high demand for oxygen and glucose (Lo & Rosenberg, 2009). The arterial system mainly comes from the internal carotid arteries, vertebral arteries and their branches (Figure 3.1). They form the circle of Willis at the base of the brain. This ring-like structure regulates and compensates for cerebral blood flow, ensuring a relatively balanced and stable blood supply to the brain areas. Arteries from the circle of Willis branch into a dense cortical arterial network. Perforating arteries then carry blood to the cerebral cortex and deep structures like the basal ganglia and thalamus. Capillaries, numerous and key in the cerebrovascular system, have walls made of endothelial cells. They form the blood–brain barrier (BBB), allowing oxygen, carbon dioxide and minimal nutrients to pass while blocking harmful substances. This maintains cerebral microenvironment stability. Pericytes around capillaries also help regulate blood flow by affecting vessel contraction and dilation (Mishra et al., 2016).

The venous system, including cortical and deep veins and the confluence of sinuses, drains metabolic waste and carbon dioxide from the brain (Rasmussen et al., 2018). The internal and external jugular veins return this blood to the heart. Beyond being a transport system, the cerebrovascular system dynamically regulates the cerebral microenvironment.

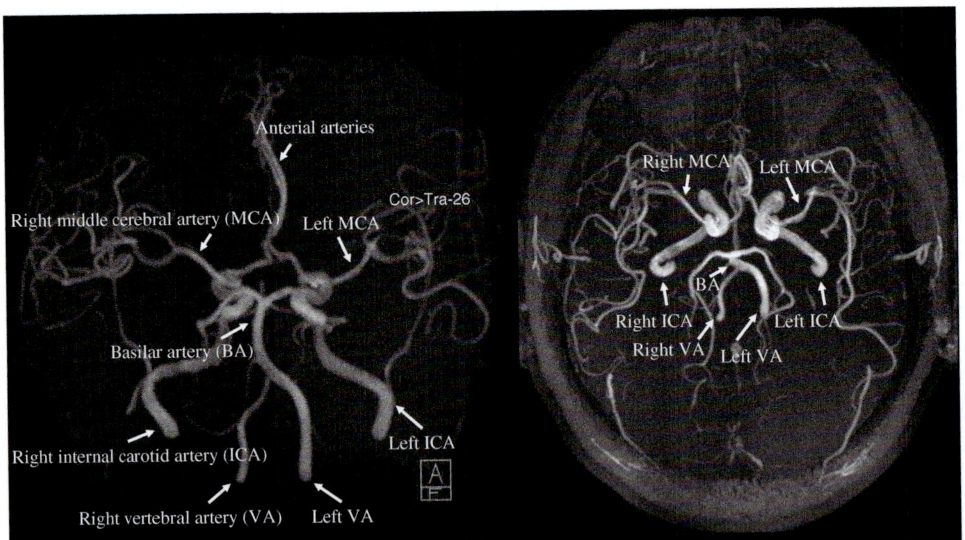

(a) Anterior view of maximum intensity projection (MIP) **(b)** Superior view of MIP

Figure 3.1 Main arteries of the cerebrovascular system (3D time-of-flight (TOF) magnetic resonance imaging (MRI); male, 64 years old).

It can adjust local blood flow via vasodilation or vasoconstriction based on metabolic activity in different brain regions. For example, during high-intensity cognitive tasks, blood vessels in active areas dilate to increase blood supply, providing more oxygen and glucose to neurons. Moreover, the cerebrovascular system aids in the circulation of cerebrospinal and interstitial fluids. It helps eliminate metabolic waste like β-amyloid (Aβ) through the brain lymphatic system, maintaining a clean and stable internal brain environment. This is crucial for preserving normal cognitive function.

3.1.3 The Evolution of Understanding the Correlation Between Cerebrovascular Disease and Dementia

In the late 19th and early 20th centuries, some neurologists and pathologists observed a connection between cerebrovascular disease and cognitive impairment (Iadecola et al., 2019; Gorelick et al., 2011). At the time, postmortem examinations revealed brain lesions in stroke patients, such as cerebral infarction and haemorrhage, and found that these patients often had cognitive decline. However, the understanding at that time was still fragmented and one-sided, without a systematic theory.

In the mid-20th century, with the gradual application of neuroimaging technologies such as computed tomography (CT) and magnetic resonance imaging (MRI), doctors could more intuitively observe cerebrovascular lesions and their effects on brain tissue in vivo (Latchaw et al., 2009). People's understanding of the relationship between cerebrovascular disease and dementia was further deepened. Research found that in addition to brain damage caused by obvious strokes, some minor cerebrovascular lesions, such as lacunar infarction and leukoaraiosis, were also related to gradual cognitive decline, prompting the gradual formation and development of the concept of vascular dementia (VaD) (Vermeer et al., 2003; Pantoni, 2010).

From the late 20th century to the early 21st century, a large number of epidemiological studies and clinical trials provided stronger support for this relationship. These studies showed that cerebrovascular disease is an important risk factor for dementia, and VaD became the second most common type of dementia after Alzheimer's disease (AD) (Gorelick et al., 2011; Nichols et al., 2022; O'Brien & Thomas, 2015). Some studies found that in elderly dementia patients, the proportion of dementia caused solely by cerebrovascular lesions and dementia co-existing with other neurodegenerative diseases was quite high. As research on the mechanisms of cognitive impairment deepened, people gradually realised that cerebrovascular lesions not only cause cognitive impairments by directly damaging brain tissue but also affect the neural microenvironment and neuron function through multiple pathways, such as chronic cerebral ischemia and hypoxia, BBB disruption and neurotransmitter changes, leading to comprehensive cognitive decline. This enriched the understanding of the correlation between cerebrovascular disease and dementia and promoted the development of related diagnostic and therapeutic research.

3.1.4 Urgency in the Context of Population Ageing

Against the backdrop of intensifying global population ageing, studying the relationship between the cerebrovascular system and cognitive function is urgently significant. As the elderly population grows, so do the incidences of cerebrovascular diseases and cognitive impairments, imposing heavy burdens on families and society. Vascular cognitive impairment

(VCI) and VaD have become major factors affecting seniors' quality of life and healthy life expectancy (Gorelick et al., 2011). Exploring their relationship aids in early identification and diagnosis of cognitive damage from cerebrovascular lesions, enabling timely interventions to slow or halt disease progression. Currently, while progress has been made in early-diagnostic technologies like high-resolution MRI and vascular ultrasound, and in treatment strategies for vascular risk factor control, many issues remain unresolved. For instance, how to precisely evaluate the potential cognitive impact of cerebrovascular lesions, develop better therapies for different VCI stages and prevent cognitive impairment via comprehensive vascular risk factor management are current research hotspots and challenges. Moreover, as research deepens on the interaction between the cerebrovascular system and neurodegenerative diseases, it may offer theoretical grounds for new therapeutic targets and strategies. This could address ageing-related cognitive challenges and enhance seniors' health and quality of life.

3.2 Neurovascular Physiology and the Neurobiological Basis of Cognition

3.2.1 Molecular Mechanisms of Neurovascular Coupling

Neurovascular coupling (NVC) refers to the tight coupling between neuronal activity and local cerebral blood flow, with complex molecular and cellular mechanisms (Lo & Rosenberg, 2009; Attwell et al., 2010; Iadecola, 2017). Nitric oxide (NO), a key signalling molecule, is synthesised by endothelial nitric oxide synthase (eNOS) in endothelial cells and neuronal nitric oxide synthase in neurons (Iadecola, 1993). It freely diffuses across cell membranes to activate guanylate cyclase in vascular smooth muscle cells, boosting cyclic guanosine monophosphate levels and triggering vasodilation to increase local cerebral blood flow. Prostaglandins (PGs) also play a role in NVC. For example, prostacyclin (PGI2) binds to prostacyclin receptors (IP receptors) on vascular smooth muscle cells, activating adenylyl cyclase, increasing cyclic AMP levels and causing vasodilation (Zonta et al., 2003; Niwa et al., 2001).

During neuronal activity, potassium ions (K^+) are released in increased amounts (Webb, 2003; Jackson, 2000). At low concentrations, K^+ causes depolarisation of vascular smooth muscle cell membranes, opening voltage-dependent calcium channels and increasing calcium ion (Ca^{2+}) influx, leading to vasoconstriction. However, when K^+ concentrations rise further, they activate potassium channels in endothelial cells, causing endothelial hyperpolarisation. This leads to the release of vasodilators like NO and PGI2, resulting in vasodilation. Glutamate, the main excitatory neurotransmitter, acts on metabotropic glutamate receptors on astrocytes after synaptic release. This activates signalling pathways like phospholipase C and protein kinase C, prompting astrocytes to release vascular-active substances such as eNOS and regulating local blood flow.

Furthermore, norepinephrine (NA) released from locus coeruleus neurons, acting on α2-adrenergic receptors (α2AR), can induce vasodilation and regulate cerebral blood flow (McBurney-Lin et al., 2019). Abnormalities in NVC mechanisms may lead to cerebral blood flow regulation disorders, thereby affecting cognitive function. For instance, in neurodegenerative diseases like AD, impaired NVC reduces cerebral blood flow, exacerbating neuronal damage and cognitive decline.

3.2.2 The Structure, Function and Role of the BBB in Maintaining Brain Homeostasis

The BBB, a key structure for maintaining brain homeostasis, is composed of brain microvascular endothelial cells, the basement membrane, pericytes and the foot processes of astrocytes (Lacoste et al., 2025). Tight junctions, a core component of the BBB, are formed by transmembrane proteins such as occludin and claudin-5, along with intracellular adaptors like ZO-1. These elements tightly link adjacent endothelial cells, restricting paracellular permeability and blocking most plasma components, pathogens and immune cells from entering the brain. Transporters are another crucial part of the BBB. For example, glucose transporter GLUT1, located on the luminal membrane of endothelial cells, transports glucose from blood to brain tissue. P-glycoprotein (P-gp), found on the abluminal membrane, effluxes various xenobiotics and endogenous substances from the brain to the blood, limiting their brain entry. The enzyme barrier, also important for the BBB (Hardebo & Owman, 1980; Joó & Várkonyi, 1969), involves endothelial cells expressing enzymes like acetylcholinesterase and monoamine oxidase. These enzymes degrade neurotransmitters and certain drugs, preventing their excessive accumulation in the brain. Through these structural and functional features, the BBB strictly regulates material exchange, maintaining brain environment stability. It selectively transports nutrients like glucose and amino acids into brain tissue and removes metabolic waste and potential neurotoxins. Moreover, the BBB has an immune surveillance function (Engelhardt & Coisne, 2011). Under normal conditions, it prevents peripheral immune cells from entering brain tissue, preserving the central nervous system's immune privilege. However, under pathological conditions like infection or inflammation, BBB permeability can increase, allowing immune cells to enter brain tissue for immune responses, which may also lead to neuroinflammation and neuronal damage.

3.2.3 The Neurogliovascular Unit: Synergy Between Astrocytes, Pericytes, Microglia and Endothelial Cells in the NVU

The neurogliovascular unit (NVU), a complex network of neurons, astrocytes, pericytes, microglia and endothelial cells, collectively regulates cerebral blood flow, BBB function, inflammation and synaptic plasticity (Lo and Rosenberg, 2009; Mishra et al., 2016; Iadecola, 2017; Zonta et al., 2003). Astrocytes, with foot processes covering vascular endothelial cells and pericytes, express proteins like AQP4 to regulate brain water balance and NVC. After neuronal activation, astrocytes release factors such as NO and PGI2 to modulate vascular smooth muscle and pericyte contraction/dilation, adjusting local cerebral blood flow. They also secrete VEGF and Ang-1 to guide BBB development and maintain its homeostasis (Abbott, 2002; Abbott et al., 2006).

Pericytes, located within capillary walls and covering nearly 100% of CNS endothelial length, interact with endothelial cells via PDGF-B signalling to preserve BBB integrity. They release cytokines and growth factors to regulate vessel contraction/dilation and cerebral blood flow. Microglia, the CNS's resident immune cells, are usually dormant but can be activated to release pro- and anti-inflammatory cytokines. Activated microglia interact with endothelial cells, astrocytes and pericytes to influence BBB permeability and cerebral blood flow (Nation et al., 2019; Haruwaka et al., 2019; Faust et al., 2025; Jha et al., 2019; Dudvarski Stankovic et al., 2016).

Endothelial cells, a key BBB component, restrict transcellular transport via tight junctions. They also express transporters and enzymes to selectively regulate material exchange. By secreting vasodilators like NO and PGI2, they maintain stable cerebral blood flow. Additionally, endothelial cells communicate with neurons, astrocytes, pericytes and microglia to coordinate NVU regulatory functions (Faust et al., 2025; Jha et al., 2019).

3.2.4 Cerebral Autoregulation: Intrinsic Mechanisms for Stable Cerebral Perfusion Pressure

Cerebral blood flow autoregulation enables cerebrovascular adjustment of cerebral blood flow within a certain arterial pressure range (50–150 mmHg), maintaining relatively stable cerebral perfusion pressure. It primarily relies on metabolic and neurogenic regulation. Metabolic regulation refers to the direct impact of cerebral metabolic activities on cerebral blood flow. For example, enhanced metabolic activity increases CO_2 and H^+ levels, triggering vasodilation to meet the brain's oxygen and nutrient needs (Mishra et al., 2016; Kontos et al., 1977). Neurogenic regulation involves neurotransmitters like norepinephrine and acetylcholine, which bind to receptors on vascular smooth muscle cells to adjust cerebrovascular tone.

This autoregulation is crucial for stable cognitive function (Ishikawa et al., 2021). It ensures sufficient brain blood supply during daily activities and physiological changes, maintaining normal neuronal activity and cognitive function. For instance, during postural changes or emotional excitement, arterial pressure fluctuates, but the autoregulation mechanism quickly stabilises cerebral perfusion pressure, preventing cognitive impairments caused by abrupt changes in cerebral blood flow (Zhang et al., 1998). Notably, impaired autoregulation is associated with various cerebrovascular diseases and cognitive disorders, such as stroke and VaD. Thus, in-depth research on cerebral blood flow autoregulation is vital for preventing and treating these conditions.

3.2.5 Cerebrovascular Reserve: Compensatory Capacity to Meet Increased Demands

Cerebrovascular reserve is the cerebrovascular system's compensatory ability to handle increased cerebral blood flow demands, involving vascular dilation and blood flow augmentation (Iadecola, 2017; Derdeyn et al., 2002; Willie et al., 2014). Its main assessment methods are CO_2 reactivity testing and acetazolamide testing. CO_2 reactivity testing assesses vascular reactivity by observing cerebral blood flow velocity changes during inhalation of different CO_2 concentrations. Normally, increased CO_2 concentration induces vasodilation and higher blood flow velocity. Reduced CO_2 reactivity indicates compromised cerebrovascular reserve. Acetazolamide, a carbonic anhydrase inhibitor, increases H^+ concentration in cerebrospinal fluid (CSF) and blood, causing vasodilation. Cerebrovascular dilation capacity can be evaluated by measuring blood flow velocity changes before and after acetazolamide administration.

Cerebrovascular reserve is crucial for stable cognitive function. During heightened cognitive activities (e.g., learning, memory, thinking), adequate cerebrovascular reserve ensures sufficient brain tissue blood supply, maintaining normal neuronal and cognitive function. When this reserve declines, as in cerebrovascular diseases, hypertension or diabetes, the brain may not meet the blood flow demands of cognitive activities, leading to cognitive decline or impairment.

In summary, the neurobiology of cerebrovascular physiology and cognition is a complex system where NVC, the BBB, the neurogliovascular unit, cerebral autoregulation and cerebrovascular reserve work together to maintain brain homeostasis and cognitive function. Investigating these mechanisms is highly significant for understanding cerebrovascular disease pathogenesis, developing new therapies and preventing cognitive impairment.

3.3 Mechanisms of Cognitive Impairment Caused by Cerebrovascular Pathological Changes

3.3.1 Chronic Cerebral Hypoperfusion: Energy Deficiency and Neuronal Dysfunction

Chronic cerebral hypoperfusion, a key mechanism linking cerebrovascular pathology to cognitive impairment, is characterised by long-term insufficient brain blood supply, restricting oxygen and energy provision (Shibata et al., 2004; Iadecola et al., 2016). Given the brain's high energy demands, met primarily through aerobic glucose metabolism to produce ATP, inadequate perfusion first impacts mitochondrial function. Mitochondria, being highly oxygen-sensitive, experience impaired function, reducing ATP production and triggering cellular dysfunction (Kim et al., 2006).

Intracellular energy deficiency impairs neuronal ion pump function, such as the Na^+/K^+-ATPase pump (Galeffi et al., 2004; Rossi et al., 2000, 2007). This leads to imbalanced ion gradients, cell depolarisation and excessive release of excitatory neurotransmitters like glutamate. Overactivation of glutamate receptors, particularly N-methyl-D-aspartate (NMDA) receptors, causes excessive Ca^{2+} influx, inducing intracellular calcium overload. Calcium overload activates calcium-dependent enzymes that break down cellular components, ultimately causing neuronal damage and death (Seeley & Woods, 2003).

Moreover, chronic cerebral hypoperfusion intensifies oxidative stress. Insufficient oxygen disrupts redox balance, increasing reactive oxygen species (ROS) generation while antioxidant enzyme activities are inadequate to neutralise them. ROS attack biomolecules, causing oxidative damage. For example, lipid peroxidation from ROS reduces membrane fluidity and increases permeability, disrupting normal cellular function (Hirata et al., 2023).

Chronic cerebral hypoperfusion also triggers inflammation. Under ischemic and hypoxic conditions, inflammatory cytokines like IL-1β and TNF-α are upregulated (Liu et al., 1994). They recruit immune cells such as microglia and peripheral immune cells to the affected area, releasing more inflammatory mediators and forming a damaging cascade. This inflammation not only directly injures neurons but also compromises the BBB, allowing more inflammatory substances and immune cells into the brain, exacerbating tissue damage (Ding et al., 2019; Tomimoto et al., 2003; Zhang et al., 2020).

Chronic cerebral hypoperfusion also disrupts neurotransmitter metabolism and transmission (Tanaka et al., 2000; Zhao et al., 2014). For instance, synthesis of dopamine and acetylcholine is reduced, and their reuptake and degradation processes are affected, leading to neurotransmitter system dysfunction. This disruption impairs neuronal signalling and network function, consequently affecting cognitive abilities. Studies show chronic cerebral hypoperfusion is closely related to various cognitive-impairment diseases like VCI and AD. In these diseases, the aforementioned pathological processes initiated by chronic cerebral hypoperfusion jointly contribute to the brain's gradual cognitive decline.

3.3.2 Microvascular Hemodynamic Changes and BBB Disruption

Changes in the microvascular hemodynamic environment are crucial in cerebrovascular pathology, especially in small vessel diseases (SVDs) (Iadecola et al., 2019; Gorelick et al., 2011; Pantoni, 2010; Ter Telgte et al., 2018; Dichgans & Leys, 2017). Normally, perivascular tissues support microvessels and stabilise blood flow. However, pathological factors like chronic inflammation, oxidative stress and ageing can weaken this support, causing microvascular tortuosity.

Microvascular tortuosity alters blood flow dynamics, leading to slower and uneven flow with changed shear stress. These changes affect endothelial cell function and structure (Dong et al., 2024; Zhang et al., 2018; Dai et al., 2004). Normally, endothelial cells form tight junctions to maintain BBB integrity, but altered haemodynamics disrupt this, increasing BBB permeability.

Slowed blood flow prolongs nutrient and oxygen exchange in microvessels but also increases contact time between harmful substances and endothelial cells, raising their damage risk. Altered shear stress affects endothelial cell cytoskeletal structure and intercellular connections, causing tight-junction protein abnormalities, widening endothelial gaps and increasing BBB permeability.

Microvascular tortuosity can also cause local ischemia and hypoxia, activating pathophysiological responses like free-radical production and inflammatory factor release. These further damage endothelial cells and exacerbate BBB disruption. Inflammatory factors (e.g., IL-1, TNF-α) bind to endothelial receptors, activate intracellular signalling, increase cell-adhesion-molecule expression and promote leukocyte adherence and transmigration, triggering neuroinflammation. This compromises brain tissue homeostasis, impairs neuronal function and leads to cognitive deficits.

In cerebral amyloid angiopathy (CAA), Aβ deposition in microvascular walls causes structural and functional abnormalities, making microvessels fragile and prone to microbleeds (Cozza et al., 2023; Yamada, 2015). CAA also disrupts the microvascular hemodynamic environment, impairing the BBB and hindering Aβ clearance, which aggravates Aβ deposition in brain tissue. This creates a vicious cycle of cognitive impairment. Thus, microvascular tortuosity and hemodynamic microenvironment changes are key mechanisms of cognitive impairment in cerebrovascular pathology. Research in this area may help find effective ways to diagnose and treat chronic cerebrovascular-related cognitive disorders.

3.3.3 BBB Disruption: Leakage, Inflammation and Neurotoxin Invasion

The BBB is a key structure of the central nervous system, composed of brain capillary endothelial cells, pericytes, astrocytic foot processes and the basement membrane. It plays a crucial role in maintaining brain homeostasis and ensuring the normal function of neurons. In cerebrovascular diseases, the disruption of the BBB is a common and important pathological process that leads to neuronal damage and cognitive impairment through various mechanisms (Lacoste et al., 2025).

When the BBB is disrupted, blood components such as plasma proteins and immune cells leak into brain tissue, triggering a series of inflammatory responses. For example, leaked fibrinogen can be converted into an active form in brain tissue, activating the complement system to produce anaphylatoxins like C3a and C5a (Merlini et al., 2019). These attract inflammatory cells such as neutrophils and monocytes to the brain tissue. These inflammatory cells release a large number of inflammatory mediators, including cytokines (such as IL-6, TNF-α), chemokines ROS and reactive nitrogen species (RNS). These mediators not only further damage endothelial cells and exacerbate BBB disruption but also directly affect neurons, causing dysfunction or even death.

Moreover, the disruption of the BBB allows neurotoxic substances in the blood to enter brain tissue. For example, heavy metal ions (such as iron and copper), metabolic waste (such as uremic toxins) and some exogenous toxins (such as environmental pollutants) are normally blocked by the BBB. However, they can easily enter the brain when the BBB is disrupted. These neurotoxic substances accumulate in the brain, interfering with the normal metabolism and function of neurons. For example, excessive accumulation of iron ions in the brain can enhance oxidative stress reactions, promote the generation of free radicals and damage biological macromolecules such as cell membranes, DNA and proteins, leading to neurodegeneration (Tran et al., 2022; Ndayisaba et al., 2019; Bergsland et al., 2019). This is closely related to the occurrence of cognitive impairment diseases such as AD.

The mechanisms of neuronal damage caused by inflammatory responses following BBB disruption are multifaceted. On one hand, ROS and RNS released by inflammatory cells lead to lipid peroxidation, protein oxidation and nitration, and DNA damage, thereby disrupting cellular structure and function. On the other hand, proteases of inflammatory cell origin, such as matrix metalloproteinases, can degrade the extracellular matrix and structures on neuronal cell membranes, including receptors and ion channels, which in turn affects signal transmission between neurons. Meanwhile, inflammatory mediators can activate microglia, causing them to transition from a resting state to an activated state (Cunningham et al., 2009). This leads to the release of more inflammatory factors and neurotoxic substances, creating an inflammatory cascade that further exacerbates neuronal damage.

The disruption of the BBB also triggers cerebral oedema (Krueger et al., 2015), which increases intracranial pressure and compresses surrounding neural tissue. This might result in mechanical injury to neurons and changes in haemodynamics, further affecting cerebral blood and oxygen supply and worsening neuronal energy metabolism disorders. Moreover, BBB disruption disrupts the homeostasis of the neurovascular unit, affecting the normal interactions between neurons and vascular endothelial cells, astrocytes and other cells.

This interferes with the NVC mechanism, preventing neuronal activity from effectively regulating local cerebral blood flow and thereby impacting normal neuronal function and neural network activity.

For example, in ischemia-reperfusion injury of the brain, BBB disruption following reperfusion allows a large number of inflammatory cells and mediators to enter brain tissue, triggering an inflammatory storm that causes extensive neuronal damage (Tang et al., 2023; Aoki et al., 2002; Zhang M. et al., 2024). This leads to irreversible brain tissue injury and severe cognitive impairment. Studies have shown that increased BBB permeability is positively correlated with the degree of cognitive impairment (Zhang et al., 2017; Montagne et al., 2015). In various cerebrovascular diseases, such as cerebral infarction, cerebral haemorrhage and CAA, BBB disruption is one of the key factors leading to cognitive decline. Therefore, in-depth research into the mechanisms of BBB disruption and its role in cerebrovascular pathology is of great significance for finding effective strategies to treat cognitive disorders.

3.3.4 Microvascular Changes and Endothelial Dysfunction: Oxidative Stress, Inflammation and Nitric Oxide Imbalance

Microvascular changes are an important component of cerebrovascular diseases, and one of the core mechanisms is endothelial dysfunction, in which oxidative stress, inflammatory responses and nitric oxide (NO) imbalance play key roles. Under normal conditions, vascular endothelial cells release various vasoactive substances, such as nitric oxide (NO) and PGI2, to maintain the balance of vascular dilation and contraction, regulate vascular permeability and inhibit platelet aggregation and leukocyte adhesion, thereby ensuring the normal operation of the microcirculation and adequate tissue perfusion.

Oxidative stress is one of the important factors leading to endothelial dysfunction (Nakagami, 2003; Taddei et al., 1998). In cerebrovascular lesions, due to various causes such as hypertension, hyperglycaemia, smoking and ageing, the production of ROS and RNS increases, or the function of the antioxidant defence system declines, leading to oxidative stress. The large amounts of ROS and RNS generated by oxidative stress can directly damage the biological macromolecules in endothelial cells, such as lipids, proteins and DNA, disrupting the structure and function of endothelial cells. For example, ROS can oxidise the unsaturated fatty acids in the endothelial cell membrane, producing lipid peroxides, which reduce the fluidity and increase the permeability of the cell membrane, thereby affecting the normal physiological functions of endothelial cells (Villa et al., 1994; Hennig, 1988). In addition, ROS can also oxidatively modify low-density lipoprotein (LDL) to form oxidised LDL (ox-LDL). After being taken up by endothelial cells, ox-LDL accumulates within the cells, triggering endoplasmic reticulum stress and mitochondrial dysfunction, further exacerbating endothelial cell damage.

Inflammatory responses and oxidative stress are intertwined, jointly promoting the development of endothelial dysfunction. Inflammatory factors such as interleukin-1 (IL-1), tumour necrosis factor-α (TNF-α) and interleukin-6 (IL-6) are elevated in cerebrovascular lesions (Liu et al., 1994; Zafar et al., 2021). These inflammatory factors can activate signal pathways such as nuclear factor-κB (NF-κB) within endothelial cells, inducing increased expression of cell adhesion molecules like vascular cell adhesion molecule-1 (VCAM-1) and intercellular adhesion molecule-1 (ICAM-1) (Chu et al., 2025; Kempe, 2005; Tozawa, 1996). This process facilitates the adhesion and migration of leukocytes to the vascular wall, triggering inflammatory responses within the vessel wall. Inflammatory cells within the vascular wall release more ROS and inflammatory mediators, creating a vicious cycle that further damages endothelial cells and exacerbates microvascular changes.

Nitric oxide (NO) plays a crucial role in maintaining endothelial function, and NO imbalance is another key factor in microvascular changes (Cyr et al., 2020; Lundberg & Weitzberg, 2022). NO is catalysed by eNOS in endothelial cells and diffuses to vascular smooth muscle cells, causing vasodilation, reducing vascular resistance and increasing blood flow. Additionally, NO has effects such as inhibiting platelet aggregation, anti-leukocyte adhesion and antioxidant stress. However, in cerebrovascular lesions, the activity and expression of eNOS are inhibited due to oxidative stress and inflammatory responses, leading to reduced NO production. Moreover, excessive ROS can react with NO to form peroxynitrite, not only neutralising the effects of NO but also causing further oxidative stress damage. This deficiency and imbalance of NO leads to dysfunction in the contraction and relaxation of microvessels, enhanced vasoconstriction, reduced blood flow perfusion, and consequently, tissue ischemia and hypoxia, thereby worsening brain tissue damage.

For example, in cerebral small vessel disease (CSVD), microvascular changes include arteriosclerosis, tortuosity of microvessels and reduced capillary density, all of which are closely related to endothelial dysfunction. Endothelial cell damage caused by oxidative stress and inflammatory responses impairs the autoregulation of blood vessels, preventing them from adjusting blood flow in a timely manner according to the metabolic needs of brain tissue, resulting in chronic cerebral hypoperfusion. Moreover, damaged endothelial cells release pro-fibrotic factors, promoting fibrosis and thickening of the vascular wall (Zeisberg et al., 2007), further exacerbating microvascular changes. Studies have shown that a decrease in NO is positively correlated with the severity of CSVD (Ji et al., 2024; Liao et al., 2021). Improving endothelial function and increasing NO production and bioavailability may become potential strategies for treating CSVD and its associated cognitive impairments.

Microvascular changes and endothelial dysfunction interact through mechanisms such as oxidative stress, inflammatory responses and nitric oxide imbalance, jointly leading to ischemia, hypoxia and inflammatory damage in brain tissue, thereby impairing cognitive function. In-depth research into these mechanisms can help develop early diagnostic and intervention measures for microvascular changes, delaying or preventing the decline of cognitive function.

3.3.5 White Matter Hyperintensities/Lesions: Myelin Damage, Axonal Conduction Impairment and Neural Network Disruption

White matter hyperintensities (WMH)/lesions are one of the important radiological features of CSVD (Figure 3.2), and they conceal a series of complex pathological and physiological processes such as myelin damage, axonal conduction impairment and neural network disruption, which have a profound impact on cognitive function (Shibata et al., 2004; Altschuler et al., 2025; Lee et al., 2016; Lim et al., 2025). White matter is primarily composed of myelin sheaths of nerve fibres, and myelin plays a crucial role in the rapid

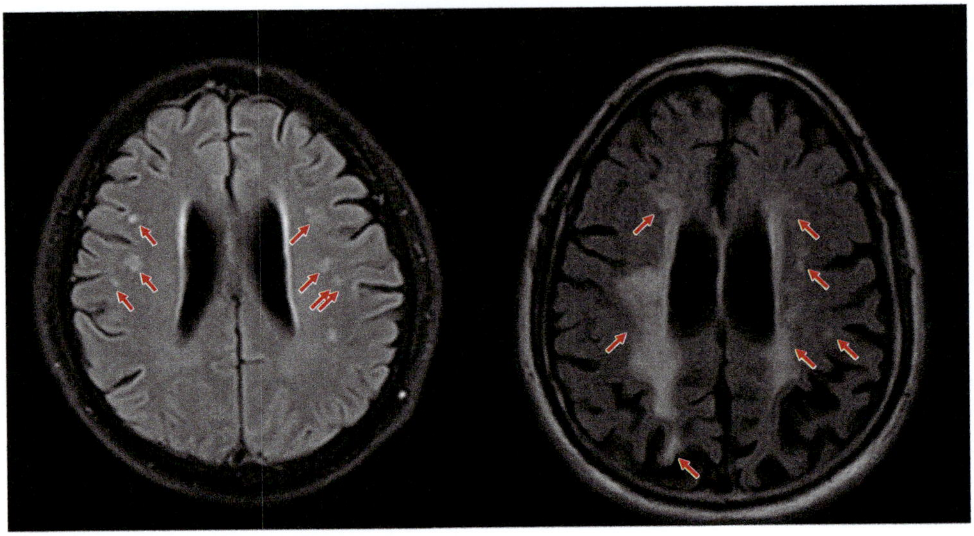

(a) Deep and isolated white matter hyperintensities (WMHs) **(b)** Fused deep and periventricular WMHs

Figure 3.2 Periventricular and deep white matter hyperintensities (red arrows) (T_2 FLAIR MRI; (a): male, 59 years old; (b): male, 80 years old).

conduction of nerve signals. Its integrity is vital for maintaining the brain's efficient information processing and cognitive function.

In SVD of the brain, structural and functional abnormalities of the microvessels lead to chronic ischemia and hypoxia in the white matter regions. The long-term state of low perfusion causes damage to oligodendrocytes in the white matter, which are key cells in the formation and maintenance of myelin (Tomimoto et al., 2003). Dysfunction of oligodendrocytes leads to reduced synthesis of myelin and decreased repair capacity. Meanwhile, the maintenance of existing myelin is also affected, resulting in demyelination and myelin damage. Myelin damage directly leads to a reduction in nerve conduction velocity, affecting the efficiency of nerve signal transmission and slowing down the exchange of information between different regions of the brain (Petrova et al., 2018). This, in turn, affects various aspects of cognitive function, such as attention, executive function and memory.

In addition, ischemia and hypoxia in white matter can trigger oxidative stress and inflammatory responses, further exacerbating damage to myelin and axons (Xu et al., 2025; Farkas et al., 2004). The large amounts of ROS generated by oxidative stress attack the lipid components of myelin, causing lipid peroxidation and disrupting the structure and function of myelin. During the inflammatory response, activated microglia and infiltrating peripheral immune cells release a variety of inflammatory factors, such as IL-1 and TNF-α. These factors not only damage oligodendrocytes but also directly act on axons, causing axonal swelling, rupture and loss. Axonal damage can severely affect the conduction of nerve signals and even lead to the interruption of connections between neurons, disrupting the integrity of neural networks.

White matter lesions also cause changes in the topological structure of neural networks (Xin et al., 2024; Zhang et al., 2014; Lin et al., 2019). The normal brain neural network has small-world characteristics and an efficient connection pattern, which allows information to be transmitted quickly and effectively between different brain regions. However, axonal damage and demyelination caused by white matter lesions can reduce the strength of connections in the neural network, increase path length and decrease network efficiency, thereby affecting the overall function and cognitive abilities of the brain. For example, studies have shown that the severity of WMH is positively correlated with the degree of cognitive impairment (Kan et al., 2020; Guo & Shi, 2022; Hu et al., 2021), especially in terms of executive function and information processing speed.

In addition, white matter lesions can also trigger secondary neuronal damage. Due to the disruption of neural networks, the normal input and output between neurons are disturbed, leading to abnormal neuronal activity. Long-term abnormal activity can cause metabolic disorders and functional impairments in neurons, further exacerbating cognitive dysfunction. Moreover, white matter lesions can also affect the metabolism and transmission of neurotransmitters, such as the synthesis and release of dopamine and norepinephrine (Mather & Harley, 2016; Li X. et al., 2024). These neurotransmitters play important roles in regulating cognitive functions, and their decreased levels can lead to a comprehensive decline in cognitive abilities.

In summary, WMH/lesions lead to cognitive impairment through multiple mechanisms, including myelin damage, axonal conduction impairment and neural network disruption. These pathological changes accumulate gradually during the development of CSVD, eventually leading to significant cognitive impairments. In-depth research into the mechanisms of white matter lesions is of great significance for the early diagnosis and intervention of cognitive impairments related to CSVD and may provide a theoretical basis for the development of new therapeutic strategies.

3.3.6 Microbleeds and Microinfarcts: Subtle Structural Damage and Cumulative Effects

Microbleeds and microinfarcts are common, subtle structural injuries in cerebrovascular diseases and play an important role in the process of chronic and subtle cognitive impairment. Microbleeds refer to haemorrhagic foci in the brain parenchyma with a diameter of less than 5 mm, usually caused by structural damage to the microvascular wall leading to the extravasation of red blood cells into the surrounding tissue. Microinfarcts (Figure 3.3), on the other hand, are small areas of infarction caused by microvascular occlusion, with a lesion diameter generally less than 3 mm. Although these tiny lesions may not cause obvious clinical symptoms individually, over time, their cumulative effects can gradually cause widespread damage to brain tissue and affect cognitive function.

Microbleeds lead to the deposition of hemosiderin and the accumulation of iron in the local brain tissue. Hemosiderin is a strong oxidant that can trigger oxidative stress reactions in brain tissue, producing large amounts of ROS (Schipper et al., 2019; Wu & Hsieh, 2022; Ryter, 2019). ROS attack biological macromolecules such as lipids, proteins and DNA in cell membranes, causing destruction of cell structure and function. In addition, the accumulation of iron can also promote the generation of free radicals, further exacerbating oxidative stress damage (Ndayisaba et al., 2019; Bergsland et al., 2019). This oxidative stress reaction not only directly damages neurons and glial cells but also destroys the integrity of

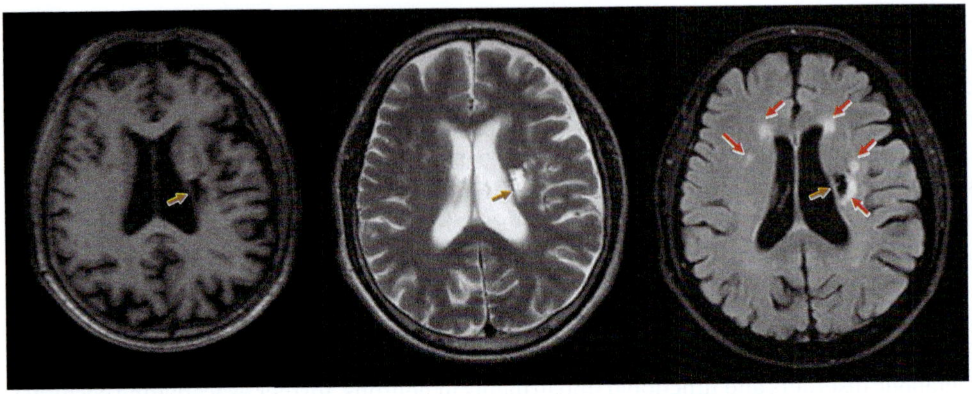

(a) T_1-weighted image **(b)** T_2-weighted image **(c)** T_2 FLAIR

Figure 3.3 Lacunal lesion (yellow arrow) and white matter hyperintensities (red arrow) (male, 74 years old).

the BBB, allowing more harmful substances to enter brain tissue, intensifying inflammatory reactions and neurotoxicity, and ultimately leading to neuronal death and cognitive decline (Tran et al., 2022).

Microinfarcts cause local brain tissue ischemia and hypoxia. Although the lesions of microinfarcts are small, they are numerous and widely distributed, especially in the white matter and subcortical regions. Long-term microinfarcts can lead to chronic ischemic state of brain tissue, affecting the metabolism and function of neurons. Neurons in the ischemic area will suffer from dysfunction due to insufficient energy supply, and ischemia will also trigger inflammatory reactions, releasing inflammatory mediators such as IL-1 and TNF-α (Liu et al., 1994). These inflammatory mediators can further damage the surrounding neural tissue and expand the scope of damage. In addition, microinfarcts can also lead to the interruption of neural network connections, affecting the efficiency of information transmission, and thereby affecting cognitive function (Yin et al., 2025; Zhang et al., 2021).

The cumulative effects of microbleeds and microinfarcts gradually lead to brain tissue atrophy and a reduction in the number of neurons. As the number of lesions increases, the affected brain tissue areas expand continuously, the degree of neural network disruption worsens, and cognitive impairment gradually intensifies. Studies have shown that the number of microbleeds and microinfarcts is positively correlated with the severity of cognitive impairment (Pantoni, 2010; Yin et al., 2025; Zhang et al., 2021), which is more evident in executive function, memory and attention. Moreover, microbleeds and microinfarcts may interact with each other, jointly promoting brain tissue damage. For example, local oxidative stress and inflammatory reactions after microbleeds increase the fragility of microvessels, making them more susceptible to microinfarcts; while the ischemia and hypoxia in the microinfarct area further exacerbate the damage to the microvascular wall, increasing the risk of microbleeds, forming a vicious cycle and further worsening cognitive impairment.

In summary, as subtle structural injuries, microbleeds and microinfarcts lead to cognitive impairment through multiple mechanisms, including oxidative stress, inflammatory reactions, neural network disruption and cumulative effects in cerebrovascular diseases. These lesions accumulate gradually in chronic cerebrovascular diseases and eventually trigger significant cognitive impairments. Therefore, early detection and intervention of microbleeds and microinfarcts are of great significance for the prevention and treatment of VCI.

3.3.7 Neuroinflammation: Microglial Activation, Cytokine Release and Synaptic Damage

Neuroinflammation is one of the important mechanisms by which cerebrovascular pathological changes impair cognitive function (Heneka et al., 2025). The activation of microglia, the release of cytokines and synaptic damage play a central role in this process. Microglia are the primary immune cells in the central nervous system, and they are normally in a resting state, monitoring the surrounding environment. When cerebrovascular lesions occur, such as ischemia, haemorrhage or disruption of the BBB, microglia are activated and transformed into a pro-inflammatory phenotype (M1 type), releasing large amounts of pro-inflammatory cytokines and inflammatory mediators, triggering neuroinflammatory responses.

In cerebrovascular diseases, damage signals such as ischemia and hypoxia, oxidative stress and cellular debris can prompt microglia to shift from a resting to an activated state (Yenari et al., 2006; Song et al., 2022). Activated microglia release pro-inflammatory cytokines such as interleukin-1β (IL-1β), TNF-α and IL-6. These cytokines not only attract more immune cells, such as T lymphocytes and monocytes, into brain tissue, exacerbating the inflammatory response, but also directly affect neurons, leading to their dysfunction and death. For instance, TNF-α can bind to receptors on the neuronal surface, activating apoptotic signalling pathways within the cell and inducing neuronal apoptosis (Livne-Bar et al., 2016; Yang et al., 2015; Dey & Snow, 2007). Meanwhile, inflammatory mediators like ROS and RNS can cause oxidative damage to neuronal cell membranes, proteins and DNA, further worsening neuronal injury.

Neuroinflammation also leads to synaptic damage, which is of significant importance in cognitive impairment. Synapses are the key structures for information transfer between neurons, and their functional integrity is crucial for cognitive functions such as learning and memory. Cytokines and inflammatory mediators released by inflammatory cells can affect the structure and function of synapses (Lee et al., 2025; Zipp et al., 2023). For example, IL-1β and TNF-α can interfere with the neurotransmitter release mechanism at presynaptic terminals, reducing neurotransmitter release, while also affecting the expression and function of postsynaptic receptors, decreasing synaptic transmission efficiency. In addition, inflammatory mediators can induce changes in synaptic structure, such as the atrophy and disappearance of dendritic spines, leading to a reduction in synaptic number and decreased connectivity of neural networks, thereby affecting cognitive function.

There is an interaction between neuroinflammation and cerebrovascular disease. On the one hand, neuroinflammation caused by cerebrovascular disease can further damage cerebral vascular endothelial cells, disrupt the BBB and lead to the entry of more inflammatory cells and mediators into brain tissue, exacerbating the inflammatory response and damage in brain tissue. On the other hand, neuroinflammation can also worsen the pathological changes in cerebral vessels, such as promoting the proliferation and migration of vascular smooth muscle cells, leading to thickening of the vessel wall and narrowing of the lumen, which further intensifies the ischemic and hypoxic state of brain tissue, creating a vicious cycle. For example, in CAA, the deposition of Aβ in the vessel wall activates microglia, triggering a neuroinflammatory response. In turn, the inflammatory response promotes further deposition of Aβ and the progression of vascular lesions, worsening cognitive impairment.

In addition, neuroinflammation can also induce changes in neuroplasticity (Calderone et al., 2024). Neuroplasticity refers to the ability of the nervous system to adjust its structure and function to adapt to environmental changes and is the basis for learning and memory. In a chronic inflammatory state, inflammatory mediators can interfere with signalling pathways related to neuroplasticity, such as the expression and function of brain-derived neurotrophic factor (BDNF), affecting synaptic plasticity and the remodelling of neural networks, leading to a decline in cognitive function. Studies have shown that inhibiting the overactivation of microglia and inflammatory responses can reduce synaptic damage and improve cognitive function, providing a potential therapeutic target for the treatment of VCI.

In summary, neuroinflammation plays an important role in cognitive impairment caused by cerebrovascular pathological changes through mechanisms such as microglial activation, cytokine release and synaptic damage. In-depth research into this mechanism can help develop therapeutic strategies targeting neuroinflammation to delay or prevent the decline of cognitive function.

3.3.8 Impaired Aβ Clearance: The Role of the Cerebrovascular System in AD Pathology

Impaired clearance of Aβ is one of the important pathological mechanisms of AD, and the cerebrovascular system plays a key role in this process (Chen et al., 2025; Tsoy et al., 2024; Huang et al., 2024). Under normal conditions, Aβ is primarily cleared through various pathways of the cerebrovascular system, including convective diffusion in CSF, clearance driven by the contraction of cerebral vascular smooth muscle cells and transport across the BBB. However, when the cerebrovascular system is affected by pathological changes, such as CAA, chronic cerebral hypoperfusion or disruption of the BBB, the clearance of Aβ is severely impacted, leading to abnormal deposition of Aβ in the brain and promoting the pathological progression of AD.

CAA is a condition caused by the deposition of Aβ in the walls of cerebral blood vessels and is commonly seen in patients with AD (Greenberg et al., 2020). CAA can damage the structure and function of blood vessels, causing the vessel walls to thicken and become more brittle, which in turn leads to the expansion of the perivascular spaces and affects the flow and convective diffusion of CSF, thereby hindering the clearance of Aβ. In addition, CAA can also disrupt the integrity of the BBB, impeding the transport of Aβ across the BBB. Under normal circumstances, endothelial cells of the BBB use specific transport proteins, such as P-glycoprotein (P-gp) and low-density lipoprotein receptor-related protein 1 (LRP1), to transport Aβ from the brain to the blood. However, when the BBB is disrupted, the function of these transport proteins is affected, reducing the efficiency of Aβ clearance and leading to the accumulation of Aβ in brain tissue.

Chronic cerebral hypoperfusion is also one of the important factors that lead to impaired Aβ clearance. A reduction in cerebral blood flow slows down the convective diffusion of Aβ in CSF and its transport across the BBB (Elbert et al., 2022), prolonging the residence time of Aβ in the brain and increasing the risk of its deposition. Moreover, hypoperfusion can cause hypoxia and metabolic disturbances in brain tissue, affecting the metabolism and clearance of Aβ. For example, mitochondrial dysfunction can lead to abnormal metabolism of Aβ within cells (D'Alessandro et al., 2025), promoting its aggregation inside cells and secretion outside cells, further exacerbating Aβ deposition.

The disruption of the BBB not only affects the transport of Aβ across the BBB but also triggers inflammatory responses, further exacerbating Aβ deposition and neurotoxicity. Inflammatory factors such as TNF-α and IL-1β can inhibit the activity of enzymes involved in Aβ clearance, such as insulin-degrading enzyme and neprilysin (NEP), which normally participate in the degradation and clearance of Aβ (Attiq et al., 2025). Moreover, inflammatory responses can promote the aggregation and fibrillation of Aβ, enhancing its neurotoxicity. Aβ deposition, in turn, activates microglia and astrocytes, leading to the release of more inflammatory mediators, creating a vicious cycle that accelerates the pathological process of AD.

Lesions in the cerebrovascular system also lead to abnormal distribution of Aβ within the brain (Verma et al., 2023; Che et al., 2025). In a healthy brain, Aβ primarily deposits around large vessels in the meninges and cortical surface. However, when the cerebrovascular system is dysfunctional, Aβ is more likely to deposit around small vessels and capillaries within the brain parenchyma, forming typical AD senile plaques. These plaques interfere with signal transmission between neurons, triggering neuroinflammation and oxidative stress responses, leading to neuronal damage and death.

Furthermore, the cerebrovascular system plays a crucial role in maintaining the stability of the brain's internal environment. Dysfunction of the cerebrovascular system affects the normal function of the neurovascular unit, preventing neurons from effectively regulating local cerebral blood flow, thereby impacting neuronal energy supply and metabolism and exacerbating Aβ production and deposition. Studies have shown that improving cerebrovascular function and promoting Aβ clearance may be potential strategies for treating AD (Chen et al., 2025; Bonnar et al., 2025; Zhao et al., 2025). For example, pharmacological interventions to enhance cerebrovascular dilation, improve the integrity of the BBB and promote the transport and degradation of Aβ may slow the pathological progression of AD and improve cognitive function. Therefore, in-depth research into the role of the cerebrovascular system in Aβ clearance impairment is of great significance for the development of effective treatments for AD.

In summary, lesions in the cerebrovascular system lead to Aβ clearance impairment through multiple mechanisms, thereby promoting the pathological process of AD and cognitive impairment. These mechanisms include vascular structural and functional abnormalities caused by CAA, clearance slowing due to chronic cerebral hypoperfusion, transport obstruction caused by BBB disruption, and exacerbated inflammatory responses. A thorough understanding of these mechanisms is of great clinical significance for elucidating the pathogenesis of AD and developing new therapeutic strategies.

3.4 Clinical Spectrum of Cerebrovascular Disease and Cognitive Impairment

3.4.1 Post-stroke Cognitive Impairment: Pathogenesis, Risk Factors, Clinical Manifestations and Prognosis

Cerebrovascular disease is one of the common causes of cognitive impairment, among which post-stroke cognitive impairment (PSCI) is an important clinical issue. Approximately one-third of stroke patients experience a decline in cognitive function after a stroke, which significantly affects the patients' quality of life and rehabilitation outcomes (Lo et al., 2022).

- **Pathogenesis**
  The pathogenesis of PSCI is complex and involves multiple aspects. First, brain tissue damage caused by stroke is the direct cause. For example, large-area cerebral infarction can destroy the cerebral cortex and subcortical structures, leading to impaired corresponding cognitive functions. Second, abnormal cerebral blood flow perfusion after stroke, including reduced local cerebral blood flow and dysfunction of cerebral vascular autoregulation, can further exacerbate cerebral ischemia and hypoxia, affecting the metabolism and function of nerve cells. In addition, inflammatory responses play an important role in PSCI. The release of inflammatory factors in the brain after stroke triggers neuroinflammation, which damages nerve cells and synapses, interferes with the synthesis and release of neurotransmitters, and thereby affects cognitive function. Moreover, post-stroke neurotransmitter disorders, such as reduced synthesis or release of neurotransmitters like acetylcholine and dopamine, can also lead to cognitive decline.

- **Risk Factors**
  Multiple risk factors are closely related to the occurrence of PSCI (Filler et al., 2024, 2025; Chau et al., 2023). Age is one of the important factors. As age increases, the risk of cerebrovascular disease rises, and the brain's compensatory capacity declines, making it easier for cognitive impairment to occur after a stroke. Vascular risk factors such as hypertension, diabetes and hyperlipidaemia not only increase the risk of stroke but also accelerate the progression of cerebrovascular disease, exacerbate brain tissue damage and promote the occurrence of cognitive impairment. In addition, low education level, limited cognitive reserve and pre-stroke mild cognitive impairment are also risk factors for PSCI.

- **Clinical Manifestations**
  The clinical manifestations of PSCI are diverse and can involve multiple cognitive domains. Common symptoms include memory decline, especially in recent memory, where patients may forget events or conversations that just occurred. There is also difficulty in concentrating, making it hard to maintain focus for extended periods and complete tasks without being easily distracted. Executive function is impaired, showing as weakened abilities in planning, organising and problem-solving, such as struggling to arrange daily activities or handle complex work tasks. Language abilities may be affected, with symptoms like word-finding difficulties, non-fluent speech or decreased comprehension. In addition to these cognitive symptoms, patients may also

experience emotional disorders, such as depression and anxiety, which can further exacerbate cognitive dysfunction.

- **Prognosis**
  The prognosis for PSCI is generally poor, with some patients gradually progressing to VaD. The prognosis is related to the severity of the stroke, the scope and degree of cognitive impairment and whether timely rehabilitation treatment is provided. Patients with mild cognitive impairment may experience some improvement or stabilisation of cognitive function if they undergo early, active rehabilitation training and cognitive exercises. However, for patients with severe cognitive impairment, the prognosis is often unfavourable, with continuous decline in cognitive function, gradual loss of self-care abilities and the need for long-term care and assistance.

3.4.2 Cognitive Impairment Related to CSVD: Imaging Markers (WMH, Lacunes, CMBs, EPVS, Brain Atrophy) and Clinical Phenotypes

CSVD is the most common cause of VCI and VaD. CSVD encompasses a variety of small vessel lesions in the brain, and its imaging markers and clinical phenotypes are characteristic (Cui et al., 2024).

- **Imaging Markers**
  - WMH: One of the most common imaging manifestations of CSVD (Jochems et al., 2025), WMH, appear as hyperintense lesions in the white matter regions on MRI (Figure 3.2), mainly located around the lateral ventricles and in the deep white matter. WMH reflect demyelination and axonal loss in the white matter caused by SVD and are associated with pathological processes such as reduced cerebral blood flow perfusion and disruption of the BBB. The extent and severity of WMH are positively correlated with the severity of cognitive impairment, especially in terms of executive function, information processing speed and memory.
  - Lacunes: These are small, round or oval, asymptomatic cerebral infarcts (Figure 3.3), usually less than 15 mm in diameter, mainly distributed in deep brain regions such as the basal ganglia, thalamus, pons and corona radiata. The number and location of lacunes are closely related to cognitive impairment. For example, multiple lacunes can affect several neural circuits, leading to cognitive damage in multiple aspects such as executive function, attention and memory.
  - Cerebral microbleeds (CMBs): Displayed as small, round, hypointense foci in the brain parenchyma on gradient echo sequences of MRI, representing microvascular ruptures and hemosiderin deposition in small vessels. CMBs are associated with CAA and hypertensive SVD, and their distribution and number in the brain have different impacts on cognitive function. For instance, CMBs in the subcortical region may have a greater impact on executive function and information processing speed, while those in the cortex are related to memory decline.
  - Enlarged perivascular spaces (EPVS): These are the enlargements of the perivascular spaces around small vessels in the brain (Figure 3.4), appearing as linear or punctate signal voids in the periventricular and deep white matter regions on MRI.

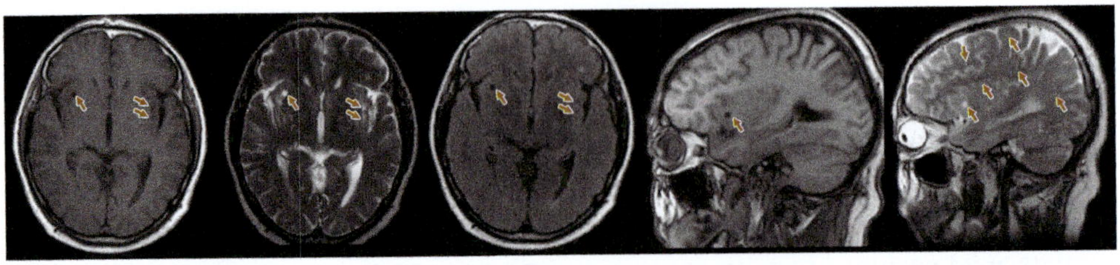

(a) T_1-weighted image (axial view) **(b)** T_2 -weighted image (axial view) **(c)** T_2 FLAIR (axial view) **(d)** T_1-weighted image (sagittal view) **(e)** T_2-weighted image (sagittal view)

Figure 3.4 Enlarged perivascular spaces (yellow arrow) (male, 57 years old).

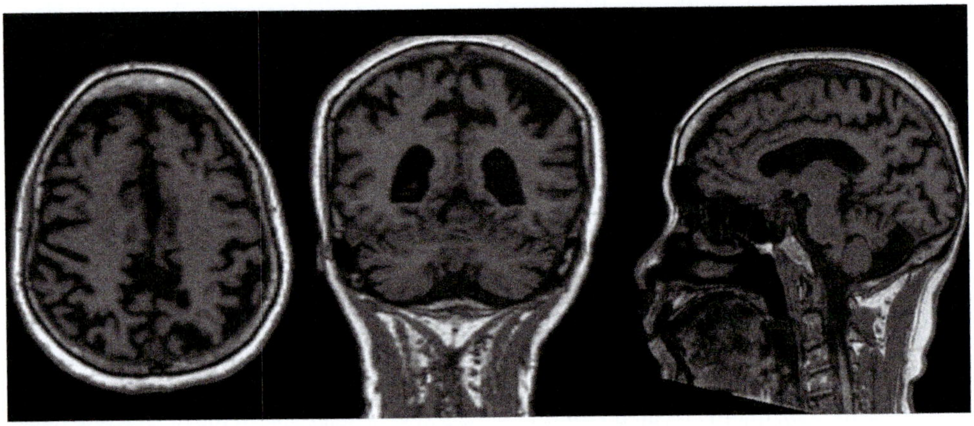

(a) T_1-weighted image (axial view) **(b)** T_1-weighted image (coronal view) **(c)** T_1-weighted image (sagittal view)

Figure 3.5 Severe brain atrophy with global cortical atrophy grade 3 (male, 69 years old).

EPVS reflect abnormal fluid exchange in the perivascular spaces and degenerative changes in brain tissue and are associated with reduced cerebral blood flow perfusion and brain atrophy. The severity of EPVS is related to the progression of cognitive impairment, particularly in executive function and attention.

- Brain atrophy: Patients with CSVD often have brain atrophy, including cortical atrophy (Figure 3.5) and ventricular enlargement (Figure 3.6). Brain atrophy reflects the loss of nerve cells and the degeneration of brain tissue and is associated with a comprehensive decline in cognitive function (Fjell et al., 2025; Morrison et al., 2022; Degraff et al., 2023), and asymmetric thinning of the cerebral cortex (Figure 3.7) across the adult lifespan is accelerated in Alzheimer's disease (Roe et al., 2021). The degree of ventricular enlargement is positively correlated with the severity of memory impairment and executive dysfunction (Apostolova et al., 2012; Lee et al., 2024).

- **Clinical Phenotypes**

Cognitive impairment related to CSVD usually presents as an insidious onset and slow progression of multi-domain cognitive damage (Jochems et al., 2025). In the early

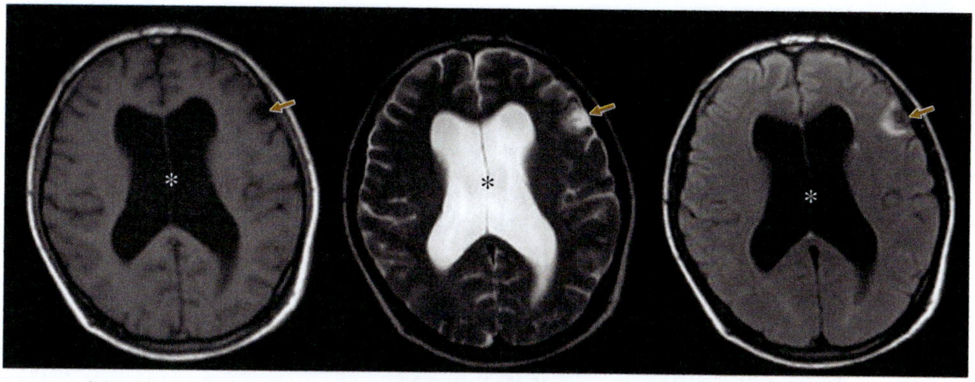

(a) T_1-weighted image **(b)** T_2-weighted image **(c)** T_2 FLAIR

Figure 3.6 Ventricular enlargement (asterisk) and infarction (yellow arrow) (male, 53 years old).

stage, patients mainly show executive dysfunction, such as difficulty in completing complex tasks and decreased ability in planning and organising, because CSVD has a greater impact on brain regions related to executive function, such as the frontal lobe and anterior cingulate gyrus. As the disease progresses, patients gradually experience memory decline, especially in recent memory, along with inattention, slowed information processing speed and impaired language function. In addition, patients may also experience emotional disorders, such as depression, anxiety and emotional blunting, which can further exacerbate cognitive impairment and affect the patients' quality of life and rehabilitation outcomes.

3.4.3 Mixed Dementia: The Coexistence and Interaction of Vascular Factors and Alzheimer's Pathology

Mixed dementia refers to the simultaneous presence of VaD and AD in a patient, a condition with complex pathophysiological mechanisms and one that is relatively common among elderly patients with dementia.

- **Coexistence of Vascular Factors and Alzheimer's Pathology**
  Among elderly patients with dementia, those with dementia caused solely by AD pathology or vascular factors are relatively rare. Most patients exhibit both types of pathological changes (Van Norden et al., 2012; Eisenmenger et al., 2023). Vascular factors include atherosclerosis, hypertension, diabetes, hyperlipidaemia and others, which can lead to cerebrovascular diseases such as cerebral infarction, cerebral haemorrhage and CSVD. These conditions, in turn, cause cerebral ischemia and hypoxia, neuronal injury and death. In contrast, AD pathology is characterised by the deposition of Aβ in the brain, the formation of neurofibrillary tangles (NFTs) and the reduction of neurotransmitters. These pathological changes are intertwined and act together on the brain, resulting in severe cognitive impairment.

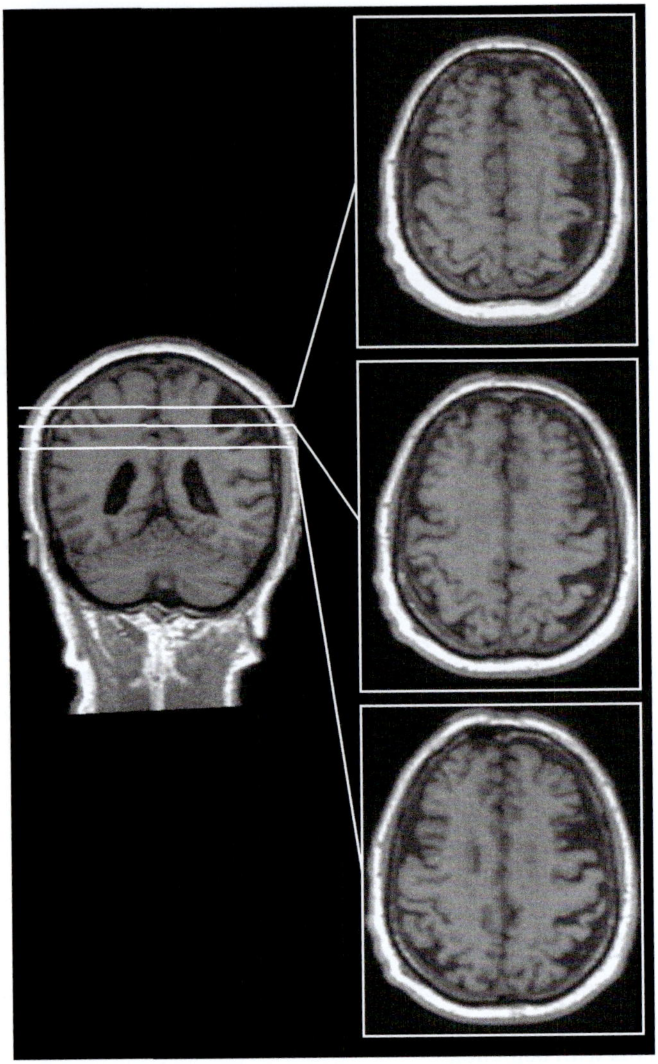

Figure 3.7 Asymmetric atrophy (male, 74 years old).

- **Interaction**
  There is a complex interplay between vascular factors and AD pathology (Mekala & Qiu, 2025; Zhang J. et al., 2024; Rius-Pérez et al., 2018). On the one hand, vascular lesions can exacerbate the pathological process of AD. For example, reduced cerebral blood flow perfusion and cerebral hypoxia caused by cerebrovascular disease can promote the production and deposition of Aβ, while also interfering with the clearance of Aβ, further increasing its accumulation in the brain. Additionally, inflammation and oxidative stress caused by vascular factors can damage neurons, promote the formation of NFTs and accelerate neuronal degeneration. On the other hand, AD pathological changes can also affect vascular function. The deposition of Aβ in the cerebral

vascular wall can lead to CAA, causing vascular wall degeneration and increased fragility, which makes microbleeds and cerebral haemorrhages more likely, thereby worsening the degree of cerebrovascular disease. Moreover, the impaired cerebral blood flow regulation in AD patients further intensifies cerebral ischemia and hypoxia, creating a vicious cycle that leads to a rapid decline in cognitive function.

- **Clinical Manifestations**
  The clinical manifestations of patients with mixed dementia are diverse and complex (Almeida et al., 2025; Custodio et al., 2017). Patients may exhibit the typical symptoms of AD, such as progressive memory decline, language disorders, disorientation and decreased abstract thinking and judgement. They may also show the characteristic manifestations of VaD, such as gait disturbances, pseudobulbar palsy and urinary and faecal incontinence. In addition, patients may experience emotional disorders, such as depression, anxiety and apathy, as well as behavioural abnormalities, such as agitation and wandering. Given that the pathological basis of mixed dementia involves multiple factors, the scope of cognitive impairment is broader, the degree is more severe and the progression is faster, leading to a relatively poor prognosis.
- **Diagnostic Challenges**
  The diagnosis of mixed dementia is relatively difficult because its clinical manifestations and imaging characteristics overlap with those of pure AD or VaD. Currently, common diagnostic methods include detailed medical history collection, neuropsychological assessment, imaging examinations (such as MRI, CT) and biomarker detection (such as the measurement of Aβ and tau protein levels in CSF). However, these examination methods have certain limitations in distinguishing mixed dementia from other types of dementia. For example, MRI can show the characteristic manifestations of cerebrovascular disease, such as WMH, lacunes and CMBs, but it cannot directly detect the pathological changes of AD. While biomarker detection can reflect the pathological characteristics of AD, it is not sensitive enough to assess vascular factors. Therefore, in clinical practice, it is necessary to comprehensively use various examination methods, combine the clinical manifestations and disease course characteristics of patients, and conduct individualised analysis and judgement to improve the diagnostic accuracy of mixed dementia.

3.4.4 Hereditary Cerebrovascular Disease and Cognitive Impairment: Such as CADASIL, CARASIL, Fabry Disease

- **CADASIL**
  CADASIL (cerebral autosomal dominant arteriopathy with subcortical infarcts and leukoencephalopathy) is an autosomal dominant hereditary cerebrovascular disease caused by mutations in the *NOTCH3* gene (Chabriat & Lesnik Oberstein, 2022; Felix-Ilemhenbhio et al., 2025). This gene encodes a transmembrane receptor that is primarily expressed in smooth muscle cells and is involved in vascular development, maintaining the integrity of the vascular wall. Mutations in the *NOTCH3* gene can lead to abnormal function of vascular smooth muscle cells, which in turn causes SVD.

Cognitive impairment in CADASIL patients typically begins in midlife, manifesting as progressive cognitive decline, mainly including executive dysfunction, memory loss and inattention. The scope and severity of cognitive impairment progressively worsen and may eventually develop into VaD. Imaging examinations show that CADASIL patients have extensive WMH in the bilateral periventricular and deep white matter, cingulate gyrus and temporal lobes, and the extent of WMH is related to the severity of cognitive impairment. In addition, patients may also present with imaging features such as lacunar infarcts, CMBs and EPVS, which further exacerbate cognitive impairment.

- **CARASIL**

  CARASIL (cerebral autosomal recessive arteriopathy with subcortical infarcts and leukoencephalopathy) is an autosomal recessive hereditary cerebrovascular disease caused by mutations in the *HTRA1* gene (Menezes Cordeiro et al., 2015; Liu et al., 2020). The *HTRA1* gene encodes a serine protease involved in the degradation and remodelling of the extracellular matrix and plays an important role in maintaining the stability of the vascular wall. Mutations in the *HTRA1* gene can lead to degeneration of elastic and collagen fibres in the vascular wall, causing the wall to thin and become more brittle, making it prone to rupture and bleeding.

  CARASIL patients often present with clinical symptoms in adolescence or early adulthood, including recurrent headaches, cognitive decline, pseudobulbar palsy and spinal cord disease. Cognitive impairment is mainly manifested as progressive memory loss, executive dysfunction and decreased language ability, with a relatively rapid progression of cognitive impairment. Imaging examinations show that CARASIL patients have WMH mainly distributed in the periventricular and deep white matter, along with brain atrophy and CMBs. In addition, patients may also present with demyelinating changes in the periventricular and deep white matter, as well as ventricular enlargement indicative of brain atrophy, which are closely related to cognitive impairment.

- **Fabry Disease**

  Fabry disease is an X-linked recessive lysosomal storage disorder caused by mutations in the alpha-galactosidase A (GLA) gene, leading to a deficiency of this enzyme and the accumulation of glycosphingolipid-rich substances in various tissues throughout the body, including vascular endothelial cells, smooth muscle cells, cardiomyocytes and renal tubular epithelial cells (Sigmundsdottir et al., 2014; Zarate & Hopkin, 2008). In the nervous system, Fabry disease can cause cerebrovascular disease, manifested by thickening of the cerebral vascular wall and narrowing of the lumen, leading to cerebral ischemia and hypoxia and subsequently causing cognitive impairment.

  Cognitive impairment in Fabry disease patients typically appears in mid-adulthood, manifesting as mild to moderate cognitive decline, mainly including memory loss, inattention and slowed information processing speed. The degree of cognitive impairment is related to the severity and duration of the disease. Imaging examinations show that patients with Fabry disease may have lesions such as WMH, lacunar infarcts and CMBs, which are mainly distributed in deep brain regions such as the basal ganglia, thalamus and pons. In addition, patients may also present with demyelinating changes in the periventricular and deep white matter, as well as brain atrophy, which reflect the degree of brain tissue damage caused by cerebrovascular disease and are closely related to cognitive impairment.

3.4.5 Other Cerebrovascular Diseases: The Potential Impact of Intracranial Artery Stenosis/Occlusion, Arteriovenous Malformations and Cavernous Hemangiomas on Cognition

- **Intracranial Artery Stenosis/Occlusion**
  Intracranial artery stenosis/occlusion is one of the important causes of cerebral ischemia and stroke (Holmstedt et al., 2013; Gutierrez et al., 2022), and its impact on cognitive function mainly depends on the location and extent of the affected vessels. For example, stenosis or occlusion of the carotid or middle cerebral artery can lead to ischemia and hypoxia in the cerebral cortex and subcortical structures, affecting functions such as language, movement and sensation (Figure 3.8). It can also involve brain regions closely related to cognitive function, such as the frontal and temporal lobes, resulting in cognitive impairments such as executive dysfunction and memory decline. In addition, intracranial artery stenosis/occlusion may cause recurrent minor

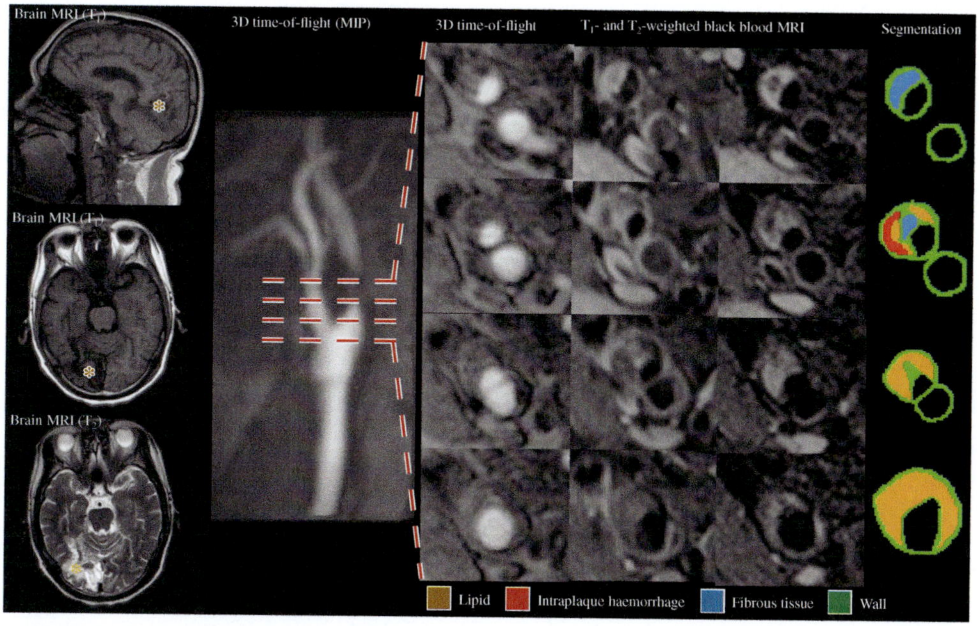

Figure 3.8 A symptomatic stroke patient (male, 68 years old, the infarction located in the right occipital lobe as indicated by the asterisk) with a vulnerable carotid atherosclerotic plaque imaged using multi-contrast, high-resolution black blood MR sequences. *Source*: JINSHAN Science & Technology.

strokes or transient ischemic attacks (TIAs), each of which can cause further damage to brain tissue and exacerbate cognitive impairment. Imaging examinations show that patients with intracranial artery stenosis/occlusion often have SVD in the brain, such as WMH and lacunar infarcts, which are closely related to the occurrence and development of cognitive impairment.

- **Arteriovenous Malformations**
  Arteriovenous malformation (AVM) is a congenital abnormality of cerebrovascular development, consisting of a tangle of tortuous vessels that lack a normal capillary bed, resulting in a direct connection between arteries and veins and the formation of a high-flow arteriovenous fistula. The impact of AVM on cognitive function is mainly related to the pathological processes it causes, such as cerebral haemorrhage, cerebral ischemia and brain oedema (Steinvorth et al., 2002; De Souza Coelho et al., 2020). When an AVM ruptures and bleeds, it can cause acute brain injury, leading to sudden neurological deficits and cognitive decline, such as memory loss, inattention and language disorders. Even in the absence of bleeding, AVM may cause ischemia and hypoxia in surrounding brain tissue due to steal phenomenon, affecting the metabolism and function of nerve cells and thereby causing cognitive impairment. In addition, recurrent seizures caused by AVM can also exacerbate brain tissue damage and further impair cognitive function.
- **Cavernous Hemangiomas**
  Cavernous hemangioma is a benign vascular lesion composed of dilated thin-walled vascular cavities that can occur in the central nervous system, including the brain, cerebellum and spinal cord. The impact of cavernous hemangioma on cognitive function mainly depends on its location and whether there is bleeding. When a cavernous hemangioma is located in brain regions closely related to cognitive function, such as the temporal and frontal lobes, it may cause cognitive impairment, such as memory decline, inattention and executive dysfunction, even without bleeding, due to mass effect and compression of surrounding brain tissue (Awad & Polster, 2019; Dulamea & Lupescu, 2024). If a cavernous hemangioma bleeds, it can cause acute brain injury and sudden cognitive decline, such as aphasia and agnosia. In addition, recurrent bleeding and iron deposition can trigger chronic inflammatory reactions and oxidative stress, further exacerbating brain tissue damage and leading to persistent cognitive impairment.

There is a close link between cerebrovascular disease and cognitive impairment, with different types of cerebrovascular disease causing specific patterns of cognitive impairment through different mechanisms. PSCI, cognitive impairment related to CSVD, mixed dementia, hereditary cerebrovascular disease, as well as other cerebrovascular diseases such as intracranial artery stenosis/occlusion, AVMs and cavernous hemangiomas, all have varying degrees of impact on cognitive function. A thorough understanding of the relationship between these diseases and cognitive impairment can help with early identification, diagnosis and intervention, thereby improving patients' cognitive function and quality of life.

3.5 Diagnosis and Assessment of VCI and Dementia

3.5.1 Evolution of Diagnostic Criteria: From Multi-infarct Dementia to the Concept of VCI

The diagnostic criteria for VCI and VaD have undergone several changes (Sachdev et al., 2014). The initial diagnostic criteria focused on the causal relationship between definite cerebrovascular lesions and dementia. For example, the NINDS-AIREN criteria proposed in the 1980s divided VaD into probable, possible and excluded categories, emphasising evidence of cerebrovascular disease, evidence of cognitive impairment and the causal relationship between the two (Román et al., 1993). However, this criterion was not sensitive enough to early and mild cognitive impairments and had a certain degree of subjectivity in practical application.

With a deeper understanding of VCI, it has been recognised that VCI is a spectrum that covers all stages from mild cognitive impairment to dementia (Pohjasvaara et al., 2000). The DSM series of criteria has also been continuously updated. DSM-IV emphasised the temporal relationship between cognitive impairment and cerebrovascular lesions and neuropsychological evidence in the diagnosis of VaD, but still had limitations. The DSM-5, released in 2013, provided a more detailed classification of vascular neurocognitive disorders, including mild and major neurocognitive disorders, and clarified diagnostic conditions, such as the need for evidence of cognitive impairment, evidence of vascular disease and a close relationship between the two, thereby improving the accuracy and applicability of the diagnosis.

In 2011, the American Heart Association/American Stroke Association (AHA/ASA) proposed the VCI classification and diagnostic criteria, which cover the entire spectrum from mild cognitive impairment to dementia, including pure vascular and mixed (coexisting with AD) conditions, providing a more comprehensive framework for clinical diagnosis. The VASCOG criteria (Sachdev et al., 2014), announced in 2014, further refined and flexibly defined the diagnosis of VCI, not only including severe VaD but also covering mild VCI. The diagnosis also took into account a variety of vascular risk factors and biomarkers. These continuous updates and improvements in criteria reflect a gradual deepening of the understanding of the complexity of VCI and have also improved the accuracy and consistency of clinical diagnosis.

3.5.2 Core Diagnostic Elements: Medical History, Neuropsychological Assessment and Neurological Examination

Medical history taking is the foundation of diagnosing VCI/VaD and requires a detailed understanding of the patient's cognitive functional changes, including the onset time, development process, and whether there are fluctuations or stepwise progressions. At the same time, it is necessary to inquire whether the patient has had a history of stroke, hypertension, diabetes, heart disease and other cerebrovascular disease risk factors, as well as a family history of dementia or other related diseases. This information helps to determine the possible causes and pathogenesis of cognitive impairment.

Neuropsychological assessment is the core part of the diagnosis and can comprehensively evaluate the patient's cognitive functional status. The core assessment areas include executive function, attention, memory, language ability and visuospatial function. Commonly used assessment tools, such as the Montreal Cognitive Assessment (MoCA) and the Addenbrooke's Cognitive Examination-Revised (ACE-III), can be used to repeatedly assess patients at different time points to observe changes in cognitive function, thereby providing an important basis for diagnosis and disease monitoring.

The neurological examination helps to identify neurological signs related to cerebrovascular lesions, such as focal neurological signs like hemiplegia, sensory impairment and ataxia, as well as some non-specific signs that may suggest cerebrovascular disease or neurodegenerative changes, such as gait abnormalities and tremors. In combination with the medical history and neuropsychological assessment results, it can further determine whether cognitive impairment is related to cerebrovascular lesions and exclude other neurological diseases that may cause cognitive impairment.

3.5.3 Neuroimaging: Structural MRI (WMH, Lacunes, CMBs, Atrophy Patterns), Functional Imaging (fMRI, PET Perfusion/Metabolism) and Vascular Imaging (MRA, CTA)

Structural MRI holds significant value in the diagnosis of VCI/VaD (Tae et al., 2025). It can clearly display characteristic manifestations of vascular brain injury such as WMH, lacunar infarcts (Lacunes) and CMBs. The extent and severity of WMH are correlated with the degree of cognitive impairment, with extensive WMH possibly indicating the severity of CSVD (Altschuler et al., 2025; Lee et al., 2016; Guo & Shi, 2022; Hu et al., 2021). The distribution and number of lacunar infarcts also help to determine the scope and extent of cerebrovascular lesions. CMB is another important marker of CSVD, with its quantity and distribution potentially related to cognitive decline. Moreover, structural MRI can assess patterns of brain atrophy, with atrophy in different locations possibly indicating different pathological processes. For example, atrophy in the frontal and parietal lobes may be associated with executive dysfunction (Eldaief et al., 2023).

Functional imaging techniques such as fMRI and PET perfusion/metabolism imaging can provide information on brain function and help understand the neurobiological mechanisms of cognitive impairment (Ko et al., 2013). fMRI can reflect neural activity in the brain during the execution of specific cognitive tasks by detecting changes in cerebral blood flow and metabolism, thereby aiding in the identification of brain regions with functional abnormalities in cognitive impairment. PET perfusion/metabolism imaging can display changes in cerebral blood flow and glucose metabolism, providing important evidence for judging the viability and functional status of brain tissue and assisting in the differential diagnosis of other types of dementia.

Vascular imaging techniques such as MRA and CTA can visually display the morphology and blood flow of cerebral vessels, aiding in the detection of vascular lesions such as arterial stenosis, plaque formation and vessel occlusion. These findings provide important clues for the diagnosis of VCI and also guide vascular intervention treatments.

3.5.4 Exploration of Biomarkers: Blood Biomarkers (Inflammatory Markers, BBB Damage Markers, Neurofilament Proteins) and CSF Biomarkers

The study of blood biomarkers offers new directions for the early diagnosis and disease monitoring of VCI/VaD. Inflammatory markers such as IL-6 and C-reactive protein (CRP) may play a significant role in the development of VCI, with changes in their levels potentially associated with cerebral vascular inflammation and cognitive decline. BBB damage markers, such as brain-type fatty acid-binding protein (BFABP) in plasma (Boneva et al., 2011), can be released into the bloodstream when the BBB is compromised. Elevated levels of these markers may indicate the activity of cerebrovascular lesions and assist in the early identification of potential VCI risks. Neurofilament light chain (NfL) (Li J. et al., 2024), a neuronal cytoskeletal protein, is released into the blood when neurons are damaged. An increase in its levels may be associated with neuronal degeneration and worsening cognitive function, making it a potential biomarker for disease progression and prognosis assessment.

CSF biomarkers also hold significant importance in the diagnosis and differential diagnosis of VCI/VaD (Custodero et al., 2022; Wallin et al., 2017). Biomarkers such as Aβ, tau protein and its phosphorylated form (P-tau) in CSF can help differentiate VCI from other types of dementia, such as AD. For example, in AD patients, CSF levels of Aβ42 are typically reduced, while tau protein and P-tau levels are elevated. In contrast, changes in these biomarkers are relatively less pronounced in VCI patients. Additionally, certain inflammatory markers and neurotrophic factors in CSF may also play a role in the pathogenesis of VCI/VaD, with changes in their levels providing valuable information for disease diagnosis and treatment.

3.5.5 Differential Diagnosis: AD, Dementia with Lewy Bodies, Frontotemporal Dementia

VCI/VaD needs to be differentiated from other types of dementia to ensure that patients receive the correct treatment and management. AD is the most common type of dementia, and there are some differences between AD and VCI/VaD in terms of clinical manifestations and pathological mechanisms. Patients with AD typically present with progressive memory impairment as the initial symptom, accompanied by gradual damage to other cognitive domains such as language and visuospatial function, and the condition progresses continuously. In contrast, cognitive impairment in patients with VCI/VaD may show stepwise or fluctuating progression, often accompanied by focal neurological signs and gait abnormalities. Imaging examinations and biomarker tests also aid in differentiation. For example, in AD patients, there is a decrease in Aβ42 and an increase in tau protein in CSF (Ye et al., 2020; Constantinides et al., 2023), while in VCI patients, the main manifestations are imaging features of cerebrovascular disease and related changes in blood or CSF biomarkers.

Dementia with Lewy bodies (DLB) is characterised by fluctuating cognitive impairment, visual hallucinations and parkinsonism (McKeith et al., 2017). Compared with VCI/VaD,

DLB patients respond better to medications such as acetylcholinesterase inhibitors but are more sensitive to neuroleptics. Neuroimaging in DLB patients may show atrophy in the basal ganglia and brainstem, but evidence of cerebrovascular disease is relatively less apparent. Frontotemporal dementia (FTD) mainly manifests as behavioural changes, language disorders and personality changes. The pattern of cognitive impairment in FTD is different from that in VCI/VaD, usually with imaging features of frontal and temporal lobe atrophy and less evidence of cerebrovascular disease.

In addition, VCI/VaD also needs to be differentiated from cognitive impairment caused by other neurological diseases such as normal pressure hydrocephalus (NPH). Patients with NPH often present with a triad of gait instability, urinary incontinence and cognitive impairment. Imaging examinations may show ventricular enlargement, and CSF shunting is effective for some patients.

Accurate differential diagnosis is crucial in the diagnostic process of VCI/VaD, as it directly affects the treatment plan and prognosis assessment for patients. By considering a comprehensive range of information, including clinical manifestations, neuroimaging features and biomarker test results, the accuracy of diagnosis can be improved, enabling the provision of individualised treatment and management strategies for patients.

3.6 Neuropsychological Characteristics and Patterns of Cognitive Impairment

3.6.1 Executive Dysfunction

Executive function is a crucial component of cognitive ability, and in patients with VCI/VaD, executive dysfunction is particularly prominent, exhibiting a unique pattern of characteristics (Cristofori et al., 2019; Logue & Gould, 2014). Executive function encompasses multiple cognitive processes, including planning, working memory, cognitive flexibility and inhibitory control, all of which are vital for solving complex problems, adapting to new environments and completing multi-step tasks in daily life.

Impaired planning ability is evident in patients with VCI/VaD (Skrobot et al., 2017). They often struggle to devise rational action plans to achieve goals. For instance, when engaging in multi-stage activities, they are unable to prearrange the sequence of each stage and the required resources, leading to disorganised actions. This deficit significantly affects their ability to handle life situations that require advance planning, such as arranging travel or managing financial affairs, severely impacting their capacity for independent living.

Working memory, a key aspect of executive function, is also significantly impaired in VCI/VaD patients. Working memory refers to the ability to store and manipulate information over a short period, such as remembering numbers during mental arithmetic or recalling key points after listening to a passage. Due to impaired working memory, patients often make mistakes when performing tasks that require simultaneous processing and memorisation of information. For example, they may find it difficult to remember the names and quantities of multiple items while shopping, affecting the efficiency and accuracy of their daily activities.

Reduced cognitive flexibility is another important manifestation of executive dysfunction in VCI/VaD patients (Moorhouse et al., 2010). Cognitive flexibility is the ability to adjust thinking and strategies flexibly in response to changes in the environment or task requirements. For example, when solving a problem, it involves being able to switch to an alternative method promptly if the initial approach fails. However, these patients typically have difficulty adapting to new rules or situations, showing rigid and stereotyped thinking. They tend to persist in using existing methods that may no longer be applicable, leading to an inability to cope effectively with complex and changing life scenarios and to make appropriate adjustments.

Deficits in inhibitory control are also common in VCI/VaD patients (Pantsiou et al., 2018). Inhibitory control is the ability to suppress irrelevant information or impulsive behaviour and focus on target tasks. Due to impaired inhibitory control, patients are easily distracted by external factors or internal impulses, leading to behaviours such as inattention and impulsive decision-making. For example, they may find it difficult to suppress distracting thoughts in situations that require quiet concentration, or they may make unnecessary purchases on impulse while shopping, affecting their social functioning and quality of daily life.

The occurrence of executive dysfunction in VCI/VaD patients is closely related to the involvement of the frontal lobes and subcortical neural circuits by cerebrovascular lesions. The frontal lobes are key brain regions for executive function, responsible for regulating higher cognitive functions such as planning, decision-making and working memory (Chayer & Freedman, 2001; Grafman & Litvan, 1999). When cerebrovascular lesions cause damage to the frontal cortex or its underlying white matter fibres, the transmission of neural signals is impeded, preventing the brain from effectively integrating and processing information to execute complex cognitive tasks. For example, ischemia and hypoxia in the frontal cortex can impair the normal physiological function of neurons, affecting their release and regulation of neurotransmitters related to planning and decision-making, thereby leading to a decline in these abilities. Additionally, subcortical white matter lesions can disrupt communication between the frontal lobes and other brain regions, weakening the overall efficacy of executive function. Moreover, subcortical structures such as the basal ganglia also play important roles in cognitive flexibility and inhibitory control (Kim, 2025; Asch et al., 2025). Damage to these structures can result in patients' inability to adjust behavioural strategies in response to environmental changes and to suppress inappropriate impulsive behaviours. These neurostructural damages and neurophysiological dysfunctions caused by cerebrovascular lesions collectively lead to comprehensive impairment of executive function in VCI/VaD patients, severely affecting their daily living abilities and quality of life. This also increases the difficulty of caring for and managing these patients. Clinical physicians need to pay high attention to this issue in the assessment and treatment process. Through targeted rehabilitation training and comprehensive measures such as pharmacological therapy, it is possible to improve patients' executive function as much as possible, enhance their ability to take care of themselves and improve their social adaptability.

3.6.2 Slowed Information Processing Speed: A Universal Characteristic and Its Impacts

Slowed information processing speed is a universal characteristic of cognitive impairment in VCI/VaD patients and has extensive and profound impacts on various aspects of their daily lives (Mahon et al., 2020; Haworth et al., 2016). Information processing speed refers to the rate at which an individual perceives, processes and reacts to internal and external information. It permeates all aspects of cognitive activities and is the foundation for efficient cognitive function.

In cognitive tasks, the slowed information processing speed in VCI/VaD patients is manifested by prolonged reaction times to both simple and complex stimuli. Whether processing visual, auditory or tactile sensory information, patients require more time to respond than healthy individuals. For example, during the Symbol Digit Modalities Test (Smith, 1968), patients often take longer to match symbols with corresponding numbers and perform the conversion. This reflects a slower encoding, retrieval and integration of information in their brains. This slowdown not only affects the efficiency of individual cognitive tasks but also, when multiple tasks need to be switched and integrated, leads to patients' inability to keep up with the normal pace, resulting in task backlogs and increased errors.

The slowed information processing speed has a significant negative impact on the daily functioning of VCI/VaD patients. In communication, patients may need more time to understand others' speech and facial expressions, impairing the fluency of conversations and leading to delayed responses. This can affect interpersonal relationships and social activities. In handling daily affairs, such as shopping, cooking and taking medication, the slow processing speed may cause patients to miss important steps, repeat operations or fail to complete tasks within the specified time, increasing errors and risks in daily life. For example, when cooking, they may fail to adjust the heat or add seasonings in time, affecting the taste and quality of the food and even posing safety hazards.

From a neurophysiological perspective, the slowed information processing speed is closely related to the damage to neural conduction pathways caused by cerebrovascular lesions. Cerebrovascular lesions lead to ischemia and hypoxia in the brain's white matter, damaging the integrity of the myelin sheath surrounding nerve fibres and slowing the conduction speed of nerve signals in axons (Li X. et al., 2024). Myelin is an insulating layer that wraps around nerve fibres and accelerates the transmission of nerve impulses. When damaged, the efficiency of nerve signal transmission drops significantly, similar to how electrical current slows when the insulation of a wire is damaged. Additionally, neuronal loss and decreased synaptic function also affect information processing speed. Neurons are the basic units of information processing, and their reduced number and abnormal function weaken the brain's ability to process information. At the same time, the neurotransmitter system is disrupted. The synthesis, release and reuptake of neurotransmitters closely related to cognitive function, such as dopamine and acetylcholine, are affected, further interfering with the transmission of nerve signals and the efficiency of information processing (Coyle, 2006; Stone et al., 2007).

Slowed information processing speed also interacts with other cognitive domains, exacerbating the overall degree of cognitive impairment. For example, slow processing speed can worsen memory deficits because patients need more time to encode and retrieve memory information, making them more susceptible to external interference and leading to the loss or error of memory information (Estelle & Voelbel, 2024). In terms of executive function, slow processing speed affects patients' efficiency and accuracy in solving complex problems, making it difficult for them to plan, organise and monitor tasks within a limited time. This interaction creates a vicious cycle, causing patients' cognitive functions to deteriorate continuously and further limiting their daily living abilities and social participation. Therefore, in the treatment and rehabilitation of VCI/VaD patients, training and intervention targeting information processing speed are of great significance. Comprehensive measures, including cognitive training, pharmacological therapy and the management of risk factors for cerebrovascular lesions, can be employed to improve patients' information processing speed, enhance their overall cognitive function and improve their quality of life.

3.6.3 Attention Deficits: Sustained, Selective and Divided Attention

Attention is the foundation of cognitive function, and attention deficits in VCI/VaD patients are significantly manifested in multiple aspects, mainly including impairments in sustained attention, selective attention and divided attention (Onandia-Hinchado et al., 2021; Loetscher et al., 2019). These deficits are interwoven and severely affect patients' cognitive processing and daily functioning.

Sustained attention refers to an individual's ability to maintain focus on a specific task or activity over a period of time. VCI/VaD patients often struggle to maintain stable attention and are prone to fluctuations and lapses in attention. For example, during activities such as prolonged reading, watching TV or attending lectures, patients may fail to stay focused from beginning to end, frequently experiencing mind-wandering and distraction. This instability in attention makes it difficult for patients to perform well in tasks that require continuous concentration, such as completing a report or engaging in a craft project. They may experience task interruptions or poor quality of completion due to an inability to sustain attention. Sustained attention deficits are associated with damage to the frontal and parietal cortices and related neural pathways caused by cerebrovascular lesions (Chayer & Freedman, 2001; Coll et al., 2024; Liu et al., 2023). These brain regions play a crucial role in regulating the alertness network and executive control network of attention. When damaged, the brain cannot effectively maintain the continuous activation and stable maintenance of attention, thereby affecting patients' daily activities and work efficiency.

Selective attention refers to an individual's ability to selectively focus on task-relevant information while ignoring irrelevant distracting information when faced with multiple stimuli. VCI/VaD patients have significant deficits in selective attention, being easily distracted by external irrelevant information and struggling to concentrate on target stimuli (Bialystok, 2025; Stevens & Bavelier, 2012). For example, in a noisy environment, patients may fail to focus on a conversation with others, being drawn to surrounding noise or other people's conversations, leading to an inability to accurately understand the content of the dialogue. In visual tasks, patients may also have difficulty distinguishing target objects from complex backgrounds. For instance, when searching for an item, they may be easily distracted by surrounding clutter, prolonging the search time. The occurrence of selective attention deficits is closely related to cerebrovascular lesions involving brain regions such as the anterior cingulate gyrus and parietal cortex (Carter et al., 1997; Bush, 2011). These regions are responsible for regulating the selection and allocation of attention. Damage to these areas results in a decreased ability of the brain to distinguish between relevant and irrelevant information, making it difficult for patients to effectively filter and focus on target information in complex environments, thus affecting the accuracy and efficiency of information processing.

Divided attention refers to an individual's ability to allocate attention to multiple tasks or activities simultaneously, such as paying attention to traffic conditions, steering the wheel and adjusting the car audio while driving. VCI/VaD patients have impaired divided attention and struggle to allocate attention reasonably among multiple tasks, often neglecting one aspect while focusing on another (Park et al., 1999; Robertson & Schmitter-Edgecombe, 2017). For example, when performing a task that requires simultaneous attention to multiple steps, patients may only focus on one step while ignoring the progress of the others, leading to difficulties in task execution. This divided attention deficit limits patients' multitasking abilities, causing significant challenges and stress when facing situations in daily life that require handling multiple tasks at the same time. Its neural mechanism is related to the impact of cerebrovascular lesions on the brain's multitasking network, which involves the coordinated activity of multiple brain regions such as the frontal lobe, parietal lobe and thalamus. When the connections and functions between these brain regions are damaged, the brain cannot effectively integrate and allocate limited attentional resources, resulting in divided attention deficits.

Attention deficits have profound impacts on the cognitive function and daily life of VCI/VaD patients. From a cognitive perspective, impaired attention affects the functioning of other cognitive domains, such as memory, language and executive function. For example, inattention can lead to insufficient encoding and storage of information during memory tasks, thereby exacerbating memory impairments. In language communication, an inability to focus on the content of a conversation can affect the accuracy of language comprehension and expression. In daily life, attention deficits pose numerous challenges for patients when engaging in various activities, such as the potential for unsafe cooking (which may lead to fires due to distraction) and increased risk of accidents while walking in traffic environments. These issues severely threaten patients' safety and quality of life. Therefore, intervention measures targeting attention deficits are crucial in clinical assessment and rehabilitation treatment. Through cognitive training, environmental adjustments and pharmacological therapy, patients can be helped to improve various aspects of attention, thereby enhancing their daily living abilities and cognitive function levels.

3.6.4 Memory Impairment: Patterns (Primarily Characterised by Retrieval Difficulties Rather Than Encoding and Storage) and Distinctions from AD

Memory impairment in VCI/VaD patients exhibits a unique pattern, primarily characterised by difficulties in memory retrieval, while the processes of encoding and storage remain relatively preserved (Lim & Alexander, 2009; Henry, 2021). This is in marked contrast to the memory impairment pattern observed in AD patients.

In VCI/VaD patients, the functions of memory encoding and storage are typically well-maintained. Memory encoding refers to the process of transforming external information into neural memory representations, while storage involves preserving these representations in the brain. These patients are often able to effectively encode new information through repetition and association when learning new material, such as remembering a list of words or the general content of a story, and can store this information for a short period. This indicates that their memory system's input and preservation mechanisms are not severely compromised. For example, during a word-learning test, VCI/VaD patients may perform similarly to healthy individuals in the learning phase, gradually remembering an increasing number of words. However, when it comes to memory retrieval, problems become apparent (Campos et al., 2010; Alfei et al., 2025; O'Sullivan et al., 2023). Memory retrieval is the ability to extract desired information from memory storage, and VCI/VaD patients experience significant difficulties in this regard. They often struggle to recall previously learned information, requiring more prompts and cues to complete recall tasks, or their recollections may be incomplete or inaccurate. This difficulty in memory retrieval may be due to cerebrovascular lesions that damage the neural network connections in the brain, preventing stored information from being smoothly extracted. For instance, in daily life, patients may remember the general plot of an event but have difficulty recalling specific details, or they may need additional cues such as location or people involved to aid their memory.

Compared to VCI/VaD, the memory impairment pattern in AD patients is characterised by early damage to memory encoding and storage, which subsequently extends to memory retrieval (Jahn, 2013). In the early stages of AD, the main issue lies in the decline of memory encoding and storage functions. This is because the pathological changes in AD are primarily concentrated in the hippocampus and its surrounding cortex, which are key areas for memory encoding and storage. The hippocampus plays a crucial role in converting short-term memory into long-term memory, and the loss of neurons and the formation of NFTs in this region lead to inefficient memory encoding and difficulty in effectively storing new information. For example, AD patients may repeatedly ask the same question because they have difficulty learning and remembering new information, failing to convert newly acquired information into stable memory traces. As the disease progresses, the ability to retrieve memories in AD patients also gradually declines. However, due to the early damage to encoding and storage, even if the retrieval stage is not further affected, the amount of information available for extraction is already significantly reduced, resulting in more severe and comprehensive memory impairments. Additionally, memory impairments in AD patients are usually accompanied by damage to other cognitive domains, such as language, visuospatial function and executive function, whereas memory impairments in VCI/VaD patients are relatively more localised, primarily focusing on retrieval difficulties, with more prominent impairments in other cognitive areas like executive function and information processing speed.

These differences in memory impairment patterns are also reflected in neuroimaging and pathological features. Brain MRI of VCI/VaD patients shows that their memory impairments are associated with subcortical white matter lesions and ischemic damage in areas such as the basal ganglia and thalamus (Wang et al., 2023; Liu et al., 2019). These regions are connected to memory-related brain areas like the hippocampus and prefrontal cortex through neural fibre tracts. When these connection pathways are damaged, it affects the transmission and integration of memory information between different brain regions, leading to difficulties in memory retrieval. In contrast, brain MRI of AD patients reveals hippocampal atrophy and thinning of the temporal lobe cortex, pathological changes that directly impair memory encoding and storage functions. PET scans show that in VCI/VaD patients, brain metabolism is primarily reduced in subcortical regions, while in AD patients, the most significant reductions in brain metabolism are seen in the temporoparietal lobes and posterior cingulate cortex (Spano et al., 2023; Perron et al., 2025), further reflecting the different neurobiological bases of memory impairments in these two diseases.

In summary, memory impairments in VCI/VaD patients are primarily characterised by difficulties in memory retrieval, with relatively preserved encoding and storage functions, which is distinctly different from the memory impairment pattern in AD patients. This difference is not only evident in clinical manifestations but also reflects the distinct neuro-pathological mechanisms and brain lesion locations of the two diseases, holding significant importance for diagnosis, differential diagnosis and the formulation of treatment strategies. In clinical practice, detailed neuropsychological assessments and neuroimaging examinations can more accurately identify the memory impairment patterns in VCI/VaD and AD, thereby providing more targeted treatment and rehabilitation plans for patients.

3.6.5 Other Domains: Visuospatial Function, Language (Fluency) and Emotional Regulation (Apathy, Depression)

In addition to the commonly affected executive function, information processing speed, attention and memory in VCI/VaD patients, other cognitive domains such as visuospatial function, language fluency and emotional regulation are also frequently impaired (Wang et al., 2024; Buening & Brown, 2018). These additional impairments further exacerbate the cognitive burden and difficulties in daily living for patients.

Visuospatial dysfunction is relatively common in VCI/VaD patients, primarily characterised by a decline in the ability to perceive and judge spatial relationships, as well as difficulties in integrating and processing visual information. Patients may struggle to accurately determine the position, distance and shape of objects. For example, they may have difficulty correctly inserting their arms into sleeves when dressing or may frequently bump into surrounding objects while walking. In visual construction tasks, such as copying geometric shapes or building with blocks, patients often exhibit difficulties and are unable to accurately replicate complete figures or structures. This reflects impaired integration and coordination of visuospatial information in their brains. The occurrence of visuospatial dysfunction is associated with cerebrovascular lesions involving the parietal and occipital cortices and related visuospatial processing pathways (Baltaretu et al., 2023; Matsuyoshi et al., 2012). These brain regions are responsible for processing visual information and spatial localisation. When damaged by ischemia or hypoxia, the connections between neurons and signal transmission are disrupted, preventing the brain from effectively processing visuospatial information and leading to visuospatial dysfunction. This impairment not only affects patients' ability to perform daily self-care activities, such as dressing and eating, but may also increase their risk of falls and accidental injuries, especially in unfamiliar or complex environments where their safety is severely threatened.

Impaired language fluency is another common manifestation in VCI/VaD patients, mainly reflected in the speed, quantity and fluency of speech expression. Patients often speak slowly, produce less language output, and have difficulty selecting words. For example, they may frequently pause during conversations, repeatedly use simple vocabulary, or fail to accurately express their thoughts. In language fluency tests, such as asking patients to list as many words as possible from a specific category (e.g., names of animals) within a set time, their scores are usually lower than those of healthy individuals. This language fluency disorder may be related to damage to the frontal and temporal cortices, which play important roles in language generation and organisation (Ghanavati et al., 2019). When cerebrovascular lesions affect these areas, the brain's language network connections are damaged, leading to obstacles in the retrieval and expression of language information. Impaired language fluency not only affects patients' communication abilities, making it difficult for them to interact effectively with others, but may also lead to social isolation and emotional problems, further reducing their quality of life and mental health.

Emotional regulation disorders are also prominent in VCI/VaD patients, mainly including symptoms such as apathy and depression (Lanctôt et al., 2023; Steffens et al., 2022). Apathy is characterised by a lack of interest and emotional response to surrounding events and a lack of enthusiasm for daily activities. For example, patients may sit on the sofa all day, unwilling to participate in family activities or go for a walk. This apathetic mood may be due to cerebrovascular lesions affecting the prefrontal cortex and limbic system of the brain, which play key roles in regulating motivation and emotions. When their functions are impaired, patients' emotional drive and reward mechanisms are suppressed, leading to a decrease in enthusiasm and motivation for life. Depressive symptoms are more complex, with patients exhibiting symptoms such as low mood, pessimism and despair, low self-esteem, sleep disturbances and reduced appetite. The occurrence of depressive mood is closely related to neurotransmitter imbalances, neuroendocrine disorders and decreased neuroplasticity caused by cerebrovascular lesions. For example, cerebrovascular lesions may lead to reduced synthesis or abnormal metabolism of neurotransmitters related to emotional regulation, such as serotonin and dopamine in the brain, thereby triggering depressive mood. Emotional regulation disorders not only severely affect patients' mental health but may also have a negative impact on cognitive function, creating a vicious cycle. For example, depressive mood may lead to a lack of enthusiasm for cognitive rehabilitation training in patients, further exacerbating cognitive decline. Conversely, cognitive decline may intensify patients' low mood, plunging them into deeper despair. Therefore, in the treatment and care of VCI/VaD patients, it is of great significance to pay attention to emotional regulation disorders and provide timely and effective interventions. Through comprehensive measures such as psychotherapy, pharmacological therapy and rehabilitation training, patients can be helped to improve their emotional state and enhance their quality of life and cognitive function levels.

3.6.6 The Relationship Between Lesion Location and Cognitive Features: The Primary Role of Frontal-Subcortical Circuit Damage

In VCI/VaD, different locations of cerebrovascular lesions are closely related to specific combinations of cognitive features, among which damage to the frontal-subcortical circuit plays a central role in cognitive impairment (Liu et al., 2023; Bonelli & Cummings, 2007; Tzovenos Starosta & Vinícius Vidor, 2016).

The frontal cortex is a key area of the brain responsible for higher cognitive functions. It is closely connected to subcortical structures such as the basal ganglia, thalamus and white matter through neural fibre tracts, forming a complex frontal-subcortical circuit.

This circuit plays a crucial role in regulating cognitive and behavioural processes such as executive function, information processing speed, attention and emotions. When cerebrovascular lesions involve the frontal cortex, neuronal damage and neurotransmitter disruption directly affect various aspects of executive function, such as planning, decision-making, working memory and cognitive flexibility. For example, ischemia and hypoxia in the frontal cortex can lead to dysfunction of the prefrontal dopamine system, which is crucial for regulating executive function and motivation. Its deficiency can result in symptoms such as decreased planning ability, rigid behaviour and lack of initiative in patients. At the same time, damage to the frontal cortex can also disrupt information exchange between it and subcortical structures, further exacerbating the degree of cognitive impairment.

The basal ganglia, as an important component of the frontal-subcortical circuit, play a significant role in motor regulation and cognitive function. The basal ganglia participate in regulating the initiation, control and coordination of movement through reciprocal neural pathways with the frontal cortex. They are also involved in reward mechanisms, decision-making and working memory in cognitive processes. When cerebrovascular lesions cause damage to the basal ganglia, in addition to causing motor disorders such as parkinsonism, they can also have a significant impact on cognitive function. Lesions in the basal ganglia may lead to symptoms such as decreased cognitive flexibility, inattention and executive dysfunction in patients. For example, the striatum in the basal ganglia plays an important role in maintaining and updating working memory. Its damage can cause patients to make mistakes when performing tasks that require simultaneous processing and updating of information, such as difficulty in switching and integrating information between different tasks during multitasking.

The thalamus is the “relay station” of the brain. It receives and integrates neural signals from the sensory and motor systems and then projects them to the corresponding areas of the cerebral cortex for further processing. Damage to the thalamus can interfere with the normal reception and transmission of information in the brain, leading to cognitive issues such as sensory processing disorders, inattention and memory impairment. For example, ischemic lesions in the thalamus may affect its efficiency in integrating and transmitting sensory information, causing delays and inaccuracies in patients’ processing of visual, auditory and other sensory information. This, in turn, affects their perception of and response to the surrounding environment. Additionally, the neural connections between the thalamus and the frontal cortex are crucial for regulating attention and alertness. Damage to these connections can lead to persistent attention deficits and distractibility in patients, affecting their performance in various cognitive tasks.

Subcortical white matter is an important neural fibre tract that connects the cerebral cortex with subcortical structures, providing pathways for information exchange between different brain regions. When cerebrovascular lesions cause white matter damage, neural signal transmission is impeded during conduction, impairing the coordination and integration functions between brain regions. This type of white matter lesion is very common in VCI/VaD patients and is one of the important factors leading to cognitive dysfunction. For example, damage to the frontal white matter can disrupt neural signal transmission between the frontal cortex and structures such as the basal ganglia and thalamus, affecting executive function and information processing speed. Patients often exhibit slow reactions and increased error rates when performing tasks that require quick responses and flexible adjustments. Moreover, white matter lesions may also trigger damage in multiple cognitive domains through the "disconnected syndrome" mechanism. This occurs when the disconnection of white matter fibres interrupts connections between different brain areas, preventing normally collaborative neural networks from functioning properly. This can lead to various cognitive impairments, such as memory decline and language disorders.

There is a certain regularity in the combination of cognitive features caused by cerebrovascular lesions in different locations. For example, lesions in the frontal cortex and basal ganglia are usually associated with executive dysfunction, bradykinesia and emotional symptoms such as apathy; thalamic lesions may cause attention deficits, memory retrieval difficulties and sensory abnormalities; and widespread white matter lesions are often accompanied by slowed information processing speed, decreased executive function and gait abnormalities. The association between lesion location and cognitive features provides important clues for clinical diagnosis and treatment. Through neuroimaging examinations such as MRI and CT, the specific location and extent of cerebrovascular lesions can be clearly identified. This allows for the prediction of the types and severity of cognitive impairments that patients may experience, providing a basis for formulating personalised treatment and rehabilitation plans.

However, it is important to note that cognitive impairments in VCI/VaD patients exhibit significant heterogeneity, which arises from the combined effects of multiple factors (Kalaria, 2016; Wei et al., 2025). First, the diversity of aetiology is one of the important reasons for the heterogeneity of cognitive impairments. Cerebrovascular lesions can be caused by atherosclerosis, hypertension, diabetes, embolism and other causes. Different causes may lead to different patterns and locations of cerebrovascular lesions, which in turn cause different cognitive damage characteristics. For example, atherosclerosis may lead to stenosis or occlusion of large cerebral vessels, involving cortical areas such as the frontal and parietal lobes, causing cognitive impairments mainly characterised by executive dysfunction and language disorders. In contrast, hypertensive SVD mainly affects subcortical white matter and the basal ganglia, resulting in cognitive decline characterised by slowed information processing speed and attention deficits. Second, the distribution range and severity of cerebrovascular lesions also affect the heterogeneity of cognitive impairments. Lesions may be confined to a specific area or widely distributed across multiple brain regions. The severity of the lesions ranges from mild white matter changes to large-scale cerebral infarctions. These differences result in the diversity and complexity of cognitive impairments. In addition, the presence of other pathological changes, such as AD pathological features (Aβ deposition, NFTs) or Lewy body formation in Lewy body dementia (LBD), can also make the manifestations of cognitive impairments more complex and diverse. VCI patients with coexisting AD pathological changes may exhibit more severe memory impairments and language function decline on the basis of existing executive dysfunction and slowed information processing speed. Their cognitive impairment pattern is different from that of pure VCI or AD.

In summary, cognitive impairments in VCI/VaD patients have multidimensional and multimodal characteristics. Their neuropsychological features are closely related to the location, range, aetiology of cerebrovascular lesions and other coexisting pathological changes. In clinical practice, a deep understanding of these features and their underlying neural mechanisms is of vital importance for accurate diagnosis, differential diagnosis and the formulation of targeted treatment and rehabilitation strategies. By comprehensively using neuropsychological assessments, neuroimaging examinations and biomarker tests, a more comprehensive understanding of patients' cognitive impairments can be achieved, thereby improving treatment outcomes and patients' quality of life and prognosis.

3.7 Intervention Strategies: Prevention, Management and Treatment

3.7.1 Primary Prevention: Control of Vascular Risk Factors

VCI, including VaD, hinges on the control of vascular risk factors for its primary prevention.

Hypertension is one of the significant risk factors for VCI. Research indicates that intensive antihypertensive therapy can reduce the risk of cognitive impairment. For instance, a secondary analysis of the SPRINT study revealed that intensive antihypertensive therapy targeting a systolic blood pressure of less than 120 mmHg can yield more pronounced cognitive benefits for participants with a higher predicted baseline risk (SPRINT Research Group, 2015; Rapp et al., 2020). Regarding the selection of antihypertensive medications, angiotensin-converting enzyme inhibitors and angiotensin receptor blockers may be more beneficial to cognition.

Individuals with diabetes have an increased risk of cognitive decline (Xue et al., 2019). For every 1% reduction in glycated haemoglobin (HbA1c), the risk of VCI decreases by 16%. GLP-1 receptor agonists, such as dulaglutide, apart from controlling blood glucose, may also exert a protective effect on cognitive function. However, some studies have shown that intensive control of blood glucose, blood pressure and lipids in diabetic patients did not result in long-term cognitive benefits.

Dyslipidaemia is closely associated with VCI (Zhao et al., 2024). Statins, which can lower low-density lipoprotein cholesterol (LDL-C), may be beneficial for the primary prevention of VCI. Smoking leads to impaired vascular endothelial function and increased inflammatory responses, thereby increasing the risk of VCI. Quitting smoking can reduce the risk of VCI, with a more significant risk reduction observed with longer smoking cessation periods.

Obesity is an independent risk factor for VCI (Zhang et al., 2025). For every 5 kg/m^2 increase in body mass index (BMI), the risk of VCI increases by 16%. Weight reduction through dietary control and exercise intervention can help prevent VCI. Patients with atrial fibrillation, due to an increased risk of thromboembolism, are prone to stroke, which in turn can lead to VCI. Anticoagulant therapy can effectively reduce the risk of VCI in patients with atrial fibrillation. A sedentary lifestyle increases the risk of VCI. It is recommended that adults engage in at least 150 minutes of moderate-intensity aerobic exercise per week, such as brisk walking or swimming.

In summary, actively controlling vascular risk factors is the most effective and fundamental strategy for preventing VCI/VaD, significantly reducing the risk of its occurrence.

3.7.2 Secondary Prevention: Post-stroke Management and Prevention of SVD Progression

The risk of VCI after stroke is significantly increased, hence the crucial importance of post-stroke management (Bangad et al., 2023).

Regarding blood pressure management, an individualised antihypertensive regimen should be formulated based on the patient's specific circumstances to maintain blood pressure within an appropriate range, thereby reducing the risk of stroke recurrence and VCI. For stroke patients with dyslipidaemia, statin therapy can lower the risk of stroke recurrence, thus preventing VCI. In terms of blood glucose control, it is essential to achieve target HbA1c levels for diabetic patients while avoiding hypoglycaemic events.

Antiplatelet therapy is one of the vital measures for secondary stroke prevention. For most non-cardioembolic stroke patients, monotherapy (e.g., aspirin) can reduce the risk of stroke recurrence. For certain specific types of stroke or high-risk patients, dual antiplatelet therapy may be more effective. Anticoagulant therapy is primarily used for cardioembolic stroke patients, such as those with atrial fibrillation, and can significantly reduce the risk of stroke recurrence and VCI.

SVD is an important pathological basis for VCI. Intensified control of risk factors, such as strict control of blood pressure, blood glucose and lipids, can delay the progression of SVD. In terms of pharmacological treatment, certain endothelial protective drugs may be beneficial for SVD-related VCI.

3.7.3 Management of Cognitive Symptoms: Cognitive Training and Rehabilitation, Non-pharmacological Treatments

Cognitive training and rehabilitation are essential non-pharmacological treatments for VCI (Liang et al., 2025). For instance, cognitive training targeting memory, attention and executive function can improve patients' cognitive abilities to a certain extent. A study showed that six months of cognitive training significantly enhanced memory and executive function in VCI patients.

Non-pharmacological treatments also include environmental adjustments and behavioural interventions. Creating a safe and familiar environment for patients can help reduce their anxiety and confusion. Behavioural interventions, such as establishing regular daily routines and engaging in appropriate activities, can improve patients' behaviour and emotional issues.

Group cognitive stimulation therapy, which involves organising patients to participate in group activities like discussions and games, can stimulate patients' thinking and social skills, thereby improving cognitive function and quality of life. Music therapy can positively impact patients' cognition and emotions by stimulating the auditory and emotional areas of the brain (Lorek et al., 2023).

3.7.4 Current Status of Pharmacological Treatment: Application of Cholinesterase Inhibitors and Memantine in VaD and Mixed Dementia

Cholinesterase inhibitors are one of the commonly used drugs for VCI treatment (Xu et al., 2021). Donepezil can mildly improve cognitive function in VCI patients (Rocca et al., 2002). Systematic reviews and Bayesian network meta-analyses have shown that 10 mg of donepezil is superior to placebo in terms of MMSE and ADAS-cog scores (Sheikh & Ammar, 2024). Rivastigmine has certain benefits for the cognitive function of VCI patients, but there are differences in the results of various studies. Galantamine can improve cognitive function in VaD patients, but its improvement in activities of daily living is similar to that of placebo.

Memantine, a glutamate receptor antagonist, can modulate glutamatergic neurotransmission and has some beneficial effects on cognitive function, behaviour and emotional problems in VCI patients, although the quality of evidence is mostly moderate to low.

Overall, cholinesterase inhibitors and memantine have certain therapeutic effects in the treatment of VaD and mixed dementia, but there are certain limitations, and further exploration of more effective pharmacological treatment regimens is needed.

3.7.5 Emerging Therapies Targeting Pathological Mechanisms: Improving Cerebral Perfusion, Protecting the BBB, Anti-inflammatory and Promoting Angiogenesis

Improving cerebral perfusion is an important target for VCI treatment. Some drugs, such as butylphthalide (Han et al., 2025), can improve cognitive function in VCI patients through mechanisms such as dilating cerebral vessels and improving cerebral blood flow. In terms of protecting the BBB, studies have shown that certain drugs may protect the BBB by reducing inflammatory responses and oxidative stress, thereby improving cognitive function.

Anti-inflammatory therapy is also one of the potential strategies for VCI treatment. Inflammatory responses play a significant role in the pathogenesis of VCI, and some anti-inflammatory drugs, such as nonsteroidal anti-inflammatory drugs (NSAIDs), may be beneficial for VCI, although the long-term safety and efficacy of their use still require further research (Karplus & Saag, 1998; Waldstein et al., 2010).

Promoting angiogenesis can positively impact VCI by increasing cerebral blood supply and improving neurological function. Some growth factors, such as vascular endothelial growth factor (VEGF), have certain potential in promoting angiogenesis but are still in the research stage.

Currently, clinical trials of emerging therapies for VCI are continuously being conducted, such as the development and clinical trials of new drugs targeting cerebral vascular endothelial dysfunction and neuroinflammation, which are expected to bring new breakthroughs in the treatment of VCI.

3.7.6 Lifestyle Interventions: Physical Exercise, Healthy Diet (e.g., MIND Diet), Cognitive and Social Activities

Physical exercise is of great significance for the prevention and treatment of VCI (Tari et al., 2025). Regular aerobic exercise, such as brisk walking or swimming for 30–60 minutes, 3–5 times a week, can improve cerebral vascular function and promote the secretion of neurotrophic factors, thereby improving cognitive function.

Healthy diets, such as the Mediterranean diet and MIND diet (Kamrani et al., 2024), which are rich in vegetables, fruits, whole grains and fish, can provide abundant antioxidants and unsaturated fatty acids, helping to maintain cerebral vascular health and slow cognitive decline (Dalile et al., 2022). A study showed that elderly people who adhered to the Mediterranean diet had a slower rate of cognitive decline and a lower risk of VCI.

Actively participating in cognitive and social activities, such as reading, playing chess and social gatherings, can stimulate the brain and delay cognitive decline (Stafford et al., 2024). Research has found that elderly people who frequently engage in cognitive and social activities have better cognitive function and a lower risk of VCI.

3.7.7 Summary of Comprehensive Management Strategies

The management of VCI needs to take into account multiple aspects, including primary prevention, secondary prevention, management of cognitive symptoms, pharmacological treatment, exploration of emerging therapies and lifestyle interventions. Actively controlling vascular risk factors is the cornerstone of preventing VCI/VaD; for post-stroke patients, rational secondary prevention measures can reduce the risk of VCI; cognitive training and rehabilitation, non-pharmacological treatments and pharmacological treatments can improve patients' cognitive function and quality of life to a certain extent; the continuous exploration and research of emerging therapies bring new hope for the treatment of VCI; and healthy lifestyle interventions run through the entire process of prevention and treatment of VCI. In the future, with the in-depth research on the pathogenesis of VCI and the conduct of more clinical trials, it is expected that comprehensive management strategies will be further optimised to improve the prognosis of VCI patients.

3.8 Challenges, Future Directions and Conclusions

3.8.1 Challenges Faced

- **Heterogeneity**
  There is significant heterogeneity in the clinical manifestations, neuroimaging changes and pathophysiological mechanisms of VaD (Cui et al., 2024; Custodero et al., 2022; Kalaria, 2016; Wei et al., 2025). In terms of clinical symptoms, patients can exhibit a wide range of cognitive impairments, such as some presenting primarily with memory decline, while others show more prominent deficits in executive function and attention, and may also have psychiatric and behavioural symptoms like emotional disorders (Loetscher et al., 2019; Bush, 2011; Park, et al., 1999; Robertson & Schmitter-Edgecombe, 2017; Lim & Alexander, 2009; Henry, 2021; Campos et al., 2010; Alfei et al., 2025; Jahn, 2013; Matsuyoshi et al., 2012). The severity and progression of symptoms vary greatly among different patients. On neuroimaging, the types, locations and extents of vascular brain injuries such as cerebral infarction, cerebral haemorrhage and white matter disease are diverse (Tomimoto et al., 2003; Ji et al., 2024; Jochems et al., 2025; Silbert et al., 2008). For example, Binswanger's disease mainly manifests as bilateral periventricular and centrum semiovale white matter lesions, while strategic infarcts cause dementia due to the location of the infarct in specific key brain regions. In terms of pathophysiological mechanisms, various complex pathways are involved, including neuronal injury, neurotransmitter changes and neuroinflammation caused by vascular brain injury, and the dominant mechanisms may differ among patients.

- **Difficulties in Early Diagnosis**
  Currently, there is a lack of specific, sensitive and convenient early diagnostic tools in clinical practice. Neuropsychological assessment scales have issues such as long testing times and susceptibility to external interference during administration, making it difficult to widely apply them for early screening in community populations. Their sensitivity for diagnosing mild VCI and early VaD is also limited. Although neuroimaging can detect cerebrovascular lesions, it is challenging to identify subtle early vascular brain injuries. Moreover, it is hard to distinguish physiological ageing from pathological changes. Although there has been some progress in the development of biomarkers, most are still in the research phase and have not been widely applied in clinical practice (Tai et al., 2025). For example, NfL protein in blood or CSF has certain diagnostic value, but its specificity is insufficient and further validation is needed (Hofmann et al., 2024; Ramani et al., 2021).

- **Insufficient Validation of Biomarkers**
  Most existing biomarkers have been studied from a single pathological process. However, VaD is the result of the combined effects of multiple factors, and a single biomarker cannot comprehensively reflect its complex pathophysiological process. Although inflammatory markers, endothelial dysfunction markers and coagulation cascade markers are related to VaD, there are differences in the results of different studies, and it is difficult to distinguish specific markers for VaD from other neurodegenerative diseases such as AD. In addition, there are differences in the detection methods of biomarkers among different laboratories, and there is a lack of standardised detection procedures and quality control standards. This leads to poor comparability and reproducibility of research results, affecting their clinical application and validation.

- **Lack of Effective Therapeutic Approaches**
  The treatment of VaD currently focuses mainly on the control of risk factors and symptomatic treatment. Although it can slow the progression of the disease, its effect on the recovery of already damaged neurological functions is limited. In terms of risk factor intervention, such as controlling hypertension, diabetes and hyperlipidaemia, although it can reduce the risk of developing VaD, the therapeutic effect on patients who have already developed the disease is not significant. There are limited options for pharmacological treatments to improve cognitive function, and their efficacy is not satisfactory. For example, cholinesterase inhibitors and NMDA receptor antagonists have certain effects (Joó and Várkonyi, 1969; Xu et al., 2021), but they are not specifically targeted at the pathophysiological mechanisms of VaD. Non-pharmacological treatments such as rehabilitation training are helpful for some patients, but there is a lack of evidence from large-scale clinical trials. Moreover, issues regarding the individualisation of treatment plans and long-term adherence need to be addressed.

- **Complexity of Mixed Pathology**
  VaD often coexists with other neurodegenerative diseases such as AD, forming a mixed pathological state, which increases the complexity of diagnosis and treatment (Almeida et al., 2025; Custodio et al., 2017). In the case of mixed pathology, vascular

and neurodegenerative factors are intertwined, jointly causing cognitive impairment. Moreover, there may be interactions and synergistic effects between their pathological mechanisms. For example, vascular lesions can promote the deposition of Aβ, which in turn can exacerbate cerebral vascular damage and neuroinflammation, making the condition more complex and severe. Currently, the diagnostic and treatment strategies for VaD in the context of mixed pathology are not clear. How to accurately identify mixed pathology and develop targeted treatment plans is an urgent problem to be solved.

3.8.2 Future Research Directions for VaD

- **Precision Typing and Biomarker Development:**
  - Typing Based on Molecular Characteristics: By delving into the molecular mechanisms of VaD, identify different molecular subtypes, such as inflammatory, vascular leakage and neurodegenerative types, to provide a basis for precision treatment. For instance, utilise single-cell sequencing technology to analyse cell subpopulations and their molecular expression characteristics in brain tissue and peripheral blood, uncovering key cell types and molecular pathways associated with different pathological processes of VaD.
  - Development of Multimodal Biomarkers: Integrate various means such as imaging, body fluid biomarkers and genetic testing to develop a comprehensive biomarker panel. For example, combine the characteristics of cerebrovascular lesions detected by MRI, neurotransmitter metabolites, inflammatory factors in blood or CSF, and genetic susceptibility genes to establish predictive models, enhancing the accuracy and specificity of early diagnosis. Machine learning and artificial intelligence technologies can play a significant role in this process, identifying biomarker combinations with diagnostic and prognostic value through the analysis and mining of large amounts of data.
- **In-depth Analysis of the Dynamic Changes of the Neurovascular Unit in Health and Disease**
  - Research on the Molecular Mechanisms of the Neurovascular Unit: Further explore the interactions and regulatory mechanisms among the components of the neurovascular unit, such as the roles of endothelial cells, pericytes, smooth muscle cells, astrocytes, neurons, etc., in maintaining cerebral blood flow regulation, BBB integrity and neurotransmitter transmission. Investigate their dynamic changes in the development of VaD, including vascular contraction and dilation dysfunction, neuroinflammation and protein deposition caused by BBB disruption, and the disruption of NVC.
  - Development of Real-time Monitoring Technologies: Develop technologies capable of real-time monitoring of the functional state of the neurovascular unit, such as those based on optical imaging and magnetic resonance spectroscopy, to achieve dynamic monitoring of indicators like cerebral blood flow, oxygen metabolism and ion concentration. This will enable the timely detection of early vascular dysfunction and neural injury, providing opportunities for early intervention.

- **Mechanistic Studies on the Interactions between Vascular and Neurodegenerative Pathologies**
  - Exploration of Shared Pathways: Investigate the shared pathological mechanisms between VaD and neurodegenerative diseases such as AD and Parkinson's disease, such as oxidative stress, mitochondrial dysfunction and autophagy dysregulation. Explore the roles and interactions of these common pathways in different types of dementia, seeking potential therapeutic targets. For example, study how cerebral ischemia and hypoxia caused by vascular damage trigger oxidative stress reactions in neurons, thereby promoting the hyperphosphorylation of tau protein and the deposition of Aβ, accelerating the progression of neurodegenerative changes.
  - Research on Cross-disease Models: Establish animal and cell models that simultaneously possess vascular and neurodegenerative pathologies. For instance, use gene editing technology to introduce both vascular risk genes and neurodegenerative-related genes into animal models to simulate the mixed pathological state of human dementia diseases, providing a basis for developing treatments targeting mixed pathologies.

- **Novel Therapeutic Strategies Targeting Specific Mechanistic Targets**
  - Drug Development: Based on the pathological mechanisms of VaD, develop new drugs targeting specific targets. For example, develop drugs targeting inflammatory pathways, such as selective inhibitors of inflammatory factor production or action, like IL-1 receptor antagonists and TNF-α inhibitors; drugs targeting angiogenesis and repair, such as VEGF and its receptor agonists, to promote cerebral angiogenesis and improve cerebral blood flow; and drugs targeting neuroprotection, such as neurotrophic factors and antioxidants, to protect neurons from vascular damage and neurotoxic substances.
  - Gene Therapy and Cell Therapy: With the development of gene editing and stem cell technologies, gene and cell therapies offer new hope for the treatment of VaD. For example, use viral vectors to introduce therapeutic genes into specific brain regions to correct pathological gene expression or regulatory abnormalities; and utilise the differentiation potential and paracrine effects of mesenchymal stem cells and neural stem cells to promote the repair and regeneration of damaged neural tissue, improving cognitive function.
  - Device Therapy: Explore the development of medical devices to improve cerebral blood flow and promote neural repair. For example, neuroregulatory technologies such as transcranial magnetic stimulation and transcranial direct current stimulation can improve cognitive function by modulating the excitability of the cerebral cortex; and implantable bioartificial adrenal devices can monitor and regulate the intracranial metabolic environment in real time, reducing vascular brain damage.

- **Large-scale Preventive Intervention Studies**
  - Intervention Studies on Vascular Risk Factors: Conduct large-scale, long-term clinical trials to assess the effects of strict control of vascular risk factors such as hypertension, diabetes, hyperlipidaemia, smoking and alcohol consumption on preventing the occurrence of VaD and delaying its progression. Further clarify the intervention thresholds, optimal treatment plans, and timing for different risk factors, and develop individualised prevention strategies. For example, compare the

protective effects of different antihypertensive drugs on the cognitive function of hypertensive patients to determine the optimal blood pressure control target value.
  - Lifestyle Intervention Studies: Investigate the role of healthy lifestyles such as the Mediterranean diet, aerobic exercise and cognitive training in preventing VaD. Through community intervention trials and other methods, observe the feasibility and effectiveness of these lifestyle interventions in different populations and promote beneficial lifestyles to reduce the incidence of VaD. Additionally, explore the synergistic effects of lifestyle interventions with other preventive measures such as pharmacological treatments to develop comprehensive prevention plans.

- **Application of Digital Health Technologies in Assessment and Management**
  - Remote Assessment and Monitoring: Utilise digital technologies such as wearable devices and smartphone applications to achieve remote real-time monitoring and assessment of patients' cognitive function, physical activity, sleep quality, blood pressure, heart rate and other physiological parameters. By establishing a data platform to integrate and analyse patients' various data, promptly detect changes in patients' conditions and potential problems, and provide a basis for doctors to offer remote medical advice and intervention measures.
  - Intelligent Diagnosis and Decision Support Systems: Develop intelligent diagnosis and decision support systems based on artificial intelligence. By learning and analysing large amounts of clinical data, assist doctors in diagnosing VaD, assessing the condition and formulating treatment plans. For example, use machine learning algorithms to analyse imaging data to improve the accuracy of identifying cerebrovascular lesions and early diagnosis capabilities; and establish clinical decision-making models to recommend personalised treatment plans and intervention measures based on patients' individual characteristics and condition states, enhancing medical quality and efficiency.

3.8.3 Conclusions on VaD

Cerebrovascular health is crucial for maintaining cognitive function. VaD, the second most common type of dementia after AD, is closely related to cerebrovascular disease. Vascular risk factors such as hypertension, diabetes, hyperlipidaemia and smoking contribute to cognitive impairment through various pathways, including damaging the cerebral vascular endothelium, disrupting the BBB and triggering neuroinflammation. These factors lead to cerebral ischemia and hypoxia as well as neuronal injury, ultimately causing cognitive dysfunction. Early identification and effective intervention of vascular risk factors are of key significance for the prevention and delay of the onset and progression of VaD.

In clinical practice, it is essential to enhance the screening and management of vascular risk factors and incorporate cerebrovascular health assessment into routine physical examinations and the diagnostic and treatment processes of patients with cognitive impairment. A comprehensive management strategy should be adopted, including pharmacological treatment, lifestyle interventions and rehabilitation training, to improve patients' vascular function and cognitive status. Meanwhile, multidisciplinary collaboration is of vital importance. A multidisciplinary team, including neurologists, cardiologists, endocrinologists and rehabilitation specialists, should work closely together to provide patients with comprehensive and personalised treatment plans.

Looking to the future, with the in-depth research on the pathogenesis of VaD and the continuous emergence of new diagnostic techniques and therapeutic methods, we anticipate that targeted interventions focusing on vascular mechanisms will effectively improve the cognitive prognosis of patients with VaD and enhance their quality of life. This will require the joint efforts of researchers, clinicians, policymakers and patient organisations worldwide. Strengthening cooperation and communication to promote the continuous development of research and clinical practice in VaD will contribute to addressing this significant public health challenge.

References

Abbott, N.J. (2002). Astrocyte–endothelial interactions and blood–brain barrier permeability*. *Journal of Anatomy* 200 (6): 629–638. https://doi.org/10.1046/j.1469-7580.2002.00064.x.

Abbott, N.J., Rönnbäck, L., and Hansson, E. (2006). Astrocyte–endothelial interactions at the blood–brain barrier. *Nature Reviews Neuroscience* 7 (1): 41–53. https://doi.org/10.1038/nrn1824.

Alfei, J.M., Miller, R.R., Ryan, T.J. et al. (2025). Rethinking memory impairments: retrieval failure. *Psychological Review*. https://doi.org/10.1037/rev0000538.

Almeida, F.C., Marx, G.A., Rohde, S.K. et al. (2025). Cognitive and neuropsychological trajectories in patients with mixed neurodegenerative pathologies. *Alzheimer's & Dementia* 21 (8): e70575. https://doi.org/10.1002/alz.70575.

Altschuler, F., Canziani, V., Fraile-Vázquez, M. et al. (2025). White matter hyperintensities and their impact in brain structure and function in Alzheimer's disease and behavioral variant frontotemporal dementia across Latin America and the United States: a cross-sectional study. *Alzheimer's Research & Therapy* 17 (1): 188. https://doi.org/10.1186/s13195-025-01832-5.

Aoki, T., Sumii, T., Mori, T. et al. (2002). Blood-brain barrier disruption and matrix metalloproteinase-9 expression during reperfusion injury: mechanical versus embolic focal ischemia in spontaneously hypertensive rats. *Stroke* 33 (11): 2711–2717. https://doi.org/10.1161/01.STR.0000033932.34467.97.

Apostolova, L.G., Green, A.E., Babakchanian, S. et al. (2012). Hippocampal atrophy and ventricular enlargement in normal aging, mild cognitive impairment (MCI), and Alzheimer disease. *Alzheimer Disease & Associated Disorders* 26 (1): 17–27. https://doi.org/10.1097/WAD.0b013e3182163b62.

Asch, N., Rahamim, N., Morozov, A. et al. (2025). Basal ganglia deep brain stimulation restores cognitive flexibility and exploration-exploitation balance disrupted by NMDA-R antagonism. *Nature Communications* 16 (1): 4963. https://doi.org/10.1038/s41467-025-60044-5.

Attiq, A., Afzal, S., Raman, H. et al. (2025). Neuroinflammation to neurodegeneration: boulevard of broken nerves. *International Immunopharmacology* 161: 115015. https://doi.org/10.1016/j.intimp.2025.115015.

Attwell, D., Buchan, A.M., Charpak, S. et al. (2010). Glial and neuronal control of brain blood flow. *Nature* 468 (7321): 232–243. https://doi.org/10.1038/nature09613.

Awad, I.A. and Polster, S.P. (2019). Cavernous angiomas: deconstructing a neurosurgical disease: JNSPG 75th anniversary invited review article. *Journal of Neurosurgery* 131 (1): 1–13. https://doi.org/10.3171/2019.3.JNS181724.

Baltaretu, B.R., Stevens, W.D., Freud, E. et al. (2023). Occipital and parietal cortex participate in a cortical network for transsaccadic discrimination of object shape and orientation. *Scientific Reports* 13 (1): 11628. https://doi.org/10.1038/s41598-023-38554-3.

Bangad, A., Abbasi, M., and de Havenon, A. (2023). Secondary ischemic stroke prevention. *Neurotherapeutics* 20 (3): 721–731. https://doi.org/10.1007/s13311-023-01352-w.

Bergsland, N., Tavazzi, E., Schweser, F. et al. (2019). Targeting iron dyshomeostasis for treatment of neurodegenerative disorders. *CNS Drugs* 33 (11): 1073–1086. https://doi.org/10.1007/s40263-019-00668-6.

Bialystok, E. (2025). Selective attention in cognitive processing revisited: a long-term re-evaluation. *Brain and Language* 261: 105518. https://doi.org/10.1016/j.bandl.2024.105518.

Bonelli, R.M. and Cummings, J.L. (2007). Frontal-subcortical circuitry and behavior. *Dialogues in Clinical Neuroscience* 9 (2): 141–151. https://doi.org/10.31887/DCNS.2007.9.2/rbonelli.

Boneva, N.B., Kikuchi, M., Minabe, Y. et al. (2011). Neuroprotective and ameliorative actions of polyunsaturated fatty acids against neuronal diseases: implication of fatty acid–binding proteins (FABP) and G protein–coupled receptor 40 (GPR40) in adult neurogenesis. *Journal of Pharmacological Sciences* 116 (2): 163–172. https://doi.org/10.1254/jphs.10R34FM.

Bonnar, O., Eyre, B., and Van Veluw, S.J. (2025). Perivascular brain clearance as a therapeutic target in cerebral amyloid angiopathy and Alzheimer's disease. *Neurotherapeutics* 22 (3): e00535. https://doi.org/10.1016/j.neurot.2025.e00535.

Buening, J. and Brown, R.D. (2018). Visuospatial cognition. In: *Neuroscience of Mathematical Cognitive Development*, 79–96. Springer International Publishing. https://doi.org/10.1007/978-3-319-76409-2_5.

Bush, G. (2011). Cingulate, frontal, and parietal cortical dysfunction in attention-deficit/hyperactivity disorder. *Biological Psychiatry* 69 (12): 1160–1167. https://doi.org/10.1016/j.biopsych.2011.01.022.

Calderone, A., Latella, D., Cardile, D. et al. (2024). The role of neuroinflammation in shaping neuroplasticity and recovery outcomes following traumatic brain injury: a systematic review. *International Journal of Molecular Sciences* 25 (21): 11708. https://doi.org/10.3390/ijms252111708.

Campos, T.F., Barroso, M.T.M., and De Lara Menezes, A.A. (2010). Encoding, storage and retrieval processes of the memory and the implications for motor practice in stroke patients. *NeuroRehabilitation* 26 (2): 135–142. https://doi.org/10.3233/NRE-2010-0545.

Carter, C.S., Mintun, M., Nichols, T. et al. (1997). Anterior cingulate gyrus dysfunction and selective attention deficits in schizophrenia: [^{15}O]H_2O PET study during single-trial Stroop task performance. *American Journal of Psychiatry* 154 (12): 1670–1675. https://doi.org/10.1176/ajp.154.12.1670.

Chabriat, H. and Lesnik Oberstein, S. (2022). Cognition, mood and behavior in CADASIL. *Cerebral Circulation - Cognition and Behavior* 3: 100043. https://doi.org/10.1016/j.cccb.2022.100043.

Chau, J.P.C., Lo, S.H.S., Zhao, J. et al. (2023). Prevalence of post-stroke cognitive impairment and associated risk factors in Chinese stroke survivors. *Journal of the Neurological Sciences* 455: 122805. https://doi.org/10.1016/j.jns.2023.122805.

Chayer, C. and Freedman, M. (2001). Frontal lobe functions. *Current Neurology and Neuroscience Reports* 1 (6): 547–552. https://doi.org/10.1007/s11910-001-0060-4.

Che, P., Cai, L., Liu, F. et al. (2025). Cerebral perfusion is correlated with cerebral metabolism and amyloid deposition in Alzheimer's disease. *Translational Psychiatry* 15 (1): 189. https://doi.org/10.1038/s41398-025-03402-7.

Chen, J., Xiang, P., Duro-Castano, A. et al. (2025). Rapid amyloid-β clearance and cognitive recovery through multivalent modulation of blood–brain barrier transport. *Signal Transduction and Targeted Therapy* 10 (1): 331. https://doi.org/10.1038/s41392-025-02426-1.

Chu, T., Wang, Y., Wang, S. et al. (2025). Kaempferol regulating macrophage foaming and atherosclerosis through Piezo1-mediated MAPK/NF-κB and Nrf2/HO-1 signaling pathway. *Journal of Advanced Research* 75: 635–650. https://doi.org/10.1016/j.jare.2024.11.016.

Coll, S.Y., Marti, E., Doganci, N. et al. (2024). The disengagement deficit after right-hemisphere damage: distinct roles of lateral frontal and parietal damage. *Brain Research Bulletin* 214: 111003. https://doi.org/10.1016/j.brainresbull.2024.111003.

Constantinides, V.C., Paraskevas, G.P., Boufidou, F. et al. (2023). CSF Aβ42 and Aβ42/Aβ40 ratio in Alzheimer's disease and frontotemporal dementias. *Diagnostics (Basel)* 13 (4): 783. https://doi.org/10.3390/diagnostics13040783.

Coyle, J.T. (2006). Glutamate and schizophrenia: beyond the dopamine hypothesis. *Cellular and Molecular Neurobiology* 26 (4-6): 363–382. https://doi.org/10.1007/s10571-006-9062-8.

Cozza, M., Amadori, L., and Boccardi, V. (2023). Exploring cerebral amyloid angiopathy: insights into pathogenesis, diagnosis, and treatment. *Journal of the Neurological Sciences* 454: 120866. https://doi.org/10.1016/j.jns.2023.120866.

Cristofori, I., Cohen-Zimerman, S., and Grafman, J. (2019). Executive functions. In: *Handbook of Clinical Neurology*, Vol. 163, 197–219. Elsevier. https://doi.org/10.1016/B978-0-12-804281-6.00011-2.

Cui, M., Jin, Z., Wang, Y. et al. (2024). Imaging, biomarkers, and vascular cognitive impairment in China: rationale and design for the VICA study. *Alzheimer's & Dementia* 20 (12): 8898–8909. https://doi.org/10.1002/alz.14352.

Cunningham, C., Campion, S., Lunnon, K. et al. (2009). Systemic inflammation induces acute behavioral and cognitive changes and accelerates neurodegenerative disease. *Biological Psychiatry* 65 (4): 304–312. https://doi.org/10.1016/j.biopsych.2008.07.024.

Custodero, C., Ciavarella, A., Panza, F. et al. (2022). Role of inflammatory markers in the diagnosis of vascular contributions to cognitive impairment and dementia: a systematic review and meta-analysis. *GeroScience* 44 (3): 1373–1392. https://doi.org/10.1007/s11357-022-00556-w.

Custodio, N., Montesinos, R., Lira, D. et al. (2017). Mixed dementia: a review of the evidence. *Dementia & Neuropsychologia* 11 (4): 364–370. https://doi.org/10.1590/1980-57642016dn11-040005.

Cyr, A.R., Huckaby, L.V., Shiva, S.S. et al. (2020). Nitric oxide and endothelial dysfunction. *Critical Care Clinics* 36 (2): 307–321. https://doi.org/10.1016/j.ccc.2019.12.009.

D'Alessandro, M.C.B., Kanaan, S., Geller, M. et al. (2025). Mitochondrial dysfunction in Alzheimer's disease. *Ageing Research Reviews* 107: 102713. https://doi.org/10.1016/j.arr.2025.102713.

Dai, G., Kaazempur-Mofrad, M.R., Natarajan, S. et al. (2004). Distinct endothelial phenotypes evoked by arterial waveforms derived from atherosclerosis-susceptible and -resistant regions of human vasculature. *Proceedings of the National Academy of Sciences of the United States of America* 101 (41): 14871–14876. https://doi.org/10.1073/pnas.0406073101.

Dalile, B., Kim, C., Challinor, A. et al. (2022). The EAT–Lancet reference diet and cognitive function across the life course. *The Lancet Planetary Health* 6 (9): e749–e759. https://doi.org/10.1016/S2542-5196(22)00123-1.

De Souza Coelho, D., Fernandes De Oliveira Santos, B., Silva Da Costa, M.D. et al. (2020). Cognitive performance in patients with cerebral arteriovenous malformation. *Journal of Neurosurgery* 132 (5): 1548–1555. https://doi.org/10.3171/2018.12.JNS181883.

Degraff, Z., Souza, G.S., Santos, N.A. et al. (2023). Brain atrophy and cognitive decline in bipolar disorder: influence of medication use, symptomatology and illness duration. *Journal of Psychiatric Research* 163: 421–429. https://doi.org/10.1016/j.jpsychires.2023.05.074.

Derdeyn, C.P., Videen, T.O., Yundt, K.D. et al. (2002). Variability of cerebral blood volume and oxygen extraction: stages of cerebral haemodynamic impairment revisited. *Brain* 125 (3): 595–607. https://doi.org/10.1093/brain/awf047.

Dey, S. and Snow, D.M. (2007). Cocaine exposure in vitro induces apoptosis in fetal locus coeruleus neurons through TNF-α-mediated induction of Bax and phosphorylated c-Jun NH_2 -terminal kinase. *Journal of Neurochemistry* 103 (2): 542–556. https://doi.org/10.1111/j.1471-4159.2007.04750.x.

Dichgans, M. and Leys, D. (2017). Vascular cognitive impairment. *Circulation Research* 120 (3): 573–591. https://doi.org/10.1161/CIRCRESAHA.116.308426.

Ding, X., Sun, X., Shen, X. et al. (2019). Propofol attenuates TNF-α-induced MMP-9 expression in human cerebral microvascular endothelial cells by inhibiting Ca^{2+}/CAMK II/ERK/NF-κB signaling pathway. *Acta Pharmacologica Sinica* 40 (10): 1303–1313. https://doi.org/10.1038/s41401-019-0258-0.

Dong, M., Chen, M., Zhang, Y. et al. (2024). Oscillatory shear stress promotes endothelial senescence and atherosclerosis via STING activation. *Biochemical and Biophysical Research Communications* 715: 149979. https://doi.org/10.1016/j.bbrc.2024.149979.

Dudvarski Stankovic, N., Teodorczyk, M., Ploen, R. et al. (2016). Microglia–blood vessel interactions: a double-edged sword in brain pathologies. *Acta Neuropathologica* 131 (3): 347–363. https://doi.org/10.1007/s00401-015-1524-y.

Dulamea, A.O. and Lupescu, I.C. (2024). Cerebral cavernous malformations – an overview on genetics, clinical aspects and therapeutic strategies. *Journal of the Neurological Sciences* 461: 123044. https://doi.org/10.1016/j.jns.2024.123044.

Eisenmenger, L.B., Peret, A., Famakin, B.M. et al. (2023). Vascular contributions to Alzheimer's disease. *Translational Research* 254: 41–53. https://doi.org/10.1016/j.trsl.2022.12.003.

Elbert, D.L., Patterson, B.W., Lucey, B.P. et al. (2022). Importance of CSF-based Aβ clearance with age in humans increases with declining efficacy of blood-brain barrier/proteolytic pathways. *Communications Biology* 5 (1): 98. https://doi.org/10.1038/s42003-022-03037-0.

Eldaief, M.C., Brickhouse, M., Katsumi, Y. et al. (2023). Atrophy in behavioural variant frontotemporal dementia spans multiple large-scale prefrontal and temporal networks. *Brain* 146 (11): 4476–4485. https://doi.org/10.1093/brain/awad167.

Engelhardt, B. and Coisne, C. (2011). Fluids and barriers of the CNS establish immune privilege by confining immune surveillance to a two-walled castle moat surrounding the CNS castle. *Fluids Barriers CNS* 8 (1): 4. https://doi.org/10.1186/2045-8118-8-4.

Estelle, M.C.P. and Voelbel, G.T. (2024). The effect of processing speed on verbal and visual memory of adults with a chronic acquired brain injury. *Brain Injury* 38 (3): 170–176. https://doi.org/10.1080/02699052.2024.2309250.

Farkas, E., Donka, G., De Vos, R.A.I. et al. (2004). Experimental cerebral hypoperfusion induces white matter injury and microglial activation in the rat brain. *Acta Neuropathologica* 108 (1): 57–64. https://doi.org/10.1007/s00401-004-0864-9.

Faust, T.E., Lee, Y.H., O'Connor, C.D. et al. (2025). Microglia-astrocyte crosstalk regulates synapse remodeling via Wnt signaling. *Cell* 188 (19): 5212–5230.e21. https://doi.org/10.1016/j.cell.2025.08.023.

Felix-Ilemhenbhio, F., Kocsy, K., Azzouz, M. et al. (2025). The role of NOTCH3 in CADASIL pathogenesis: insights into novel therapies. *Brain Research* 1863: 149754. https://doi.org/10.1016/j.brainres.2025.149754.

Filler, J., Georgakis, M.K., and Dichgans, M. (2024). Risk factors for cognitive impairment and dementia after stroke: a systematic review and meta-analysis. *The Lancet Healthy Longevity* 5 (1): e31–e44. https://doi.org/10.1016/S2666-7568(23)00217-9.

Filler, J., Georgakis, M.K., Janowitz, D. et al. (2025). Risk factors for dementia and cognitive impairment within 5 years after stroke: a prospective multicentre cohort study. *The Lancet Regional Health – Europe* 56: 101428. https://doi.org/10.1016/j.lanepe.2025.101428.

Fjell, A.M., Rogeberg, O., Sørensen, Ø. et al. (2025). Reevaluating the role of education on cognitive decline and brain aging in longitudinal cohorts across 33 Western countries. *Nature Medicine* 31 (9): 2967–2976. https://doi.org/10.1038/s41591-025-03828-y.

Friederici, A.D. (2011). The brain basis of language processing: from structure to function. *Physiological Reviews* 91 (4): 1357–1392. https://doi.org/10.1152/physrev.00006.2011.

Galeffi, F., Sah, R., Pond, B.B. et al. (2004). Changes in intracellular chloride after oxygen–glucose deprivation of the adult hippocampal slice: effect of diazepam. *Journal of Neuroscience* 24 (18): 4478–4488. https://doi.org/10.1523/JNEUROSCI.0755-04.2004.

Ghanavati, E., Salehinejad, M.A., Nejati, V. et al. (2019). Differential role of prefrontal, temporal and parietal cortices in verbal and figural fluency: implications for the supramodal contribution of executive functions. *Scientific Reports* 9 (1): 3700. https://doi.org/10.1038/s41598-019-40273-7.

Gorelick, P.B., Scuteri, A., Black, S.E. et al. (2011). Vascular contributions to cognitive impairment and dementia: a statement for healthcare professionals from the American Heart Association/American Stroke Association. *Stroke* 42 (9): 2672–2713. https://doi.org/10.1161/STR.0b013e3182299496.

Grafman, J. and Litvan, I. (1999). Importance of deficits in executive functions. *The Lancet* 354 (9194): 1921–1923. https://doi.org/10.1016/S0140-6736(99)90438-5.

Greenberg, S.M., Bacskai, B.J., Hernandez-Guillamon, M. et al. (2020). Cerebral amyloid angiopathy and Alzheimer disease — one peptide, two pathways. *Nature Reviews Neurology* 16 (1): 30–42. https://doi.org/10.1038/s41582-019-0281-2.

Guo, W. and Shi, J. (2022). White matter hyperintensities volume and cognition: a meta-analysis. *Frontiers in Aging Neuroscience* 14: 949763. https://doi.org/10.3389/fnagi.2022.949763.

Gutierrez, J., Turan, T.N., Hoh, B.L. et al. (2022). Intracranial atherosclerotic stenosis: risk factors, diagnosis, and treatment. *The Lancet Neurology* 21 (4): 355–368. https://doi.org/10.1016/S1474-4422(21)00376-8.

Han, Q., Zhao, X., Shao, Y. et al. (2025). Efficacy of butylphthalide in preventing cognitive decline in ischaemic stroke survivors: a 12-month prospective following-up study. *Stroke and Vascular Neurology* 10 (5): 560–568. https://doi.org/10.1136/svn-2024-003611.

Hardebo, J.E. and Owman, C. (1980). Barrier mechanisms for neurotransmitter monoamines and their precursors at the blood-brain interface. *Annals of Neurology* 8 (1): 1–11. https://doi.org/10.1002/ana.410080102.

Haruwaka, K., Ikegami, A., Tachibana, Y. et al. (2019). Dual microglia effects on blood brain barrier permeability induced by systemic inflammation. *Nature Communications* 10 (1): 5816. https://doi.org/10.1038/s41467-019-13812-z.

Haworth, J., Phillips, M., Newson, M. et al. (2016). Measuring information processing speed in mild cognitive impairment: clinical versus research dichotomy. *Journal of Alzheimer's Disease* 51 (1): 263–275. https://doi.org/10.3233/JAD-150791.

Heneka, M.T., Van Der Flier, W.M., Jessen, F. et al. (2025). Neuroinflammation in Alzheimer disease. *Nature Reviews Immunology* 25 (5): 321–352. https://doi.org/10.1038/s41577-024-01104-7.

Hennig, B. (1988). Lipid peroxidation and endothelial cell injury: implications in atherosclerosis. *Free Radical Biology and Medicine* 4 (2): 99–106. https://doi.org/10.1016/0891-5849(88)90070-6.

Henry, J.D. (2021). Prospective memory impairment in neurological disorders: implications and management. *Nature Reviews Neurology* 17 (5): 297–307. https://doi.org/10.1038/s41582-021-00472-1.

Hirata, Y., Cai, R., Volchuk, A. et al. (2023). Lipid peroxidation increases membrane tension, Piezo1 gating, and cation permeability to execute ferroptosis. *Current Biology* 33 (7): 1282–1294.e5. https://doi.org/10.1016/j.cub.2023.02.060.

Hofmann, A., Häsler, L.M., Lambert, M. et al. (2024). Comparative neurofilament light chain trajectories in CSF and plasma in autosomal dominant Alzheimer's disease. *Nature Communications* 15 (1): 9982. https://doi.org/10.1038/s41467-024-52937-8.

Holmstedt, C.A., Turan, T.N., and Chimowitz, M.I. (2013). Atherosclerotic intracranial arterial stenosis: risk factors, diagnosis, and treatment. *The Lancet Neurology* 12 (11): 1106–1114. https://doi.org/10.1016/S1474-4422(13)70195-9.

Hu, H.Y., Ou, Y.N., Shen, X.N. et al. (2021). White matter hyperintensities and risks of cognitive impairment and dementia: a systematic review and meta-analysis of 36 prospective studies. *Neuroscience & Biobehavioral Reviews* 120: 16–27. https://doi.org/10.1016/j.neubiorev.2020.11.007.

Huang, X., Fowler, C., Li, Y. et al. (2024). Clearance and transport of amyloid β by peripheral monocytes correlate with Alzheimer's disease progression. *Nature Communications* 15 (1): 7998. https://doi.org/10.1038/s41467-024-52396-1.

Iadecola, C. (1993). Regulation of the cerebral microcirculation during neural activity: is nitric oxide the missing link? *Trends in Neurosciences* 16 (6): 206–214. https://doi.org/10.1016/0166-2236(93)90156-G.

Iadecola, C. (2017). The neurovascular unit coming of age: a journey through neurovascular coupling in health and disease. *Neuron* 96 (1): 17–42. https://doi.org/10.1016/j.neuron.2017.07.030.

Iadecola, C., Duering, M., Hachinski, V. et al. (2019). Vascular cognitive impairment and dementia. *Journal of the American College of Cardiology* 73 (25): 3326–3344. https://doi.org/10.1016/j.jacc.2019.04.034.

Iadecola, C., Yaffe, K., Biller, J. et al. (2016). Impact of hypertension on cognitive function: a scientific statement from the American Heart Association. *Hypertension* 68 (6). https://doi.org/10.1161/HYP.0000000000000053.

Ishikawa, J., Seino, S., Kitamura, A. et al. (2021). The relationship between blood pressure and cognitive function. *International Journal of Cardiology Cardiovascular Risk and Prevention* 10: 200104. https://doi.org/10.1016/j.ijcrp.2021.200104.

Jackson, W.F. (2000). Ion channels and vascular tone. *Hypertension* 35 (1): 173–178. https://doi.org/10.1161/01.HYP.35.1.173.

Jahn, H. (2013). Memory loss in Alzheimer's disease. *Dialogues in Clinical Neuroscience* 15 (4): 445–454. https://doi.org/10.31887/DCNS.2013.15.4/hjahn.

Jha, M.K., Jo, M., Kim, J.H. et al. (2019). Microglia-astrocyte crosstalk: an intimate molecular conversation. *Neuroscientist* 25 (3): 227–240. https://doi.org/10.1177/1073858418783959.

Ji, X., Zhang, X., Zhang, J. et al. (2024). Association between plasma trimethylamine N-oxide and cerebral white matter hyperintensity: a cross-sectional study. *Frontiers in Aging Neuroscience* 16: 1498502. https://doi.org/10.3389/fnagi.2024.1498502.

Jochems, A.C.C., Muñoz Maniega, S., Clancy, U. et al. (2025). Longitudinal cognitive changes in cerebral small vessel disease: the effect of white matter hyperintensity regression and progression. *Neurology* 104 (4): e213323. https://doi.org/10.1212/WNL.0000000000213323.

Joó, F. and Várkonyi, T. (1969). Correlation between the cholinesterase activity of capillaries and the blood-brain barrier in the rat. *Acta Biologica Academiae Scientiarum Hungaricae* 20 (4): 359–372.

Kalaria, R.N. (2016). Neuropathological diagnosis of vascular cognitive impairment and vascular dementia with implications for Alzheimer's disease. *Acta Neuropathologica* 131 (5): 659–685. https://doi.org/10.1007/s00401-016-1571-z.

Kamrani, F., Kachouei, A.A., Sobhani, S.R. et al. (2024). Nourishing the mind: how the EAT-Lancet reference diet (ELD) and MIND diet impact stress, anxiety, and depression. *BMC Psychiatry* 24 (1): 709. https://doi.org/10.1186/s12888-024-06165-5.

Kan, C.N., Gyanwali, B., Hilal, S. et al. (2020). Neuropsychiatric correlates of small vessel disease progression in incident cognitive decline: independent and interactive effects. *Journal of Alzheimer's Disease* 73 (3): 1053–1062. https://doi.org/10.3233/JAD-190999.

Karplus, T.M. and Saag, K.G. (1998). Nonsteroidal anti-inflammatory drugs and cognitive function: do they have a beneficial or deleterious effect? *Drug Safety* 19 (6): 427–433. https://doi.org/10.2165/00002018-199819060-00001.

Kempe, S. (2005). NF- B controls the global pro-inflammatory response in endothelial cells: evidence for the regulation of a pro-atherogenic program. *Nucleic Acids Research* 33 (16): 5308–5319. https://doi.org/10.1093/nar/gki836.

Kim, H.F. (2025). Principal axis of primate basal ganglia in processing cognitive flexibility and habitual stability. *Brain* 148 (12): 4198–4209. https://doi.org/10.1093/brain/awaf299.

Kim, J., Tchernyshyov, I., Semenza, G.L. et al. (2006). HIF-1-mediated expression of pyruvate dehydrogenase kinase: a metabolic switch required for cellular adaptation to hypoxia. *Cell Metabolism* 3 (3): 177–185. https://doi.org/10.1016/j.cmet.2006.02.002.

Ko, J.H., Tang, C.C., and Eidelberg, D. (2013). Brain stimulation and functional imaging with fMRI and PET. In: *Handbook of Clinical Neurology*, Vol. 116, 77–95. Elsevier. https://doi.org/10.1016/B978-0-444-53497-2.00008-5.

Kontos, H.A., Wei, E.P., Raper, A.J. et al. (1977). Local mechanism of CO_2 action of cat pial arterioles. *Stroke* 8 (2): 226–229. https://doi.org/10.1161/01.STR.8.2.226.

Krueger, M., Bechmann, I., Immig, K. et al. (2015). Blood—brain barrier breakdown involves four distinct stages of vascular damage in various models of experimental focal cerebral ischemia. *Journal of Cerebral Blood Flow & Metabolism* 35 (2): 292–303. https://doi.org/10.1038/jcbfm.2014.199.

Lacoste, B., Prat, A., Freitas-Andrade, M. et al. (2025). The blood–brain barrier: composition, properties, and roles in brain health. *Cold Spring Harbor Perspectives in Biology* 17 (5): a041422. https://doi.org/10.1101/cshperspect.a041422.

Lanctôt, K.L., Ismail, Z., Bawa, K.K. et al. (2023). Distinguishing apathy from depression: a review differentiating the behavioral, neuroanatomic, and treatment-related aspects of apathy from depression in neurocognitive disorders. *International Journal of Geriatric Psychiatry* 38 (2): e5882. https://doi.org/10.1002/gps.5882.

Latchaw, R.E., Alberts, M.J., Lev, M.H. et al. (2009). Recommendations for imaging of acute ischemic stroke: a scientific statement from the American Heart Association. *Stroke* 40 (11): 3646–3678. https://doi.org/10.1161/STROKEAHA.108.192616.

Lee, B., Kwon, J.T., Jeong, Y. et al. (2025). Inflammatory and anti-inflammatory cytokines bidirectionally modulate amygdala circuits regulating anxiety. *Cell* 188 (8): 2190–2202.e15. https://doi.org/10.1016/j.cell.2025.03.005.

Lee, J.S., Heo, D.Y., Choi, K.H. et al. (2024). Impact of the ventricle size on Alzheimer's disease progression: a retrospective longitudinal study. *Dementia and Neurocognitive Disorders* 23 (2): 95–106. https://doi.org/10.12779/dnd.2024.23.2.95.

Lee, S., Viqar, F., Zimmerman, M.E. et al. (2016). White matter hyperintensities are a core feature of Alzheimer's disease: evidence from the dominantly inherited Alzheimer network. *Annals of Neurology* 79 (6): 929–939. https://doi.org/10.1002/ana.24647.

Li, J., Yang, M., Wei, R. et al. (2024). The predictive ability of blood neurofilament light chain in predicting cognitive decline in the Alzheimer's disease continuum: a systematic review and meta-analysis. *Journal of Alzheimer's Disease* 97 (4): 1589–1620. https://doi.org/10.3233/JAD-231080.

Li, X., Wang, Z., Zhang, H. et al. (2024). Tract-specific white matter hyperintensities disrupt brain networks and associated with cognitive impairment in mild traumatic brain injury. *Human Brain Mapping* 45 (17): e70050. https://doi.org/10.1002/hbm.70050.

Liang, L.B., Wang, S., Li, K.P. et al. (2025). Comparative efficacy of cognitive training modalities in cognitive impairment: a systematic review and network meta-analysis. *The Journal of Prevention of Alzheimer's Disease* 12 (7): 100207. https://doi.org/10.1016/j.tjpad.2025.100207.

Liao, F.F., Lin, G., Chen, X. et al. (2021). Endothelial nitric oxide synthase–deficient mice. *The American Journal of Pathology* 191 (11): 1932–1945. https://doi.org/10.1016/j.ajpath.2021.02.022.

Lim, C. and Alexander, M.P. (2009). Stroke and episodic memory disorders. *Neuropsychologia* 47 (14): 3045–3058. https://doi.org/10.1016/j.neuropsychologia.2009.08.002.

Lim, C., Lee, H., Moon, Y. et al. (2025). Volume and permeability of white matter hyperintensity on cognition: a DCE imaging study of an older cohort with and without cognitive impairment. *Magnetic Resonance Imaging* 61 (5): 2260–2270. https://doi.org/10.1002/jmri.29631.

Lin, S.Y., Lin, C.P., Hsieh, T.J. et al. (2019). Multiparametric graph theoretical analysis reveals altered structural and functional network topology in Alzheimer's disease. *NeuroImage: Clinical* 22: 101680. https://doi.org/10.1016/j.nicl.2019.101680.

Liu, G., Tan, X., Dang, C. et al. (2019). Regional shape abnormalities in thalamus and verbal memory impairment after subcortical infarction. *Neurorehabilitation and Neural Repair* 33 (6): 476–485. https://doi.org/10.1177/1545968319846121.

Liu, J., Bayle, D.J., Spagna, A. et al. (2023). Fronto-parietal networks shape human conscious report through attention gain and reorienting. *Communications Biology* 6 (1): 730. https://doi.org/10.1038/s42003-023-05108-2.

Liu, J.Y., Zhu, Y.C., Zhou, L.X. et al. (2020). HTRA1-related autosomal dominant cerebral small vessel disease. *Chinese Medical Journal* 134 (2): 178–184. https://doi.org/10.1097/CM9.0000000000001176.

Liu, T., Clark, R.K., McDonnell, P.C. et al. (1994). Tumor necrosis factor-alpha expression in ischemic neurons. *Stroke* 25 (7): 1481–1488. https://doi.org/10.1161/01.STR.25.7.1481.

Livne-Bar, I., Lam, S., Chan, D. et al. (2016). Pharmacologic inhibition of reactive gliosis blocks TNF-α-mediated neuronal apoptosis. *Cell Death & Disease* 7 (9): e2386. https://doi.org/10.1038/cddis.2016.277.

Lo, E.H. and Rosenberg, G.A. (2009). The neurovascular unit in health and disease: introduction. *Stroke* 40 (3_suppl_1). https://doi.org/10.1161/STROKEAHA.108.534404.

Lo, J.W., Crawford, J.D., Desmond, D.W. et al. (2022). Long-term cognitive decline after stroke: an individual participant data meta-analysis. *Stroke* 53 (4): 1318–1327. https://doi.org/10.1161/STROKEAHA.121.035796.

Loetscher, T., Potter, K.J., Wong, D. et al. (2019). Cognitive rehabilitation for attention deficits following stroke. *Cochrane Database of Systematic Reviews*. https://doi.org/10.1002/14651858.CD002842.pub3.

Logue, S.F. and Gould, T.J. (2014). The neural and genetic basis of executive function: attention, cognitive flexibility, and response inhibition. *Pharmacology Biochemistry and Behavior* 123: 45–54. https://doi.org/10.1016/j.pbb.2013.08.007.

Lorek, M., Bąk, D., Kwiecień-Jaguś, K. et al. (2023). The effect of music as a non-pharmacological intervention on the physiological, psychological, and social response of patients in an intensive care unit. *Healthcare (Basel)* 11 (12): 1687. https://doi.org/10.3390/healthcare11121687.

Lundberg, J.O. and Weitzberg, E. (2022). Nitric oxide signaling in health and disease. *Cell* 185 (16): 2853–2878. https://doi.org/10.1016/j.cell.2022.06.010.

Mahon, S., Faulkner, J., Barker-Collo, S. et al. (2020). Slowed information processing speed at four years poststroke: evidence and predictors from a population-based follow-up study. *Journal of Stroke and Cerebrovascular Diseases* 29 (2): 104513. https://doi.org/10.1016/j.jstrokecerebrovasdis.2019.104513.

Mather, M. and Harley, C.W. (2016). The locus coeruleus: essential for maintaining cognitive function and the aging brain. *Trends in Cognitive Sciences* 20 (3): 214–226. https://doi.org/10.1016/j.tics.2016.01.001.

Matsuyoshi, D., Ikeda, T., Sawamoto, N. et al. (2012). Differential roles for parietal and occipital cortices in visual working memory. *PLoS One* 7 (6): e38623. https://doi.org/10.1371/journal.pone.0038623.

McBurney-Lin, J., Lu, J., Zuo, Y. et al. (2019). Locus coeruleus-norepinephrine modulation of sensory processing and perception: a focused review. *Neuroscience & Biobehavioral Reviews* 105: 190–199. https://doi.org/10.1016/j.neubiorev.2019.06.009.

McKeith, I.G., Boeve, B.F., Dickson, D.W. et al. (2017). Diagnosis and management of dementia with Lewy bodies: fourth consensus report of the DLB Consortium. *Neurology* 89 (1): 88–100. https://doi.org/10.1212/WNL.0000000000004058.

Mekala, A. and Qiu, H. (2025). Interplay between vascular dysfunction and neurodegenerative pathology: new insights into molecular mechanisms and management. *Biomolecules* 15 (5): 712. https://doi.org/10.3390/biom15050712.

Menezes Cordeiro, I., Nzwalo, H., Sá, F. et al. (2015). Shifting the CARASIL paradigm: report of a non-Asian family and literature review. *Stroke* 46 (4): 1110–1112. https://doi.org/10.1161/STROKEAHA.114.006735.

Merlini, M., Rafalski, V.A., Rios Coronado, P.E. et al. (2019). Fibrinogen induces microglia-mediated spine elimination and cognitive impairment in an Alzheimer's disease model. *Neuron* 101 (6): 1099–1108.e6. https://doi.org/10.1016/j.neuron.2019.01.014.

Mishra, A., Reynolds, J.P., Chen, Y. et al. (2016). Astrocytes mediate neurovascular signaling to capillary pericytes but not to arterioles. *Nature Neuroscience* 19 (12): 1619–1627. https://doi.org/10.1038/nn.4428.

Montagne, A., Barnes, S.R., Sweeney, M.D. et al. (2015). Blood-brain barrier breakdown in the aging human hippocampus. *Neuron* 85 (2): 296–302. https://doi.org/10.1016/j.neuron.2014.12.032.

Moorhouse, P., Song, X., Rockwood, K. et al. (2010). Executive dysfunction in vascular cognitive impairment in the consortium to investigate vascular impairment of cognition study. *Journal of the Neurological Sciences* 288 (1-2): 142–146. https://doi.org/10.1016/j.jns.2009.09.017.

Morrison, C., Dadar, M., Shafiee, N. et al. (2022). Regional brain atrophy and cognitive decline depend on definition of subjective cognitive decline. *NeuroImage: Clinical* 33: 102923. https://doi.org/10.1016/j.nicl.2021.102923.

Nakagami, H. (2003). NADPH oxidase-derived superoxide anion mediates angiotensin II-induced cardiac hypertrophy. *Journal of Molecular and Cellular Cardiology* 35 (7): 851–859. https://doi.org/10.1016/S0022-2828(03)00145-7.

Nation, D.A., Sweeney, M.D., Montagne, A. et al. (2019). Blood–brain barrier breakdown is an early biomarker of human cognitive dysfunction. *Nature Medicine* 25 (2): 270–276. https://doi.org/10.1038/s41591-018-0297-y.

Ndayisaba, A., Kaindlstorfer, C., and Wenning, G.K. (2019). Iron in neurodegeneration – cause or consequence? *Frontiers in Neuroscience* 13: 180. https://doi.org/10.3389/fnins.2019.00180.

Nichols, E., Steinmetz, J.D., Vollset, S.E. et al. (2022). Estimation of the global prevalence of dementia in 2019 and forecasted prevalence in 2050: an analysis for the Global Burden of Disease Study 2019. *The Lancet Public Health* 7 (2): e105–e125. https://doi.org/10.1016/S2468-2667(21)00249-8.

Niwa, K., Haensel, C., Ross, M.E. et al. (2001). Cyclooxygenase-1 participates in selected vasodilator responses of the cerebral circulation. *Circulation Research* 88 (6): 600–608. https://doi.org/10.1161/01.RES.88.6.600.

O'Brien, J.T. and Thomas, A. (2015). Vascular dementia. *The Lancet* 386 (10004): 1698–1706. https://doi.org/10.1016/S0140-6736(15)00463-8.

O'Sullivan, M.J., Li, X., Galligan, D. et al. (2023). Cognitive recovery after stroke: memory. *Stroke* 54 (1): 44–54. https://doi.org/10.1161/STROKEAHA.122.041497.

Onandia-Hinchado, I., Pardo-Palenzuela, N., and Diaz-Orueta, U. (2021). Cognitive characterization of adult attention deficit hyperactivity disorder by domains: a systematic review. *Journal of Neural Transmission* 128 (7): 893–937. https://doi.org/10.1007/s00702-021-02302-6.

Pantoni, L. (2010). Cerebral small vessel disease: from pathogenesis and clinical characteristics to therapeutic challenges. *The Lancet Neurology* 9 (7): 689–701. https://doi.org/10.1016/S1474-4422(10)70104-6.

Pantsiou, K., Sfakianaki, O., Papaliagkas, V. et al. (2018). Inhibitory control, task/rule switching, and cognitive planning in vascular dementia: are there any differences from vascular aging? *Frontiers in Aging Neuroscience* 10: 330. https://doi.org/10.3389/fnagi.2018.00330.

Park, N.W., Moscovitch, M., and Robertson, I.H. (1999). Divided attention impairments after traumatic brain injury. *Neuropsychologia* 37 (10): 1119–1133. https://doi.org/10.1016/S0028-3932(99)00034-2.

Perron, J., Scramstad, C., and Ko, J.H. (2025). Brain metabolic imaging-based model identifies cognitive stability in prodromal Alzheimer's disease. *Scientific Reports* 15 (1): 17187. https://doi.org/10.1038/s41598-025-02039-2.

Petrova, N., Carassiti, D., Altmann, D.R. et al. (2018). Axonal loss in the multiple sclerosis spinal cord revisited. *Brain Pathology* 28 (3): 334–348. https://doi.org/10.1111/bpa.12516.

Pohjasvaara, T., Mäntylä, R., Ylikoski, R. et al. (2000). Comparison of different clinical criteria (DSM-III, ADDTC, ICD-10, NINDS-AIREN, DSM-IV) for the diagnosis of vascular dementia. *Stroke* 31 (12): 2952–2957. https://doi.org/10.1161/01.STR.31.12.2952.

Ramani, S., Berard, J.A., and Walker, L.A.S. (2021). The relationship between neurofilament light chain and cognition in neurological disorders: a scoping review. *Journal of the Neurological Sciences* 420: 117229. https://doi.org/10.1016/j.jns.2020.117229.

Rapp, S.R., Gaussoin, S.A., Sachs, B.C. et al. (2020). Effects of intensive versus standard blood pressure control on domain-specific cognitive function: a substudy of the SPRINT randomised controlled trial. *The Lancet Neurology* 19 (11): 899–907. https://doi.org/10.1016/S1474-4422(20)30319-7.

Rasmussen, M.K., Mestre, H., and Nedergaard, M. (2018). The glymphatic pathway in neurological disorders. *The Lancet Neurology* 17 (11): 1016–1024. https://doi.org/10.1016/S1474-4422(18)30318-1.

Rius-Pérez, S., Tormos, A.M., Pérez, S. et al. (2018). Vascular pathology: cause or effect in Alzheimer disease? *Neurología (English Edition)* 33 (2): 112–120. https://doi.org/10.1016/j.nrleng.2015.07.008.

Robertson, K. and Schmitter-Edgecombe, M. (2017). Focused and divided attention abilities in the acute phase of recovery from moderate to severe traumatic brain injury. *Brain Injury* 31 (8): 1069–1076. https://doi.org/10.1080/02699052.2017.1296192.

Rocca, P., Cocuzza, E., Marchiaro, L. et al. (2002). Donepezil in the treatment of Alzheimer's disease. *Progress in Neuro-Psychopharmacology and Biological Psychiatry* 26 (2): 369–373. https://doi.org/10.1016/S0278-5846(01)00283-4.

Roe, J.M., Vidal-Piñeiro, D., Sørensen, Ø. et al. (2021). Asymmetric thinning of the cerebral cortex across the adult lifespan is accelerated in Alzheimer's disease. *Nature Communications* 12 (1): 721. https://doi.org/10.1038/s41467-021-21057-y.

Román, G.C., Tatemichi, T.K., Erkinjuntti, T. et al. (1993). Vascular dementia: diagnostic criteria for research studies. Report of the NINDS-AIREN International Workshop*. *Neurology* 43 (2): 250. https://doi.org/10.1212/WNL.43.2.250.

Rossi, D.J., Brady, J.D., and Mohr, C. (2007). Astrocyte metabolism and signaling during brain ischemia. *Nature Neuroscience* 10 (11): 1377–1386. https://doi.org/10.1038/nn2004.

Rossi, D.J., Oshima, T., and Attwell, D. (2000). Glutamate release in severe brain ischaemia is mainly by reversed uptake. *Nature* 403 (6767): 316–321. https://doi.org/10.1038/35002090.

Ryter, S.W. (2019). Heme oxygenase-1/carbon monoxide as modulators of autophagy and inflammation. *Archives of Biochemistry and Biophysics* 678: 108186. https://doi.org/10.1016/j.abb.2019.108186.

Sachdev, P., Kalaria, R., O'Brien, J. et al. (2014). Diagnostic criteria for vascular cognitive disorders: a VASCOG statement. *Alzheimer Disease & Associated Disorders* 28 (3): 206–218. https://doi.org/10.1097/WAD.0000000000000034.

Schipper, H.M., Song, W., Tavitian, A. et al. (2019). The sinister face of heme oxygenase-1 in brain aging and disease. *Progress in Neurobiology* 172: 40–70. https://doi.org/10.1016/j.pneurobio.2018.06.008.

Seeley, R.J. and Woods, S.C. (2003). Monitoring of stored and available fuel by the CNS: implications for obesity. *Nature Reviews Neuroscience* 4 (11): 901–909. https://doi.org/10.1038/nrn1245.

Sheikh, M. and Ammar, M. (2024). Efficacy of 5 and 10 mg donepezil in improving cognitive function in patients with dementia: a systematic review and meta-analysis. *Frontiers in Neuroscience* 18: 1398952. https://doi.org/10.3389/fnins.2024.1398952.

Shibata, M., Ohtani, R., Ihara, M. et al. (2004). White matter lesions and glial activation in a novel mouse model of chronic cerebral hypoperfusion. *Stroke* 35 (11): 2598–2603. https://doi.org/10.1161/01.STR.0000143725.19053.60.

Sigmundsdottir, L., Tchan, M.C., Knopman, A.A. et al. (2014). Cognitive and psychological functioning in Fabry disease. *Archives of Clinical Neuropsychology* 29 (7): 642–650. https://doi.org/10.1093/arclin/acu047.

Silbert, L.C., Nelson, C., Howieson, D.B. et al. (2008). Impact of white matter hyperintensity volume progression on rate of cognitive and motor decline. *Neurology* 71 (2): 108–113. https://doi.org/10.1212/01.wnl.0000316799.86917.37.

Skrobot, O.A., O'Brien, J., Black, S. et al. (2017). The vascular impairment of cognition classification consensus study. *Alzheimer's & Dementia* 13 (6): 624–633. https://doi.org/10.1016/j.jalz.2016.10.007.

Smith, A. (1968). The symbol-digit modalities test: a neuropsychologic test of learning and other cerebral disorders. In: *Learning Disorders*, 83–91. Special Child Publications.

Song, S., Yu, L., Hasan, M.N. et al. (2022). Elevated microglial oxidative phosphorylation and phagocytosis stimulate post-stroke brain remodeling and cognitive function recovery in mice. *Communications Biology* 5 (1): 35. https://doi.org/10.1038/s42003-021-02984-4.

Spano, M., Roytman, M., Aboian, M. et al. (2023). Brain PET imaging: approach to cognitive impairment and dementia. *PET Clinics* 18 (1): 103–113. https://doi.org/10.1016/j.cpet.2022.09.006.

Squire, L.R. and Wixted, J.T. (2011). The cognitive neuroscience of human memory since H.M. *Annual Review of Neuroscience* 34 (1): 259–288. https://doi.org/10.1146/annurev-neuro-061010-113720.

Stafford, J., Dekhtyar, S., Welmer, A.K. et al. (2024). Social health and subsequent cognitive functioning in people aged 50 years and older: examining the mediating roles of depressive symptoms and inflammatory biomarkers in two European longitudinal studies. *The Lancet Healthy Longevity* 5 (5): e356–e369. https://doi.org/10.1016/S2666-7568(24)00046-1.

Steffens, D.C., Fahed, M., Manning, K.J. et al. (2022). The neurobiology of apathy in depression and neurocognitive impairment in older adults: a review of epidemiological, clinical, neuropsychological and biological research. *Translational Psychiatry* 12 (1): 525. https://doi.org/10.1038/s41398-022-02292-3.

Steinvorth, S., Wenz, F., Wildermuth, S. et al. (2002). Cognitive function in patients with cerebral arteriovenous malformations after radiosurgery: prospective long-term follow-up. *International Journal of Radiation Oncology, Biology, Physics* 54 (5): 1430–1437. https://doi.org/10.1016/S0360-3016(02)03800-2.

Stevens, C. and Bavelier, D. (2012). The role of selective attention on academic foundations: a cognitive neuroscience perspective. *Developmental Cognitive Neuroscience* 2 (Suppl 1): S30–S48. https://doi.org/10.1016/j.dcn.2011.11.001.

Stone, J.M., Morrison, P.D., and Pilowsky, L.S. (2007). Review: glutamate and dopamine dysregulation in schizophrenia — a synthesis and selective review. *Journal of Psychopharmacology* 21 (4): 440–452. https://doi.org/10.1177/0269881106073126.

Taddei, S., Virdis, A., Ghiadoni, L. et al. (1998). Vitamin C improves endothelium-dependent vasodilation by restoring nitric oxide activity in essential hypertension. *Circulation* 97 (22): 2222–2229. https://doi.org/10.1161/01.CIR.97.22.2222.

Tae, W.S., Ham, B.J., Pyun, S.B. et al. (2025). Current clinical applications of structural MRI in neurological disorders. *Journal of Clinical Neurology* 21 (4): 277. https://doi.org/10.3988/jcn.2025.0185.

Tai, X.Y., Toniolo, S., Llewellyn, D.J. et al. (2025). Detection of cognitive deficits years prior to clinical diagnosis across neurological conditions. *Brain Communications* 7 (5): fcaf307. https://doi.org/10.1093/braincomms/fcaf307.

Tanaka, K., Wada, N., and Ogawa, N. (2000). Chronic cerebral hypoperfusion induces transient reversible monoaminergic changes in the rat brain. *Neurochemical Research* 25 (2): 313–320. https://doi.org/10.1023/A:1007544124794.

Tang, N., Gong, X.R., Huang, H. et al. (2023). Activated neutrophil-derived exosomes contribute to blood–brain barrier damage and hemorrhagic transformation after cerebral ischemia/reperfusion. *Brain Research* 1810: 148374. https://doi.org/10.1016/j.brainres.2023.148374.

Tari, A.R., Walker, T.L., Huuha, A.M. et al. (2025). Neuroprotective mechanisms of exercise and the importance of fitness for healthy brain ageing. *The Lancet* 405 (10484): 1093–1118. https://doi.org/10.1016/S0140-6736(25)00184-9.

Ter Telgte, A., Van Leijsen, E.M.C., Wiegertjes, K. et al. (2018). Cerebral small vessel disease: from a focal to a global perspective. *Nature Reviews Neurology* 14 (7): 387–398. https://doi.org/10.1038/s41582-018-0014-y.

The SPRINT Research Group (2015). A randomized trial of intensive versus standard blood-pressure control. *The New England Journal of Medicine* 373 (22): 2103–2116. https://doi.org/10.1056/NEJMoa1511939.

Tomimoto, H., Ihara, M., Wakita, H. et al. (2003). Chronic cerebral hypoperfusion induces white matter lesions and loss of oligodendroglia with DNA fragmentation in the rat. *Acta Neuropathologica* 106 (6): 527–534. https://doi.org/10.1007/s00401-003-0749-3.

Tozawa, K. (1996). Activation of nuclera factor-kappa B and control of the expression of cell adhesion molecules in human prostate cancer cells. *Japanese Journal of Urology* 87 (9): 1082–1091. https://doi.org/10.5980/jpnjurol1989.87.1082.

Tran, D., DiGiacomo, P., Born, D.E. et al. (2022). Iron and Alzheimer's disease: from pathology to imaging. *Frontiers in Human Neuroscience* 16: 838692. https://doi.org/10.3389/fnhum.2022.838692.

Tsoy, A., Umbayev, B., Kassenova, A. et al. (2024). Pathology of amyloid-β (Aβ) peptide peripheral clearance in Alzheimer's disease. *International Journal of Molecular Sciences* 25 (20): 10964. https://doi.org/10.3390/ijms252010964.

Tzovenos Starosta, R. and Vinícius Vidor, M. (2016). The frontal-subcortical syndrome. *Journal of Alzheimer's Disease & Parkinsonism* 6 (4). https://doi.org/10.4172/2161-0460.1000247.

Van Norden, A.G.W., Van Dijk, E.J., De Laat, K.F. et al. (2012). Dementia: Alzheimer pathology and vascular factors: from mutually exclusive to interaction. *Biochimica et Biophysica Acta (BBA) - Molecular Basis of Disease* 1822 (3): 340–349. https://doi.org/10.1016/j.bbadis.2011.07.003.

Verma, N., Velmurugan, G.V., Winford, E. et al. (2023). Aβ efflux impairment and inflammation linked to cerebrovascular accumulation of amyloid-forming amylin secreted from pancreas. *Communications Biology* 6 (1): 2. https://doi.org/10.1038/s42003-022-04398-2.

Vermeer, S.E., Prins, N.D., Den Heijer, T. et al. (2003). Silent brain infarcts and the risk of dementia and cognitive decline. *New England Journal of Medicine* 348 (13): 1215–1222. https://doi.org/10.1056/NEJMoa022066.

Villa, L.M., Salas, E., Darley-Usmar, V.M. et al. (1994). Peroxynitrite induces both vasodilatation and impaired vascular relaxation in the isolated perfused rat heart. *Proceedings of the National Academy of Sciences of the United States of America* 91 (26): 12383–12387. https://doi.org/10.1073/pnas.91.26.12383.

Waldstein, S.R., Wendell, C.R., Seliger, S.L. et al. (2010). Nonsteroidal anti-inflammatory drugs, aspirin, and cognitive function in the Baltimore Longitudinal Study of aging. *Journal of the American Geriatrics Society* 58 (1): 38–43. https://doi.org/10.1111/j.1532-5415.2009.02618.x.

Wallin, A., Kapaki, E., Boban, M. et al. (2017). Biochemical markers in vascular cognitive impairment associated with subcortical small vessel disease - a consensus report. *BMC Neurology* 17 (1): 102. https://doi.org/10.1186/s12883-017-0877-3.

Wang, H., Yu, M., Ren, J. et al. (2023). Neuroanatomical correlates of cognitive impairment following basal ganglia-thalamic post-hemorrhagic stroke: uncovering network-wide alterations in hemispheric gray matter asymmetry. *Brain Research* 1820: 148559. https://doi.org/10.1016/j.brainres.2023.148559.

Wang, J., Wang, Y., Cai, X. et al. (2024). A review: visuospatial dysfunction in patients with the cerebral small vessel disease. *Neuroscience* 552: 47–53. https://doi.org/10.1016/j.neuroscience.2024.06.007.

Webb, R.C. (2003). Smooth muscle contraction and relaxation. *Advances in Physiology Education* 27 (4): 201–206. https://doi.org/10.1152/advan.00025.2003.

Wei, X., Xiong, R., Xu, P. et al. (2025). Revealing heterogeneity in mild cognitive impairment based on individualized structural covariance network. *Alzheimer's Research & Therapy* 17 (1): 106. https://doi.org/10.1186/s13195-025-01752-4.

Willie, C.K., Tzeng, Y., Fisher, J.A. et al. (2014). Integrative regulation of human brain blood flow. *The Journal of Physiology* 592 (5): 841–859. https://doi.org/10.1113/jphysiol.2013.268953.

Wu, Y.H. and Hsieh, H.L. (2022). Roles of heme oxygenase-1 in neuroinflammation and brain disorders. *Antioxidants (Basel)* 11 (5): 923. https://doi.org/10.3390/antiox11050923.

Xin, H., Liang, C., Fu, Y. et al. (2024). Disrupted brain structural networks associated with depression and cognitive dysfunction in cerebral small vessel disease with microbleeds. *Progress in Neuro-Psychopharmacology and Biological Psychiatry* 131: 110944. https://doi.org/10.1016/j.pnpbp.2024.110944.

Xu, H., Garcia-Ptacek, S., Jönsson, L. et al. (2021). Long-term effects of cholinesterase inhibitors on cognitive decline and mortality. *Neurology* 96 (17). https://doi.org/10.1212/WNL.0000000000011832.

Xu, W., Huang, Y., and Zhou, R. (2025). NLRP3 inflammasome in neuroinflammation and central nervous system diseases. *Cellular & Molecular Immunology* 22 (4): 341–355. https://doi.org/10.1038/s41423-025-01275-w.

Xue, M., Xu, W., Ou, Y.N. et al. (2019). Diabetes mellitus and risks of cognitive impairment and dementia: a systematic review and meta-analysis of 144 prospective studies. *Ageing Research Reviews* 55: 100944. https://doi.org/10.1016/j.arr.2019.100944.

Yamada, M. (2015). Cerebral amyloid angiopathy: emerging concepts. *Journal of Stroke* 17 (1): 17–30. https://doi.org/10.5853/jos.2015.17.1.17.

Yang, J., Wang, C., Nie, X. et al. (2015). Perfluorooctane sulfonate mediates microglial activation and secretion of TNF-α through Ca^{2+}-dependent PKC-NF-κB signaling. *International Immunopharmacology* 28 (1): 52–60. https://doi.org/10.1016/j.intimp.2015.05.019.

Ye, L.Q., Li, X.Y., Zhang, Y.B. et al. (2020). The discriminative capacity of CSF β-amyloid 42 and Tau in neurodegenerative diseases in the Chinese population. *Journal of the Neurological Sciences* 412: 116756. https://doi.org/10.1016/j.jns.2020.116756.

Yenari, M.A., Xu, L., Tang, X.N. et al. (2006). Microglia potentiate damage to blood–brain barrier constituents: improvement by minocycline in vivo and in vitro. *Stroke* 37 (4): 1087–1093. https://doi.org/10.1161/01.STR.0000206281.77178.ac.

Yin, J., Zhou, X., Chen, Y. et al. (2025). Cerebral microbleeds linked to structural network disruption and cognitive impairment in white matter hyperintensities. *Neuroscience* 564: 31–40. https://doi.org/10.1016/j.neuroscience.2024.11.027.

Zafar, A., Farooqui, M., Ikram, A. et al. (2021). Cytokines, brain proteins, and growth factors in acute stroke patients: a pilot study. *Surgical Neurology International* 12: 366. https://doi.org/10.25259/SNI_569_2021.

Zarate, Y.A. and Hopkin, R.J. (2008). Fabry's disease. *The Lancet* 372 (9647): 1427–1435. https://doi.org/10.1016/S0140-6736(08)61589-5.

Zeisberg, E.M., Tarnavski, O., Zeisberg, M. et al. (2007). Endothelial-to-mesenchymal transition contributes to cardiac fibrosis. *Nature Medicine* 13 (8): 952–961. https://doi.org/10.1038/nm1613.

Zhang, B., Zhang, X., Zhang, F. et al. (2014). Characterizing topological patterns in amnestic mild cognitive impairment by quantitative water diffusivity. *Journal of Alzheimer's Disease* 43 (2): 687–697. https://doi.org/10.3233/JAD-140882.

Zhang, C.E., Wong, S.M., Van De Haar, H.J. et al. (2017). Blood–brain barrier leakage is more widespread in patients with cerebral small vessel disease. *Neurology* 88 (5): 426–432. https://doi.org/10.1212/WNL.0000000000003556.

Zhang, D., Shen, C., Chen, N. et al. (2025). Long-term obesity impacts brain morphology, functional connectivity and cognition in adults. *Nature Mental Health* 3 (4): 466–478. https://doi.org/10.1038/s44220-025-00396-5.

Zhang, J., Zhang, Y., Wang, J. et al. (2024). Recent advances in Alzheimer's disease: mechanisms, clinical trials and new drug development strategies. *Signal Transduction and Targeted Therapy* 9 (1): 211. https://doi.org/10.1038/s41392-024-01911-3.

Zhang, J.X., Qu, X.L., Chu, P. et al. (2018). Low shear stress induces vascular eNOS uncoupling via autophagy-mediated eNOS phosphorylation. *Biochimica et Biophysica Acta (BBA) - Molecular Cell Research* 1865 (5): 709–720. https://doi.org/10.1016/j.bbamcr.2018.02.005.

Zhang, L., Biessels, G.J., Hilal, S. et al. (2021). Cerebral microinfarcts affect brain structural network topology in cognitively impaired patients. *Journal of Cerebral Blood Flow & Metabolism* 41 (1): 105–115. https://doi.org/10.1177/0271678X20902187.

Zhang, M., Liu, Q., Meng, H. et al. (2024). Ischemia-reperfusion injury: molecular mechanisms and therapeutic targets. *Signal Transduction and Targeted Therapy* 9 (1): 12. https://doi.org/10.1038/s41392-023-01688-x.

Zhang, R., Zuckerman, J.H., Giller, C.A. et al. (1998). Transfer function analysis of dynamic cerebral autoregulation in humans. *American Journal of Physiology-Heart and Circulatory Physiology* 274 (1): H233–H241. https://doi.org/10.1152/ajpheart.1998.274.1.H233.

Zhang, W., Zhu, L., An, C. et al. (2020). The blood brain barrier in cerebral ischemic injury – disruption and repair. *Brain Hemorrhages* 1 (1): 34–53. https://doi.org/10.1016/j.hest.2019.12.004.

Zhao, D., Wang, J., Zhu, Y. et al. (2025). Targeting the glymphatic system: Aβ accumulation and phototherapy strategies across different stages of Alzheimer's disease. *Translational Neurodegeneration* 14 (1): 49. https://doi.org/10.1186/s40035-025-00510-8.

Zhao, Y., Gu, J.H., Dai, C.L. et al. (2014). Chronic cerebral hypoperfusion causes decrease of O-GlcNAcylation, hyperphosphorylation of tau and behavioral deficits in mice. *Frontiers in Aging Neuroscience* 6. https://doi.org/10.3389/fnagi.2014.00010.

Zhao, Y., Zhang, H., Cheng, J. et al. (2024). Association between dyslipidaemia and cognitive impairment: a meta-analysis of cohort and case-control studies. *Journal of Integrative Neuroscience* 23 (2): 40. https://doi.org/10.31083/j.jin2302040.

Zipp, F., Bittner, S., and Schafer, D.P. (2023). Cytokines as emerging regulators of central nervous system synapses. *Immunity* 56 (5): 914–925. https://doi.org/10.1016/j.immuni.2023.04.011.

Zonta, M., Angulo, M.C., Gobbo, S. et al. (2003). Neuron-to-astrocyte signaling is central to the dynamic control of brain microcirculation. *Nature Neuroscience* 6 (1): 43–50. https://doi.org/10.1038/nn980.

4

Multilayer Networks in Neuroimaging

Vesna Vuksanovic

Health Data Science, Swansea University Medical School, Swansea University, Sketty, UK

Corresponding author: vesna.vuksanovic@swansea.ac.uk

4.1 Introduction

Network science provides a powerful tool for unravelling complex connectivity patterns in the brain (see Bassett & Sporns, 2017; Stam, 2014 for review) and plays a major role in mapping its structural and functional interactions (see Avena-Koenigsberger et al., 2018). Until recently, most of the research on brain networks was based on traditional approaches of graph theory, where a typical network represents pair-wise relationships (links) between network units (nodes) (Rubinov & Sporns, 2010). While such a network representation has enabled years of progress in understanding connectivity patterns in the healthy and diseased human brain, recent work suggests that the description of the brain as a (monolayered) network may still be an oversimplification. Traditional representations of the brain as a single-layer network provide only limited representation of its structural and functional complexity.

Complex networks of interactions include a variety of time-varying and dynamic connections across brain sub-systems (e.g., metabolic, cellular or regional) (Patow et al., 2024; Ritter et al., 2013). The connectivity architecture of such connections exhibits an extraordinary level of 'multiplexity' – i.e., different types of interactions among its interacting units that span multiple scales – spatial, temporal and topological. This has prompted a question whether the full information about brain networks can be better studied using the multilayer networks approach (Vaiana & Muldoon, 2020). Although long present in many research fields to address diverse topics such as ecology (Pilosof et al., 2017), transportation (Kurant & Thiran, 2006), animal behaviour and biology (see De Domenico, Sasai, & Arenas, 2016; Finn et al., 2019), or protein interactions (Ou-Yang et al., 2014), studies using multilayer methods for analysis of brain networks are still few (De Domenico, 2017).

Here, I will discuss major advantages, but also challenges, in utilising the multilayer methods in analysing complex patterns of interactions in the brain. Particular attention will be paid to their potential for applications in exploring structure–function relationships, modelling brain diseases and integrating multimodal, multiscale, datasets in neuroimaging. The main focus will be on the so-called large-scale, or whole-brain, networks based on experimental recordings using magnetic resonance imaging (MRI), including positron emission tomography (PET) and functional magnetic resonance imaging (fMRI), which allowed for the in vivo characterisation of the neural correlates underlying sensory, cognitive or behavioural tasks.

4.2 Multilayer Networks: The Theoretical Background

Intuitively, a multilayer network can be described as a 'network of networks' or an 'interconnected network' – a network comprised of multiple interconnected layers, each characterising a different aspect of the same system (see Figure 4.1). Distinct layers of connectivity emerge both in nature and in human-made systems. For example, in communication networks, one can distinguish between email exchanges, instant messaging and phone calls, which together form multiple layers of interpersonal interaction using digital media. Other examples include transportation (Kurant & Thiran, 2006), ecology (Pilosof et al., 2017), biology (Finn et al., 2019) and numerous other areas (see Boccaletti et al., 2014 for review). In neuroscience, structural, functional and molecular/metabolic connections all create complex connections to shape brain dynamics (Boccaletti et al., 2014).

The potential of multilayer networks to represent complex systems more accurately than it was previously possible has led to an increasing interest in physics of these networks (De Domenico et al., 2013). One significant application in physics emerges from the study of spreading processes (De Domenico et al., 2016), which have applications, e.g., in modelling traffic flows in transportation networks (Aleta et al., 2017) or the transmission of information and diseases in social networks (Wang et al., 2015). When spreading processes are coupled within a multilayered network approach to their analysis, this coupling can lead to some intriguing phenomena, which cannot be modelled within the single-layer network framework. In multilayer networks, the onset and dynamics of one process can depend on the onset of the other one, such as the emergence of critical points in phase diagrams, revealing distinct regimes where the criticality of the dynamics becomes intertwined or independent (Granell et al., 2013). Furthermore, multilayer structures can foster cooperative behaviour in structured populations (Li et al., 2020). By providing a platform for interactions across different layers,

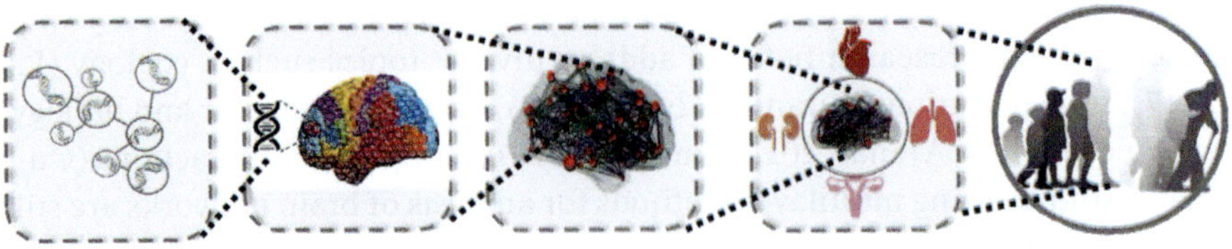

Figure 4.1 Brain as a complex system at scales of organisation and interactions across different systems and modalities.

multilayer networks offer a novel mechanism for promoting cooperation even in complex environments (Duh et al., 2019).

4.2.1 Monolayer Brain Networks

Graph theory provides one of the main mathematical frameworks for studying the structural and functional organisation of the human brain. In this approach, the brain is represented as a network (or graph) consisting of nodes and edges (links). Nodes and edges are the fundamental building blocks of any network, and how they are defined is critical for constructing realistic models of brain connectivity (Butts, 2009; Zalesky et al., 2010). However, one of the major challenges in large-scale MRI-based brain network analysis is deciding how to define meaningful nodes. Approaches vary across studies, from using the native resolution of MRI (voxel-wise resolution) (van den Heuvel et al., 2008) to applying validated parcellations of the cortex based on anatomical or functional landmarks (Desikan et al., 2018; Tzourio-Mazoyer et al., 2002), or even random parcellations to ensure equally sized nodes (Fornito et al., 2010; Hagmann et al., 2008). More data-driven strategies include defining nodes based on connectivity patterns (Glasser et al., 2016), applying statistical decomposition methods such as independent component analysis (Power et al., 2011) or relying on predefined regions derived from large-scale meta-analyses (Dosenbach et al., 2010).

Most studies of functional and anatomical MRI networks define nodes using validated parcellations (brain atlases), as these allow biologically meaningful interpretation. Their main advantage is that they are informed by measures of brain structure and function, making them well suited for testing specific hypotheses about network organisation. A limitation, however, is that such parcellations are not always easily transferable across different imaging modalities. Despite this, results from different studies show that measures of network topology are largely consistent across parcellation schemes and MRI modalities. Basic characteristics of brain network organisation – such as node degree (the number of edges connected to a node), clustering (the number of triangles in the network) and path length (the average number of edges between two nodes) – are consistent and remain stable when parcellations with the same number of nodes are compared (Chen et al., 2008; Meunier et al., 2009; Vuksanović, 2019). Similarly, modular organisation, which is another major characteristic of brain organisation, has been shown to be consistent across both anatomical and functional parcellations as well as across imaging modalities (Chen et al., 2008; Meunier et al., 2009; Vuksanović, 2019).

In functional and morphological brain networks, edges are typically defined using an association matrix that captures statistical relationships – such as cross-correlation or mutual information – between nodal features. This matrix encodes all possible pairwise associations, either between regional morphological characteristics or between time series of neural activity. For analysing network topology, these association matrices can be represented in two ways: as binary networks, where edges indicate only the presence or absence of associations, or as weighted networks, where edges reflect the strength of associations. Different strategies exist for analysing these two types of networks. Binary networks are often studied across a range of thresholds to ensure the robustness of topological properties (Bassett & Bullmore, 2009) and to account for noise or weak, spurious associations

(Vuksanović et al., 2019). Although the choice of threshold is somewhat arbitrary, it is usually guided by how much the resulting network deviates from a random (null-model) organisation (Rubinov & Bassett, 2011) and by the presence of small-world or scale-free features (Bassett et al., 2006). Weighted networks, in contrast, directly incorporate the strength of associations between nodes. In functional networks, edge weights represent temporal dependencies between time series, quantified through measures such as correlation, coherence or synchrony (Power et al., 2011). In morphological networks, edges are weighted by statistical associations between regional properties such as cortical thickness, surface area, volume or curvature (Bassett et al., 2008; Seidlitz et al., 2018; Vuksanović et al., 2019), often estimated across groups of individuals.

Although neither functional nor morphological (anatomical) correlation networks are based on direct neural (axonal) connections between cortical regions, both show strong association with the underlying structural network (Honey et al., 2007). Many studies have examined how functional interactions reflect local anatomical organisation and axonal connectivity linking these regions (Honey et al., 2007). However, the organisation of functional and morphological networks, and the way they interact, points to a complex, many-to-one mapping between structure and function (Friston & Dolan, 2010; Park & Friston, 2013). Multilayer network approaches offer a promising framework for studying the basis of these complex structure–function relationships. By integrating multiple layers of connectivity, they can capture the complex associations across functional and structural networks and potentially provide a unified model. In particular, multilayer networks allow the mapping of numerous functional corticocortical connections across scales, thereby offering deeper insights into the interdependence of brain structure and function.

Structural brain networks are usually derived from diffusion MRI (dMRI) data by tracking the diffusion of water molecules along white matter tracts (Hagmann et al., 2008). dMRI exploits the anisotropic diffusion of water to reconstruct images that reveal the orientation and connectivity of white matter pathways. Similar to brain functional networks, nodes typically correspond to distinct anatomical regions of the brain, defined based on standardised brain atlases, while links represent the white matter pathways that connect these regions. Over the last decade, structural brain networks have been used to study normal brain development (Hagmann et al., 2012; Huang et al., 2015), as well as to identify alterations in the integrity of these networks associated with neurological disorders (Griffa et al., 2013).

4.2.2 Structural Representation of Multilayer Networks

This section focuses on a mathematical framework for the structural and organisational representation of multilayer networks that allow for the detection of network properties that are uncovered by traditional single-layer network analyses. The definitions of nodes and edges in multilayer networks differ from those in traditional networks. For example, a node can be present in all layers of the network, but it does not have to be. Such an organisation allows for more creative approaches when defining network nodes and edges in the network. Within multilayer networks, three main types of edges can be identified: intra-layer edges that link nodes within the same layer, inter-layer edges that connect counterpart nodes across layers, and inter-layer edges that join nodes representing different units. Importantly, intra-layer and inter-layer edges encode relationships in fundamentally different ways, highlighting their distinct functions (Vaiana & Muldoon, 2020). For example, in a multimodal transportation system, intra-layer edges represent connections within the same mode of transport (e.g., shipping routes between ports), while inter-layer edges link different modes (e.g., a port connected to a nearby rail terminal). Similarly, in ecological networks, intra-layer edges could denote trophic interactions such as predation within one species group, while inter-layer edges capture cross-layer dependencies like pollination links between plants and insects. In neuroscience, intra-layer edges may correspond to functional connections within one imaging modality (e.g., fMRI-based correlations), whereas inter-layer edges link across modalities (e.g., functional activity mapped to diffusion-based structural pathways). Depending on the balance between intra- and inter-layer connections, a multilayer network may behave like a set of relatively independent systems with weak coupling between layers, or like a tightly integrated network in which the layers are nearly indistinguishable. In some cases, these systems even undergo transitions between such regimes.

Mathematically, interactions in a multilayer network are represented by the elements $m_{i\alpha}^{j\beta}$ of a fourth-order tensor M, referred to as the multilayer adjacency tensor. This tensor specifies the relationship between node i in layer α and node j in layer β, where $i,j \in 1,2,\ldots,N$ and $\alpha,\beta \in 1,2,\ldots,M$. Here, N denotes the total number of nodes in the network, while M corresponds to the number of layers. By formulating connectivity in terms of a tensor, it becomes possible to define new measures that capture the structural complexity of multilayer networks (Presigny & De Vico Fallani, 2022). Within this general framework, each node may belong to one or more layers, and edges can represent pairwise interactions across all possible combinations of nodes and layers. (This formulation can be further extended to include hyperedges that connect more than two nodes.) Thus, a node u in layer α may be connected to any node v in any layer β. However, such generalisation requires caution, since directly transferring concepts from monolayer networks can lead to misleading or even invalid interpretations. The compact representation of multilayer networks provided by tensors enables greater abstraction and has facilitated the advancement of mathematical frameworks for understanding complex systems (Kivelä et al., 2014). In the exploration of multilayer networks' structural properties, efforts have been dedicated to simplifying the structural complexity of multilayer networks to reveal mesoscale structures, such as densely connected communities of nodes, quantifying triadic relations like clustering, transitivity or motifs (Battiston et al., 2017), and identifying key nodes based on various measures of importance (De Domenico, Sasai, & Arenas, 2016; Yuvaraj et al., 2021).

An alternative approach to the tensor representation involves using sets of adjacency matrices to generalise concepts from monolayer to multilayer networks. This approach is based on the familiarity and the convenience of flattening adjacency tensors into matrices, also known as 'supra-adjacency matrices', for computational purposes (Presigny & De Vico Fallani, 2022). The approach is used most frequently in neuroscience. In this case, the element of the supra-adjacency matrix A, $a_{ij}^{\alpha\beta} = a_{ji}^{\alpha\beta}$, describes interactions of node i in layer α to node j in layer β of a given multilayer network. Network metrics, derived from the matrix A, have their equivalents in monolayer networks. For example, the equivalent of the *node degree* is the so-called overlapping degree or node strength, which sums the weighted degrees of node i across all layers. Some other popular network metrics in neuroscience are the *multilayer participation index* (Battiston et al., 2017), which quantifies nodal participation in interactions across layers; *triangles*, i.e., triads of interconnected nodes based on the probability of forming triangles by means of links belonging to two different layers (Cozzo et al., 2015). At the mesoscale level of network topology, one can calculate *motifs* (see Sporns & Kötter, 2004); *modularity*, which captures network modules (communities) across all the layers simultaneously (Mucha et al., 2010); in temporal networks, one can also extract *flexibility*, which quantifies how often a node changes its modular allegiance across layers (Bassett et al., 2011). Several other metrics have also been used in neuroscience (see Presigny & De Vico Fallani, 2022), depending on the research question.

The structural properties of multilayer networks are strongly determined by how their layers are interconnected to form an integrated system. Inter-layer edges provide the links between layers, encoding both structural and dynamical relationships. Their inclusion – or absence – can give rise to rich structural and dynamical behaviours. For instance, in multimodal transportation systems, where layers represent different modes of travel, the weight of an inter-layer edge may correspond to the time or cost required to switch from one mode to another (Alessandretti et al., 2023). In brain networks, inter-layer connections can link structural and functional layers, reflecting how anatomical pathways constrain functional interactions across regions. These interdependencies illustrate how inter-layer coupling shapes overall system organisation. A more detailed discussion of the dynamics of multilayer networks is provided in Section 4.2.3.

4.2.3 Dynamics of Multilayer Networks

Similar to the existence of multiple topologies, there are two types of dynamical processes in multilayer networks: (i) single dynamical processes unfolding on top of the interconnected structure of the multilayer network and (ii) mixed or coupled dynamics, where multiple processes interact across layers through inter-layer connections. Most research on the dynamics of multilayer networks explores the role of inter-layer connections in terms of interaction strength between layers and inter-layer communities. However, the behaviour of single-layer processes is shaped by both intra-layer topology – which is similar to traditional network analysis – and inter-layer topology, representing interactions between nodes across layers (De Domenico et al., 2016). Such behaviour has led to the discovery of many interesting phenomena related to network dynamics.

The study of diffusion, the simplest dynamical processes, in multilayer networks has uncovered an unexpected phenomenon: diffusion can manifest more rapidly in a multiplex network than in any of its constituent layers examined in isolation (see De Domenico et al., 2016). Another example of discrete dynamical processes is the random walk, commonly employed to model Markovian dynamics on monolayer networks and have offered valuable insights into the dynamics of spread in a network (De Domenico et al., 2016). In a random walk, a discretised form of diffusion, a walker traverses between nodes via available connections. However, in a multilayer network, these connections extend to layer-switching through inter-layer edges, a transition absent in monolayer networks. This enrichment of random-walk dynamics through layer-switching highlights a key physical insight: 'navigability'. Navigability, defined as the mean fraction of nodes visited by a random walker within a finite time, reveals an intriguing aspect of the interplay between multilayer structure and dynamic processes. Similar to continuous diffusion, navigability in multilayer networks can surpass that of an aggregated network of layers. This phenomenon indicates that multilayer networks exhibit greater resilience to uniformly random failures compared to their individual layers (Solé-Ribalta et al., 2016). This resilience emerges directly from the interplay between the multilayer structure and the underlying dynamical processes. Phenomena such as navigability can shed light on the network's resilience and efficiency in the context of various perturbations and attacks, which have already been explored in monolayer networks for modelling disease spread (Boguna et al., 2009; De Domenico et al., 2014). Another significant phenomenon to consider is congestion, which represents the delicate balance between flow across network structures and the capacity of those structures to accommodate such flow. While congestion in networks has been a subject of analysis in the physics literature for many years (Arenas et al., 2003), its study in the context of multilayer networks is a relatively recent development (Manfredi et al., 2018; Solé-Ribalta et al., 2016).

The coupled dynamics of multilayer networks consider two fundamental effects: firstly, mutual excitation, where one process facilitates the spread of another (e.g., one disease amplifying the infection of another); and secondly, mutual inhibition, where one process hinders the spread of the other (e.g., disease A inhibiting the infection of disease B, or the dissemination of awareness about a disease curbing its spread) (Raj & Powell, 2018; Wang et al., 2015). The interaction between mutual excitation and inhibition yields a complex dynamic, which makes such interplay an ideal candidate for modelling brain dynamics.

4.3 Multilayer Networks in Neuroimaging

Brain (structural and functional) interactions span across different levels of organisation and scales. Each of such levels can be described by distinct layers of connectivity and activity (function), representing an ideal foundation for multilayer network modelling. For example, different network layers can represent functional networks, that is, the co-activation of multiple brain regions involved in specific cognitive tasks or affected by diseases. Similarly, structural multilayers can map the physical connections, i.e., temporal 'connectivity' of brain units. Layers can also describe multiple features that interact at different levels of organisation, thus bridging between different levels of information and properties. An example of this type of modelling are computational models that integrate information at the neuronal level to predict whole-brain dynamics (Breakspear et al., 2010; Deco & Kringelbach, 2016; Jirsa et al., 2010).

Another important feature of multilayer networks in neuroimaging is in retaining information about individual brain networks. Traditionally, networks derived from neuroimaging data were often aggregated across participants into a single-layer representation, typically by averaging. While convenient, such aggregation discards important variability across individuals. By contrast, multilayer network methods, together with statistical tools designed to operate across layers, retain the full multidimensional structure of the data, enabling richer analyses that are not possible with aggregated single-layer networks. For example, De Domenico, Sasai and Arenas (2016) demonstrated that a multilayer representation of human brain networks yields higher classification accuracy when distinguishing

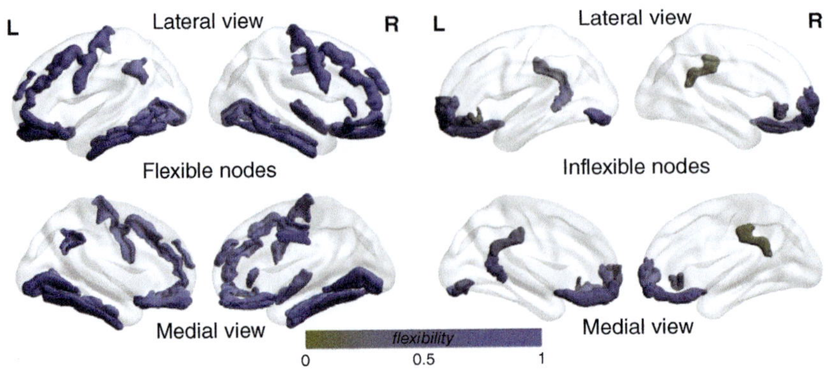

Figure 4.2 Brain 'flexible' hubs in ageing revealed by multilayer network analysis. *Source*: Vuksanović (2022)/Oxford University Press/CC BY 4.0.

between healthy individuals and patients with schizophrenia compared to single-layer or averaged network models. Similarly, Vuksanović (2022) showed that multilayer brain networks provide valuable insights into how network organisation changes with ageing (see Figure 4.2) and how it relates to variations in cognitive measures such as IQ.

The first topic in neuroimaging, which used a multilayer approach even before a formal mathematical framework was developed, was to consider time-varying functional networks from fMRI data (Braun et al., 2015). In this case, the resulting multilayer network is a collection of adjacency matrices each corresponding to a snapshot of brain activity over time. Mapping the changes in topology of these networks over time, have shown their biological relevance (Bassett et al., 2011). Similarly, in electroencephalography (EEG) and magnetoencephalography (MEG) research communities, there is an increasing number of studies that use the so-called frequency-based decomposition of multilayer networks. This method is based on building the multilayer functional brain using the EEG/MEG signal from different brain regions, decomposed in the frequency domain (see Figure 4.3). To build a multilayer network, each layer is represented by functional connectivity within different frequency bands (see Brookes et al., 2016; Tewarie et al., 2016). A more recent study has shown that the multilayer network approach provides high classification accuracy and sensitivity of subjects with cognitive impairments and Alzheimer's disease (Echegoyen et al., 2021). Moreover, by integrating data from different imaging modalities, such as fMRI, dMRI and MEG/EEG, the multilayer network approach can be utilised to construct generative models that capture the dynamic interplay between brain units across multiple *layers* of brain complexity (Mandke et al., 2018; Puxeddu et al., 2020, 2021). The advantage of network theory is that it is independent of (imaging) modality, i.e., the same mathematical framework can be applied independently of the modality from which the data come. This modality-invariant characteristic opens numerous possibilities for studying interactions between different brain systems and offers a distinct perspective on brain connectivity from a single-network approach based on networks analysed using data from a single modality. However, despite the obvious promise of these approaches, a key issue in the application of network theory is the comparison between groups or conditions. Similar to the monolayer network approach, it is difficult to distinguish between

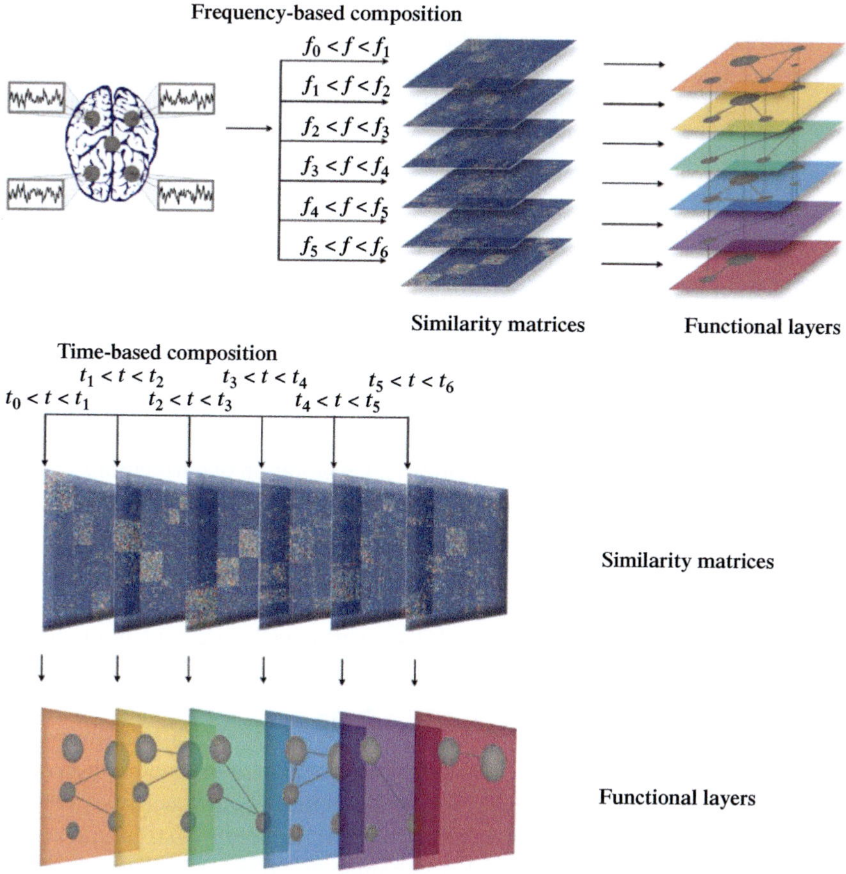

Figure 4.3 Building the multilayer functional brain. Top panels: Brain activity is measured across different regions, and the signals are decomposed in the frequency domain. The frequency domain consists of (possibly overlapping) frequency bands, and, for each band, coherence – or other similarity descriptors – is measured between all pairs of regions. A similarity matrix is built for each frequency domain, and statistical analysis of significance is used to map each matrix into a functional network, constituting a functional layer of the overall multilayer system. Bottom panels: In this case, signals are decomposed in the time domain, which consists of (possibly overlapping) consecutive temporal snapshots. A similarity matrix is calculated for each snapshot, and the corresponding functional layer is built. *Source:* De Domenico (2017)/Oxford University Press/CC BY 4.0.

alterations in the underlying brain connectivity caused by experimental manipulation or a disease process and those resulting from computational choices such as network thresholding (Van Wijk et al., 2010). The choices can potentially influence false positives or false negatives when comparing brain networks across individuals and/or groups. Multilayer brain networks are not exempt from these biases and need a careful consideration when comparing their organisation and metrics (Brookes et al., 2016; De Domenico et al., 2016; Guillon et al., 2017).

Neuroimaging models of brain dynamics and their multilayer network representation can provide a better understanding of the brain's functional connectivity. In the context of functional connectivity (network), it is important to address the concept of correlations. In single-layer networks, it has been observed that there are correlations in the properties of connected nodes. For example, the degree of a node can exhibit either positive or negative correlations with the degrees of its neighbouring nodes. Positive correlations suggest that hubs tend to connect preferentially with other hubs, while negative correlations indicate a preference for hubs to connect with low-degree nodes. However, in multiplex networks, the notion of correlations becomes much richer compared to single-layer networks (Nicosia & Latora, 2015). While it is still possible to examine standard degree–degree correlations within each layer of the network, it becomes more intriguing to define correlations in a truly multiplex context. For instance, to investigate how a certain property of a node in one layer correlates with the same or other properties of the same node in another layer. A study by Nicosia and Latora (2015) provides a comprehensive examination of the correlations of the nodal properties in multiplex networks within the framework of neuroimaging models of brain dynamics. To achieve this, the authors follow a systematic approach commonly employed in complex network analysis: (i) empirical exploration of correlations in real multiplex brain networks derived from structural and functional MRI data, (ii) introduction of various measures to characterise and quantify these correlations, offering insights into the organisation of the brain's multilayer network and (iii) proposal of a series of models to replicate the observed correlations in real brain networks or assess their significance, thus contributing to our understanding of the underlying principles governing brain dynamics.

4.3.1 Structure–Function Relationship in the Brain

Functional networks – representing distributed patterns of correlated/coherent neural activity across the brain – have long been an important topic in neuroscience. Their dynamic properties – especially when compared to the relatively static nature of structural networks – are considered critical for supporting higher cognitive functions. The interaction between structural connectivity and emergent functional activity, together with insights from multimodal neuroimaging and frequency band analyses, highlights the importance of adopting a multilayer network framework.

Despite the innate relationship between structural and functional brain networks, their direct comparisons and simultaneous analysis pose challenges (Deco & Kringelbach, 2016; Vuksanović & Hövel, 2014). While theoretical studies have made progress in understanding the relationship between brain structure and dynamics (Deco et al., 2011), it is less clear how structural network facilitates the emergence of functional network. The correspondence between functional and structural brain networks remains an active area of research (Suárez et al., 2020). Given that functional networks are constrained by invariant structural connections, functional integration during various cognitive tasks necessarily implies a many-to-one function–structure mapping (Friston & Dolan, 2010). In this context, modelling brain dynamics and organisational representation as a multilayer network from structural and functional brain networks constructed through MRI data offers a comprehensive approach to their *simultaneous* analysis.

The non-trivial relationship between structural and functional connectivity poses many challenges for making inferences about brain development, cognition, diseases, learning or ageing-related changes (Puxeddu et al., 2020; Vuksanovic, 2021). A growing body of research addresses these challenges through the prediction of functional connectivity from structural characteristics (see some early works based on the single-layer approach, e.g., Hagmann et al., 2008; Honey et al., 2007; Rubinov & Bassett, 2011). The methods used for these predictions are based on the application of statistical tools (Vázquez-Rodríguez et al., 2019), information flow framework for neural communication (Deco et al., 2009) or considering the coupling between structure and function through non-linear dynamics (Breakspear et al., 2010; Deco et al., 2009; Ritter et al., 2013), or examining the effects of brain lesions on functional organisation and behaviour (Ghajari et al., 2017).

Beyond predicting one from the other, structural and functional networks can be jointly analysed to investigate shared organisational properties, for example, their modular structure and inter-relationships. The tensor representation of multilayer networks provides a robust mathematical framework, enabling the extension of traditional complex network analysis techniques. Within the framework, methods such as detecting modular super-units and identifying central nodes are instrumental in understanding the organisational structure of the human brain (Sporns, 2013; Sporns & Betzel, 2016; Vuksanović, 2022). Modularity is a common organisational characteristic of complex networks, including those of the human brain. In this context, networks can be partitioned into modules by grouping subset of nodes that are more densely connected to each other than the rest of the network. Brain organisation into functional modules mirrors the underlying structural (axonal) pathways (Honey et al., 2007). However, while these functional modules generally align with local anatomical organisation, they also include long-range interactions linking spatially distant brain regions (Fries, 2005; Vuksanović & Hövel, 2014, 2015). Such functional modular topology of brain networks, as derived from MRIs, has been consistently demonstrated across different cortical parcellation schemes (Bertolero et al., 2015; Meunier et al., 2009; Vuksanović, 2019). Beyond functional connectivity, modularity has also been widely examined in brain morphology, particularly in the context of evolutionary and developmental processes (Melo et al., 2016). Considering both aspects together within a multilayer framework allows for the integration of structural, morphological and functional modularity, providing a more complete account of how brain networks are organised and how their modules support cognition across scales. Algorithms designed for this task typically optimise the division of the network such that inter-module connections remain relatively sparse, allowing the detection of overlapping or distinct communities (Fortunato & Hric, 2016; Newman, 2006). Identifying modules can provide insight into inter-nodal connections that enable specialised functions with some degree of independence from the rest of the system. At the same time, characterising interactions between modules helps reveal how coordinated activity supports specific brain functions (Fortunato & Hric, 2016). For example, a large-scale meta-analysis of over 1,000 fMRI studies has shown the presence of functional modules associated with distinct cognitive processes (Crossley et al., 2013).

Recent neuroimaging studies suggest the modular organisation of cortical morphology across regional thickness (Vuksanović, 2019; Vuksanović et al., 2019), surface area (Sanabria-Diaz et al., 2010) or volume (Bassett et al., 2008). There is consistency in the organisation of these networks, whether they are based on correlating these features across individuals within one group (Sanabria-Diaz et al., 2010; Váša et al., 2017; Vuksanović, 2019) or correlating regional features of an individual brain (Seidlitz et al., 2018). The brain's modular, yet integrated, functional organisation lowers the wiring cost (i.e., the average length and number of connections) of the network (Bassett et al., 2009), thus potentially lowering metabolic costs (Betzel et al., 2017) while providing more efficient information processing (Sporns & Betzel, 2016). More importantly, modularity, as mapped by large-scale brain fMRI networks, is cognitively and behaviourally relevant; for example, it correlates with variations in working memory (Yamashita et al., 2015). This property is evident in both anatomical and functional networks across different spatial and temporal scales (Bassett & Sporns, 2017).

The brain's organisation into functional modules mirrors the underlying structural (axonal) pathways (Honey et al., 2007). However, while these functional modules generally align with local anatomical organisation, they also include long-range interactions linking spatially distant brain regions (Fries, 2005; Vuksanović & Hövel, 2014, 2015). Such a functional modular topology of brain networks, as derived from MRIs, has been consistently demonstrated across different cortical parcellation schemes (Bertolero et al., 2015; Meunier et al., 2009; Vuksanović, 2019). Beyond functional connectivity, modularity has also been widely examined in brain morphology, particularly in the context of evolutionary and developmental processes (Melo et al., 2016). Considering both aspects together within a multilayer framework allows for the integration of structural, morphological and functional modularity, providing a more complete account of how brain networks are organised and how their modules support cognition across scales.

4.4 Multimodal, Multiscale Neuroimaging Data: A Multilayer Network Approach

In the past decades, several experimental measurements, based on structural and functional MRI, diffusion tensor imaging (DTI), PET imaging, or EEG and MEG, have been used for non-invasive reconstructions and recordings of brain structures and activity. The existence of such large and multimodal data sets, providing snapshots of the brain at scales, has prompted the appearance of quantitative models that involve explorations of interactions between multiple modalities in health (see Bassett and Sporns, 2017) and disease (see Franzmeier et al., 2019; Khan et al., 2020). The advantage of such an approach is to allow for multiple elements or interactions of a networked system to be analysed under a single model. Before I address multilayer networks from multimodal neuroimaging data, it should be noted that multilayer network models have been used for over a decade now to study the covariance structure of functional brain data by analysing time-varying functional networks from fMRI data (see Bassett et al., 2011; Mucha et al., 2010). A similar approach has also been used to establish nodal modular allegiances across networks representing individual brains using structural and anatomical MRI (Puxeddu et al., 2022; Vuksanović, 2022).

Recently, a smaller number of studies have begun investigating multilayer network models of the brain across multiple imaging modalities (De Domenico, 2017; Puxeddu et al., 2022; Vaiana & Muldoon, 2020). A first such attempt has been made using fMRI and DTI data by constructing layers from the fMRI and DTI networks of a single subject (Battiston et al., 2017; Crofts et al., 2022, 2016). In such models, the fMRI layer represents the fMRI network, where the nodes are brain regions and edges are statistical associations between regional activities. The DTI layer represents the weights of connections of the white matter tracts between (the same) brain regions. These studies highlight the complex relationships between structural connectivity and brain dynamics. The structural communities identified in a multilayer framework differ from those typically obtained from single-layer structural networks, indicating that multilayer approaches may provide a more suitable framework for analysing multimodal brain networks.

A comprehensive understanding of the brain requires an integrated approach across various scales and levels of organisation. This involves, for example, the interplay between genes and synapses, as well as the relationship between the structure and dynamics of the entire brain, which ultimately shape diverse behaviour and cognitive functions (Presigny & De Vico Fallani, 2022). Multiscale brain modelling poses significant challenges, partly due to the difficulty of simultaneously accessing information across multiple scales and levels (Goubran et al., 2019; Valk et al., 2017). While there has been some progress in understanding the role of specific microcircuits in generating macroscale brain activity through biologically inspired computational models (see Breakspear et al., 2010; Deco et al., 2011; Jirsa et al., 2010), a thorough characterisation of how changes at one scale or level impact others remains elusive. The theoretical advancements of multilayer networks offer a robust framework for analysing and modelling interconnected systems with interactions within and between different layers of information. Recent progress includes characterising multiscale brain organisation in terms of structure–function relationships (Schirner et al., 2018), oscillation frequencies (Deco & Kringelbach, 2016) and temporal dynamics (Betzel & Bassett, 2017). In addition, multilayer network properties have been used to identify multimodal network-based biomarkers of brain pathologies in Alzheimer's disease (Guillon et al., 2017).

I have presented some conceptual frameworks, tools and results that provide insights into how brain systems can be represented via the intra- and inter-layer network properties. However, research in this field is active, and many issues remain to be addressed to ultimately characterise the multiscale, multilevel brain organisation (De Domenico, 2017; Vaiana & Muldoon, 2020; Vuksanović, 2022). Finally, each imaging modality offers a unique perspective on brain function or structure, and data fusion leverages the strengths of each modality and their inter-relationships in a joint analysis. However, most studies favour only one data type or fail to combine modalities in an integrated manner, thereby missing significant changes that are only partially detected by individual modalities. Conversely, multimodal fusion provides a more comprehensive depiction of altered brain patterns and connectivity. This approach has demonstrated increasing utility in addressing both scientifically intriguing and clinically relevant questions, offering a more holistic understanding of brain dynamics.

4.5 Application of Multilayer Networks in Modelling Neurodegenerative Disorders

The application of monolayer network approaches has been a promising strategy for unravelling the complex role of molecular, cellular and circuit-level changes associated with brain disorders (see Fan et al., 2023; Stam, 2014). In this section, I will explore the potential of multilayer network analysis in elucidating the pathophysiology, progression and potential therapeutic targets of neurodegenerative, neurological and psychiatric disorders. As emphasised earlier, research on these questions (i) spans multiple spatial and temporal scales, (ii) relies on multiple experimental modalities and (iii) often requires comparisons across individuals or groups. A multilayer network provides a powerful framework for integrating these aspects in a coherent and consistent way. Its main advantage lies in the ability to link network nodes (e.g., brain regions) across different modes of analysis, resulting in a unified structure that preserves the full richness of the data. There are two main pathways for the multilayer network approach to improve our understanding of neurodegeneration – cellular heterogeneity and connectomics.

A common underlying characteristic of brain disorders, such as mental disorders (Richetto & Meyer, 2021), epilepsy (Guerrini et al., 2003) or dementia-causing disorders (Dujardin et al., 2020; Rademakers et al., 2012), is molecular heterogeneity. For example, neurodegenerative disorders that cause dementia are, without exception, characterised by molecular heterogeneity. Such heterogeneity involves, for example, aberrant protein aggregation (Dujardin et al., 2020), synaptic dysfunction (Dujardin et al., 2020) or neuronal loss (West et al., 1994). While traditional single-layer network models often overlook the multifaceted nature of neurodegeneration by focusing on a single imaging modality, the multilayer networks can provide an integrated framework for diverse molecular and cellular sources of neurodegeneration. By representing different molecular layers (i.e., sources of heterogeneity) as interconnected networks, the multilayer models can capture the dynamic interactions across the scales involved in neurodegeneration. For example, integrating gene expression data with multimodal imaging data (Rubido et al., 2024; Thompson et al., 2014) may lead to uncovering disease-associated connectome failures, protein dysregulation and/or protein aggregation pathways (Raj & Powell, 2018).

In addition, there is increasing evidence from multimodal studies that individuals with neurodegenerative disorders exhibit distinct morphological changes (Lieberman et al., 2001; Wurthmann et al., 1995), connectivity patterns (Franzmeier et al., 2019; Vuksanović, 2019) and functional alterations (Franzmeier et al., 2019; Wang et al., 2020). These changes are not easily discernible through separate unimodal analyses, which are typically performed in the majority of neuroimaging experiments. Thus, the application of multilayer network analyses to these multimodal characteristics could facilitate the identification of biomarkers for brain disorders, expediting differential diagnosis and leading to more appropriate treatments and improved outcomes for patients. For example, mild cognitive impairment (MCI) is challenging to diagnose due to its subtle and often insignificant symptoms of cognitive impairment. Some studies have highlighted the potential of combining structural and functional data with multimodal classification techniques to enhance the accuracy and early detection of brain abnormalities in MCI (Fan et al., 2008).

Another aspect of the multilayer network approach that could be utilised in brain disorders involves focusing on the spatial and temporal dynamics of these networks in the context of disease (see Stam, 2014). In general, disease can impact brain networks through two mechanisms. Firstly, disease can directly affect the nodes of the network, leading to either localised or widespread functional changes. This impairment can propagate along neural connections to other brain regions (Franzmeier et al., 2020, 2019), thereby amplifying the dysfunction. Secondly, certain diseases mainly target neural connections themselves, such as demyelination or axonal injury (Ettle et al., 2016). This disruption in connectivity results in anomalous information processing and widespread functional impairments, reflecting the underlying aberrant network dynamics (Raj et al., 2012; Raj & Powell, 2018; Rubido et al., 2024). For example, utilising a multilayer network approach to identify the type of disease dynamics can reveal pathological signatures across different neurodegenerative diseases, revealing convergent dysregulation of brain subsystems involved in these processes. However, clustering of multilayer networks, especially using information on higher-order interactions of system entities, still remains in its infancy, partly because higher-order network metrics are often the key in such multilayer network applications (Yuvaraj et al., 2021). Developing optimal partitioning of critical topological features can help isolate unhealthy components of the networks targeted by disease and identify multiple brain regions affected by trauma or mental disorders.

4.6 Conclusion and Future Directions

In this chapter, I have presented novel conceptual insights, tools and results that offer new perspectives to model brain systems and interactions between and within them, using the multilayer network approach. Multilayer networks offer a simple yet flexible data structure that enables quantitative analysis of complex neuroimaging data while preserving information and organisational features that are often lost in traditional statistical or single-layer network approaches. By retaining this additional information and combining it with novel network measures, the multilayer approach has revealed aspects of brain structure and function that were previously hidden from analyses using more traditional approaches. Furthermore, multilayer network approaches offer a comprehensive framework for unravelling the complexity of mechanisms and processes in neurodegenerative disorders, which can facilitate biomarker discovery and improve therapeutic strategies. By integrating diverse data sources and capturing spatial and temporal dynamics, multilayer network analysis holds significant promise for advancing our understanding of neurodegeneration and ultimately improving patient outcomes. Continued interdisciplinary collaborations and technological advancements will be essential to translate multilayer network findings into clinical practice. In the coming years, it will be important to understand how the complexity of the brain across multiple levels can be combined with multilayer network methods to build a new generation of network-based models that integrate different scales and modalities of brain organisation.

Research on multilayer networks in neuroimaging is ongoing, and numerous questions remain to be addressed to fully characterise the multiscale, multilevel organisation of the brain. For example, despite the promise of multilayer network analysis, several challenges remain in applying these approaches to neuroimaging data. Integration of heterogeneous data sources, standardisation of analytical methods and validation of findings across independent cohorts are critical features of the process. Furthermore, the development of scalable computational tools and platforms for multilayer network analysis is needed to facilitate widespread adoption in the research community. I conclude this chapter by highlighting some key directions in the advancement of multilayer networks in neuroimaging, particularly pertinent for addressing these challenges.

References

Alessandretti, L., Natera Orozco, L.G., Saberi, M. et al. (2023). Multimodal urban mobility and multilayer transport networks. *Environment and Planning B: Urban Analytics and City Science* 50 (8): 2038–2070.

Aleta, A., Meloni, S. and Moreno, Y. (2017). A multilayer perspective for the analysis of urban transportation systems. *Scientific Reports* 7 (1): 44359.

Arenas, A., Cabrales, A., Díaz-Guilera, A. et al. (2003). Search and congestion in complex networks. *Statistical Mechanics of Complex Networks* 175–194. Lecture Notes in Physics, vol 625. Springer, Berlin, Heidelberg. https://doi.org/10.1007/978-3-540-44943-0_11

Avena-Koenigsberger, A., Misic, B. and Sporns, O. (2018). Communication dynamics in complex brain networks. *Nature Reviews Neuroscience* 19 (1): 17–33.

Bassett, D.S. and Bullmore, E.T. (2009). Human brain networks in health and disease. *Current Opinion in Neurology* 22 (4): 340–347.

Bassett, D.S. and Sporns, O. (2017). Network neuroscience. *Nature Neuroscience* 20 (3): 353.

Bassett, D.S., Bullmore, E.T., Verchinski, B.A. et al. (2008). Hierarchical organization of human cortical networks in health and schizophrenia. *The Journal of Neuroscience* 28 (37): 9239–9248.

Bassett, D.S., Meyer-Lindenberg, A., Achard, S. et al. (2006). Adaptive reconfiguration of fractal small-world human brain functional networks. *Proceedings of the National Academy of Sciences of the United States of America* 103: 19518–19523.

Bassett, D.S., Meyer-Lindenberg, A., Weinberger, D.R. et al. (2009). Cognitive fitness of cost-efficient brain functional networks. *Proceedings of the National Academy of Sciences of the United States of America* 106 (28): 11747–11752.

Bassett, D.S., Wymbs, N.F., Porter, M.A. et al. (2011). Dynamic reconfiguration of human brain networks during learning. *Proceedings of the National Academy of Sciences* 108 (18): 7641–7646.

Battiston, F., Nicosia, V., Chavez, M. et al. (2017). Multilayer motif analysis of brain networks. *Chaos: An Interdisciplinary Journal of Nonlinear Science* 27 (4): 047404.

Bertolero, M.A., Yeo, B.T. and D'Esposito, M. (2015). The modular and integrative functional architecture of the human brain. *Proceedings of the National Academy of Sciences* 112 (49): E6798–E6807.

Betzel, R.F. and Bassett, D.S. (2017). Multi-scale brain networks. *Neuroimage* 160: 73–83.

Betzel, R.F., Medaglia, J.D., Papadopoulos, L. et al. (2017). The modular organization of human anatomical brain networks: accounting for the cost of wiring. *Network Neuroscience* 1 (1): 42–68.

Boccaletti, S., Bianconi, G., Criado, R. et al. (2014). The structure and dynamics of multilayer networks. *Physics Reports* 544 (1): 1–122.

Boguna, M., Krioukov, D. and Claffy, K.C. (2009). Navigability of complex networks. *Nature Physics* 5 (1): 74–80.

Braun, U., Schäfer, A., Walter, H. et al. (2015). Dynamic reconfiguration of frontal brain networks during executive cognition in humans. *Proceedings of the National Academy of Sciences* 112 (37): 11678–11683.

Breakspear, M., Heitmann, S. and Daffertshofer, A. (2010). Generative models of cortical oscillations: neurobiological implications of the Kuramoto model. *Frontiers in Human Neuroscience* 11 (4): 190.

Brookes, M.J., Tewarie, P.K., Hunt, B.A. et al. (2016). A multi-layer network approach to meg connectivity analysis. *Neuroimage* 132: 425–438.

Butts, C.T. (2009). Revisiting the foundations of network analysis. *Science* 325 (5939): 414–416.

Chen, Z.J., He, Y., Rosa-Neto, P. et al. (2008). Revealing modular architecture of human brain structural networks by using cortical thickness from MRI. *Cerebral Cortex* 18 (10): 2374–2381.

Cozzo, E., Kivelä, M., De Domenico, M. et al. (2015). Structure of triadic relations in multiplex networks. *New Journal of Physics* 17 (7): 073029.

Crofts, J.J., Forrester, M. and O'Dea, R.D. (2016). Structure-function clustering in multiplex brain networks. *Europhysics Letters* 116 (1): 18003.

Crofts, J.J., Forrester, M., Coombes, S. et al. (2022). Structure-function clustering in weighted brain networks. *Scientific Reports* 12 (1): 16793.

Crossley, N.A., Mechelli, A., Vértes, P.E. et al. (2013). Cognitive relevance of the community structure of the human brain functional coactivation network. *Proceedings of the National Academy of Sciences* 110 (28): 11583–11588.

Deco, G. and Kringelbach, M.L. (2016). Metastability and coherence: extending the communication through coherence hypothesis using a whole-brain computational perspective. *Trends in Neurosciences* 39 (3): 125–135.

Deco, G., Jirsa, V.K. and McIntosh, A.R. (2011). Emerging concepts for the dynamical organization of resting-state activity in the brain. *Nature Reviews Neuroscience* 12 (1): 43–56.

Deco, G., Jirsa, V., McIntosh, A.R. et al. (2009). Key role of coupling, delay, and noise in resting brain fluctuations. *Proceedings of the National Academy of Sciences of the United States of America* 106 (25): 10302–10307.

De Domenico, M. (2017). Multilayer modeling and analysis of human brain networks. *Giga Science* 6 (5): gix004.

De Domenico, M., Granell, C., Porter, M.A. et al. (2016). The physics of spreading processes in multilayer networks. *Nature Physics* 12 (10): 901–906.

De Domenico, M., Sasai, S. and Arenas, A. (2016). Mapping multiplex hubs in human functional brain networks. *Frontiers in Neuroscience* 10: 205256.

De Domenico, M., Solé-Ribalta, A., Cozzo, E. et al. (2013). Mathematical formulation of multilayer networks. *Physical Review X* 3 (4): 041022.

De Domenico, M., Solé-Ribalta, A., Gómez, S. et al. (2014). Navigability of interconnected networks under random failures. *Proceedings of the National Academy of Sciences* 111 (23): 8351–8356.

Desikan, R., Segonne, F., Fischl, B. et al. (2018). Release notes nda 2017. *Neuroimage* 31: 968–980.

Dosenbach, N.U., Nardos, B., Cohen, A.L. et al. (2010). Prediction of individual brain maturity using fMRI. *Science* 329 (5997): 1358–1361.

Duh, M., Gosak, M., Slavinec, M. et al. (2019). Assortativity provides a narrow margin for enhanced cooperation on multilayer networks. *New Journal of Physics* 21 (12): 123016.

Dujardin, S., Commins, C., Lathuiliere, A. et al. (2020). Tau molecular diversity contributes to clinical heterogeneity in Alzheimer's disease. *Nature Medicine* 26 (8): 1256–1263.

Echegoyen, I., López-Sanz, D., Maestú, F. et al. (2021). From single layer to multilayer networks in mild cognitive impairment and Alzheimer's disease. *Journal of Physics: Complexity* 2 (4): 045020.

Ettle, B., Schlachetzki, J.C. and Winkler, J. (2016). Oligodendroglia and myelin in neurodegenerative diseases: more than just bystanders? *Molecular Neurobiology* 53: 3046–3062.

Fan, Y., Batmanghelich, N., Clark, C.M. et al. (2008). Spatial patterns of brain atrophy in MCI patients, identified via high-dimensional pattern classification, predict subsequent cognitive decline. *Neuroimage* 39 (4): 1731–1743.

Fan, Y.-S., Xu, Y., Bayrak, Ş. et al. (2023). Macroscale thalamic functional organization disturbances and underlying core cytoarchitecture in early-onset schizophrenia. *Schizophrenia Bulletin* 49 (5): 1375–1386.

Finn, K.R., Silk, M.J., Porter, M.A. et al. (2019). The use of multilayer network analysis in animal behaviour. *Animal Behaviour* 149: 7–22.

Fornito, A., Zalesky, A. and Bullmore, E.T. (2010). Network scaling effects in graph analytic studies of human resting-state fMRI data. *Journal of Integrative Neuroscience* 17 (4): 22.

Fortunato, S. and Hric, D. (2016). Community detection in networks: a user guide. *Physics Reports* 659: 1–44.

Franzmeier, N., Neitzel, J., Rubinski, A. et al. (2020). Functional brain architecture is associated with the rate of tau accumulation in Alzheimer's disease. *Nature Communications* 11 (1): 1–17.

Franzmeier, N., Rubinski, A., Neitzel, J. et al. (2019, 02). Functional connectivity associated with tau levels in ageing, Alzheimer's, and small vessel disease. *Brain*, 142 (4): 1093–1107. https://doi.org/10.1093/brain/awz026.

Fries, P. (2005). A mechanism for cognitive dynamics: neuronal communication through neuronal coherence. *Trends in Cognitive Sciences* 9: 474–480.

Friston, K.J. and Dolan, R.J. (2010). Computational and dynamic models in neuroimaging. *Neuroimage* 52 (3): 752–765.

Ghajari, M., Hellyer, P.J. and Sharp, D.J. (2017). Computational modelling of traumatic brain injury predicts the location of chronic traumatic encephalopathy pathology. *Brain* 140 (2): 333–343.

Glasser, M.F., Coalson, T.S., Robinson, E.C. et al. (2016). A multi-modal parcellation of human cerebral cortex. *Nature* 536 (7615): 171.

Goubran, M., Leuze, C., Hsueh, B. et al. (2019). Multimodal image registration and connectivity analysis for integration of connectomic data from microscopy to MRI. *Nature Communications* 10 (1): 5504.

Granell, C., Gómez, S. and Arenas, A. (2013). Dynamical interplay between awareness and epidemic spreading in multiplex networks. *Physical Review Letters* 111 (12): 128701.

Griffa, A., Baumann, P.S., Thiran, J.-P. et al. (2013). Structural connectomics in brain diseases. *Neuroimage* 80: 515–526.

Guerrini, R., Casari, G. and Marini, C. (2003). The genetic and molecular basis of epilepsy. *Trends in Molecular Medicine* 9 (7): 300–306.

Guillon, J., Attal, Y., Colliot, O. et al. (2017). Loss of brain inter-frequency hubs in Alzheimer's disease. *Scientific Reports* 7 (1): 10879.

Hagmann, P., Cammoun, L., Gigandet, X. et al. (2008). Mapping the structural core of human cerebral cortex. *PLoS Biology* 6 (7): e159.

Hagmann, P., Grant, P.E. and Fair, D.A. (2012). MR connectomics: a conceptual framework for studying the developing brain. *Frontiers in Systems Neuroscience* 6: 43.

Honey, C.J., Kötter, R., Breakspear, M. et al. (2007). Network structure of cerebral cortex shapes functional connectivity on multiple time scales. *Proceedings of the National Academy of Sciences of the United States of America* 104 (24): 10240–10245.

Huang, H., Shu, N., Mishra, V. et al. (2015). Development of human brain structural networks through infancy and childhood. *Cerebral Cortex* 25 (5): 1389–1404.

Jirsa, V.K., Sporns, O., Breakspear, M. et al. (2010). Towards the virtual brain: network modeling of the intact and the damaged brain. *Archives Italiennes de Biologie* 148 (3): 189–205.

Khan, A.F., Palomero-Gallagher, N., Zilles, K. et al. (2020). Whole brain generative model identifies neurotransmitter alterations underlying Alzheimer's disease progression: neuroimaging/multi-modal comparisons. *Alzheimer's & Dementia* 16: e041193.

Kivelä, M., Arenas, A., Barthelemy, M. et al. (2014). Multilayer networks. *Journal of Complex Networks* 2 (3): 203–271.

Kurant, M. and Thiran, P. (2006). Layered complex networks. *Physical Review Letters* 96 (13): 138701.

Li, A., Zhou, L., Su, Q. et al. (2020). Evolution of cooperation on temporal networks. *Nature Communications* 11 (1): 2259.

Lieberman, J., Chakos, M., Wu, H. et al. (2001). Longitudinal study of brain morphology in first episode schizophrenia. *Biological Psychiatry* 49 (6): 487–499.

Mandke, K., Meier, J., Brookes, M.J. et al. (2018). Comparing multilayer brain networks between groups: introducing graph metrics and recommendations. *NeuroImage* 166, 371–384.

Manfredi, S., Di Tucci, E. and Latora, V. (2018). Mobility and congestion in dynamical multilayer networks with finite storage capacity. *Physical Review Letters* 120 (6): 068301.

Melo, D., Porto, A., Cheverud, J.M. et al. (2016). Modularity: genes, development, and evolution. *Annual Review of Ecology, Evolution, and Systematics* 47: 463–486.

Meunier, D., Achard, S., Morcom, A. et al. (2009). Age-related changes in modular organization of human brain functional networks. *Neuroimage* 44 (3): 715–723.

Mucha, P.J., Richardson, T., Macon, K. et al. (2010). Community structure in time-dependent, multiscale, and multiplex networks. *Science* 328 (5980): 876–878.

Newman, M. E. (2006). Modularity and community structure in networks. *Proceedings of the National Academy of Sciences* 103 (23): 8577–8582.

Nicosia, V. and Latora, V. (2015). Measuring and modeling correlations in multiplex networks. *Physical Review E* 92 (3): 032805.

Ou-Yang, L., Dai, D.-Q., Li, X.-L. et al. (2014). Detecting temporal protein complexes from dynamic protein-protein interaction networks. *BMC Bioinformatics* 15: 1–14.

Park, H.-J. and Friston, K. (2013). Structural and functional brain networks: from connections to cognition. *Science* 342 (6158): 1238411.

Patow, G., Martin, I., Sanz Perl, Y. et al. (2024). Whole-brain modelling: an essential tool for understanding brain dynamics. *Nature Reviews Methods Primers* 4 (1): 53.

Pilosof, S., Porter, M.A., Pascual, M. et al. (2017). The multilayer nature of ecological networks. *Nature Ecology & Evolution* 1 (4): 0101.

Power, J.D., Cohen, A.L., Nelson, S.M. et al. (2011). Functional network organization of the human brain. *Neuron* 72 (4): 665–678.

Presigny, C. and De Vico Fallani, F. (2022). Colloquium: multiscale modeling of brain network organization. *Reviews of Modern Physics* 94 (3): 031002.

Puxeddu, M.G., Petti, M. and Astolfi, L. (2021). A comprehensive analysis of multilayer community detection algorithms for application to EEG-based brain networks. *Frontiers in Systems Neuroscience* 15: 624183.

Puxeddu, M.G., Faskowitz, J., Betzel, R.F. et al. (2020). The modular organization of brain cortical connectivity across the human lifespan. *NeuroImage* 218: 116974.

Puxeddu, M.G., Faskowitz, J., Sporns, O. et al. (2022). Multi-modal and multi-subject modular organization of human brain networks. *NeuroImage* 264: 119673.

Rademakers, R., Neumann, M. and Mackenzie, I.R. (2012). Advances in understanding the molecular basis of frontotemporal dementia. *Nature Reviews Neurology* 8 (8): 423–434.

Raj, A. and Powell, F. (2018). Models of network spread and network degeneration in brain disorders. *Biological Psychiatry: Cognitive Neuroscience and Neuroimaging* 3 (9): 788–797.

Raj, A., Kuceyeski, A. and Weiner, M. (2012). A network diffusion model of disease progression in dementia. *Neuron* 73 (6): 1204–1215.

Richetto, J. and Meyer, U. (2021). Epigenetic modifications in schizophrenia and related disorders: molecular scars of environmental exposures and source of phenotypic variability. *Biological Psychiatry* 89 (3): 215–226.

Ritter, P., Schirner, M., McIntosh, A.R. et al. (2013). The virtual brain integrates computational modeling and multimodal neuroimaging. *Brain Connectivity* 3 (2): 121–145.

Rubido, N., Riedel, G. and Vuksanović, V. (2024). Genetic basis of anatomical asymmetry and aberrant dynamic functional networks in Alzheimer's disease. *Brain Communications* 6 (1): fcad320.

Rubinov, M. and Bassett, D.S. (2011). Emerging evidence of connectomic abnormalities in schizophrenia. *The Journal of Neuroscience* 31 (17): 6263–6265.

Rubinov, M. and Sporns, O. (2010). Complex network measures of brain connectivity: uses and interpretations. *Neuroimage* 52 (3): 1059–1069.

Sanabria-Diaz, G., Melie-García, L., Iturria-Medina, Y. et al. (2010). Surface area and cortical thickness descriptors reveal different attributes of the structural human brain networks. *Neuroimage* 50 (4): 1497–1510.

Schirner, M., McIntosh, A.R., Jirsa, V. et al. (2018). Inferring multi-scale neural mechanisms with brain network modelling. *elife* 7: e28927.

Seidlitz, J., Váša, F., Shinn, M. et al. (2018). Morphometric similarity networks detect microscale cortical organization and predict inter-individual cognitive variation. *Neuron* 97 (1): 231–247.

Solé-Ribalta, A., Gómez, S. and Arenas, A. (2016). Congestion induced by the structure of multiplex networks. *Physical Review Letters* 116 (10): 108701.

Sporns, O. (2013). Structure and function of complex brain networks. *Dialogues in Clinical Neuroscience* 15 (3): 247.

Sporns, O. and Betzel, R.F. (2016). Modular brain networks. *Annual Review of Psychology* 67: 613–640.

Sporns, O. and Kötter, R. (2004). Motifs in brain networks. *PLoS Biology* 2 (11): e369.

Stam, C.J. (2014). Modern network science of neurological disorders. *Nature Reviews Neuroscience* 15 (10): 683–695.

Suárez, L.E., Markello, R.D., Betzel, R.F. et al. (2020). Linking structure and function in macroscale brain networks. *Trends in Cognitive Sciences* 24 (4): 302–315.

Tewarie, P., Hillebrand, A., van Dijk, B.W. et al. (2016). Integrating cross-frequency and within band functional networks in resting-state MEG: a multi-layer network approach. *Neuroimage* 142: 324–336.

Thompson, P.M., Stein, J.L., Medland, S.E. et al. (2014). The enigma consortium: large-scale collaborative analyses of neuroimaging and genetic data. *Brain Imaging and Behavior* 8: 153–182.

Tzourio-Mazoyer, B., Landeau, D., Papathanassiou, F. et al. (2002). Automated anatomical labeling of activations in SPM using a macroscopic anatomical parcellation of the MNI MRI single-subject brain. *NeuroImage* 15: 273–289.

Vaiana, M. and Muldoon, S.F. (2020). Multilayer brain networks. *Journal of Nonlinear Science* 30 (5): 2147–2169.

Valk, S.L., Bernhardt, B.C., Trautwein, F.-M. et al. (2017). Structural plasticity of the social brain: differential change after socio-affective and cognitive mental training. *Science Advances* 3 (10): e1700489.

van den Heuvel, M.P., Stam, C.J., Boersma, M. et al. (2008). Small-world and scale-free organization of voxel-based resting-state functional connectivity in the human brain. *Neuroimage* 43 (3): 528–539.

Van Wijk, B.C., Stam, C.J. and Daffertshofer, A. (2010). Comparing brain networks of different size and connectivity density using graph theory. *PLoS One* 5 (10): e13701.

Váša, F., Seidlitz, J., Romero-Garcia, R. et al. (2017). Adolescent tuning of association cortex in human structural brain networks. *Cerebral Cortex* 28 (1): 281–294.

Vázquez-Rodríguez, B., Suárez, L.E., Markello, R.D. et al. (2019). Gradients of structure–function tethering across neocortex. *Proceedings of the National Academy of Sciences* 116 (42): 21219–21227.

Vuksanović, V. (2019). Cortical thickness and functional networks modules by cortical lobes. *Neuroscience* 423: 172–176. https://doi.org/10.1016/j.neuroscience.2019.10.038.

Vuksanović, V. (2021). Brain morphological and functional networks: implications for neurodegeneration. *Physics of Biological Oscillators: New Insights into Non-Equilibrium and Non-Autonomous Systems* 329. pp 329–342

Vuksanović, V. (2022). Brain morphometric similarity and flexibility. *Cerebral Cortex Communications* 3 (3): tgac024.

Vuksanović, V. and Hövel, P. (2014). Functional connectivity of distant cortical regions: role of remote synchronization and symmetry in interactions. *NeuroImage* 97: 1–8.

Vuksanović, V. and Hövel, P. (2015). Dynamic changes in network synchrony reveal resting-state functional networks. *Chaos: An Interdisciplinary Journal of Nonlinear Science* 25 (2): 023116.

Vuksanović, V., Staff, R.T., Ahearn, T. et al. (2019). Cortical thickness and surface area networks in healthy aging, Alzheimer's disease and behavioral variant fronto-temporal dementia. *International Journal of Neural Systems* 29 (06): 1850055. https://doi.org/10.1142/S0129065718500557.

Wang, Z., Andrews, M.A., Wu, Z.-X. et al. (2015). Coupled disease–behavior dynamics on complex networks: a review. *Physics of Life Reviews* 15: 1–29.

Wang, X., Huang, W., Su, L. et al. (2020). Neuroimaging advances regarding subjective cognitive decline in preclinical Alzheimer's disease. *Molecular Neurodegeneration* 15: 1–27.

West, M.J., Coleman, P.D., Flood, D.G. et al. (1994). Differences in the pattern of hippocampal neuronal loss in normal ageing and Alzheimer's disease. *The Lancet* 344 (8925): 769–772.

Wurthmann, C., Bogerts, B. and Falkai, P. (1995). Brain morphology assessed by computed tomography in patients with geriatric depression, patients with degenerative dementia, and normal control subjects. *Psychiatry Research: Neuroimaging* 61 (2): 103–111.

Yamashita, M., Kawato, M. and Imamizu, H. (2015). Predicting learning plateau of working memory from whole-brain intrinsic network connectivity patterns. *Scientific Reports* 5 (1): 1–8.

Yuvaraj, M., Dey, A.K., Lyubchich, V. et al. (2021). Topological clustering of multilayer networks. *Proceedings of the National Academy of Sciences* 118 (21): e2019994118.

Zalesky, A., Fornito, A., Harding, I.H. et al. (2010). Whole-brain anatomical networks: does the choice of nodes matter? *NeuroImage* 50 (3): 970–983.

5

Artificial Intelligence–inspired Subtype Analysis for Brain Imaging

Yong Liu and Kun Zhao

School of Artificial Intelligence, Beijing University of Posts and Telecommunications, Beijing, China

5.1 Introduction

There is growing clinical evidence that brain disorders are heterogeneous in phenotype, genetics and neuropathology (Habes et al., 2020; Hong et al., 2020). This heterogeneity often leads to misdiagnosis or delayed diagnosis under current clinical diagnostic criteria, potentially hindering treatment and clinical trials in disease treatment development (Drummond et al., 2017). Brain-based subtyping studies utilise brain magnetic resonance imaging and data-driven methods to discover novel subtypes of diseases, providing a new perspective on understanding disease heterogeneity in brain disorders. Substantial novel subtypes were uncovered using data-driven methods based on neuroimaging data in neurodegenerative diseases, such as Alzheimer's disease (AD) (Habes et al., 2020), and mental disorders, such as major depressive disorder (MDD) (Drysdale et al., 2017), autism spectrum disorder (Hong et al., 2020) and schizophrenia (Chand et al., 2020), contributing much to the future individualised diagnosis and treatment (Ten Kate et al., 2018).

Distinct underlying neuropathology and atrophy patterns uncovered by subtyping studies in AD show the novel disease mechanisms and phenotypes that are different from the traditional theory (Vogel et al., 2021), aiding in developing more effective targeted treatments based on subtypes (Mann et al., 2018). Typical AD is mainly characterised by memory loss, associated with atrophy in the hippocampus and temporoparietal lobe (Dubois et al., 2007), whereas some atypical AD subtypes show atrophy in other regions rather than the hippocampus (Ferreira et al., 2020). Moreover, subtyping in mental disorders showed more advantages in treatment. Four distinct subtypes of MDD identified by functional connectivity showed different responses to the repetitive transcranial magnetic stimulation treatment (Drysdale et al., 2017). Subtypes of post-traumatic stress disorder showed different brain networks and different responses to the psychotherapy treatment (Zhang et al., 2021b).

Heterogeneity uncovering aims to disentangle the novel stratification within a population of one disease, which is a typical latent representation and data structure mining problem.

The unsupervised or semi-supervised methods that can disentangle the latent representation were applied to the subtyping with the availability of big data. This chapter summarises the principles of 10 methods and shows the example codes, including hierarchical, K-means, K-medians, nonnegative matrix factorization (NMF), Gaussian mixture model (GMM), spectral clustering, unsupervised random forest (URF), Latent Dirichlet Allocation (LDA), CHIMERA and HYDRA (Ten Kate et al., 2018; Lenhart et al., 2021; Salmanpour et al., 2022; Whitwell et al., 2018; Ezzati et al., 2020; Poulakis et al., 2018; Zhang et al., 2021a; Dong et al., 2017; Varol et al., 2017).

5.2 Clustering Methods

5.2.1 Distance-based

5.2.1.1 Hierarchical

In hierarchical clustering methods, clusters are formed by iteratively dividing the patterns using a top-down or bottom-up approach. There are two primary forms of hierarchical clustering methods: agglomerative and divisive hierarchical clustering.

- **Agglomerative Hierarchical Clustering**: This method follows a bottom-up approach. It starts with each object as its own individual cluster and then progressively merges these small, atomic clusters into larger and larger clusters. This process continues until all objects are grouped into a single cluster or until certain termination conditions are met.
- **Divisive Hierarchical Clustering**: In contrast, this method follows a top-down approach. It begins with all objects in a single cluster and then iteratively breaks this cluster into smaller clusters. This process continues until each object forms its own cluster or until certain termination conditions are satisfied.

The hierarchical clustering methods could be further grouped into three categories based on similarity measures or linkages: single linkage, complete linkage and average linkage.

Steps of Agglomerative Clustering:

1. Make each point a separate cluster.
2. Until the clustering is satisfactory:
3. Merge the two clusters with the smallest inter-cluster distance.
4. End.

Steps of Divisive Clustering:

1. Construct a single cluster containing all points.
2. Until the clustering is satisfactory:
3. Split the cluster that yields the two components with the largest inter-cluster distance.
4. End.

The common criticism for classical hierarchical algorithms is that they lack robustness and are sensitive to noise and outliers. Other disadvantages of hierarchical algorithms include the tendency to form spherical shapes and reversal phenomenon, in which the normal hierarchical structure is distorted.

The code of this algorithm is as follows:

```
from sklearn import cluster
class HierarchicalClusterer:
  def __init__(self, cluster_num=2):
    self.cluster_num = int(cluster_num)
    self.labels_ = None
    self.model = None
  def fit(self, X):
    self.model = cluster.AgglomerativeClustering( n_clusters=self.cluster_num)
    self.model.fit(X)
    self.labels_ = self.model.labels_
    return self
hc = HierarchicalClusterer(cluster_num=2)
hc.fit(X)  # X: data matrix (n_samples, n_features)
labels = hc.labels_
```

5.2.1.2 K-means

Regarding partitional approaches, the K-means algorithm has been widely used by researchers. This method requires input parameters such as the number of groups (k) and a distance metric. Initially, each data point is associated with one of the k clusters according to its distance to the centroids (cluster centres) of each cluster. The objective function J is given as follows:

$$\text{Minimise } J = \sum_{j=1}^{k} \sum_{i=1}^{n} \left\|x_i^{(j)} - c_j\right\|^2 \tag{5.1}$$

where $\left\|x_i^{(j)} - c_j\right\|^2$ is the chosen distance measure between a data point $x_i^{(j)}$ and the cluster centre c_j. The steps of K-means algorithm are as follows:

1. Initialisation: Take K distinct points (patterns) randomly. These points represent initial group centroids.
2. Assign each object to the group that has the closest centroid.
3. When all objects have been assigned, recalculate the positions of the K centroids.
4. Repeat Steps 2 and 3 until the centroids no longer move. This produces a separation of the objects into groups from which the metric to be minimised can be calculated.

The final classification can strongly depend on the choice of the number of centroids. Additionally, K-means is not particularly recommended for cases where the clusters do not have a convex distribution or have very different sizes. Moreover, the K-means algorithm is sensitive to the initial seed selection.

The code of this algorithm is as follows:

```
from sklearn.cluster import KMeans
class KmeansClusterer:
  def __init__(self):
    self.cluster_num = 2
    self.labels_ = []
  def fit(self, X, cluster_num):
    self.x_data = X
    self.cluster_num = int(cluster_num)
    #self.y_data=X["true_label"]
    random_state = np.random.RandomState(seed=None)
    kmeans = KMeans(n_clusters=self.cluster_num,random_state=random_state)
    self.labels_= kmeans.fit_predict(self.x_data)
cluster = KmeansClusterer()
cluster.fit(X)
label = cluster.labels_
```

5.2.1.3 K-medians

The K-medians approach to clustering data attempts to minimise the 1-norm distances between each point and its closest cluster centre. This minimisation of distances is obtained by setting the centre of each cluster to be the median of all points in that cluster.

Given a set of points, the K-medians algorithm attempts to create k disjoint clusters that minimise the following equation. This means that the centre of each cluster minimises this objective function.

$$Q\left(\{\pi_j\}_{j=1}^{K}\right) = \sum_{j=1}^{K} \sum_{x \in \pi_j} \|x - c_j\|_1 \tag{5.2}$$

The code of this algorithm is as follows:

```
import rpy2.robjects as robjects
from rpy2.robjects.packages import importr
from rpy2.robjects import numpy2ri
class KMedianClusterer:
  def __init__(self):
    self.cluster_num = 2
    self.labels_ = []
  def fit(self,X,cluster_num):
    kmedians = importr('flexclust')
```

```
    numpy2ri.activate()
    r = robjects.r
    r.assign("x", X)
    r.assign("k_num", cluster_num)
    r('result <- kcca(x, k_num, family=kccaFamily("kmedians"))')
    r('assignments <- clusters(result)')
    self.labels_ = np.array(r('assignments'))-1
    numpy2ri.deactivate()
cluster = KMedianClusterer()
cluster.fit(X)
label = cluster.labels_
```

5.2.2 Model-based

5.2.2.1 Gaussian Mixture Model

Model-based methods can be regarded as a general framework for estimating the maximum likelihood of the parameters of an underlying distribution for a given dataset. A well-known instance of model-based methods is the expectation–maximization (EM) algorithm. Most commonly, it is assumed that the data from each class can be modelled by multivariate normal distributions. Therefore, the distribution observed for the entire dataset can be seen as a mixture of such normal distributions. A maximum likelihood approach is then applied to find the most probable parameters of the normal distributions for each class.

The EM approach for clustering is particularly suitable when the dataset includes missing values. However, the clusters obtained using this method may strongly depend on the initial conditions. Additionally, the algorithm may fail to identify very small clusters.

The procedure of EM algorithm is as follows:

1. Initialise the parameters Θ^{old}.
2. E steps: Evaluate $p(Y|X, \Theta^{\text{old}})$.
3. M steps: Re-estimate the parameters $\Theta^{\text{new}} = \text{argmax}_{\Theta} L(\Theta)$.
4. Check for convergence. If the convergence criterion is not satisfied, let $\Theta^{\text{old}} \leftarrow \Theta^{\text{new}}$ and return to Step 2.

The code of this algorithm is as follows:

```
from sklearn.mixture import GaussianMixture as GMM
class GMMClusterer:
  def __init__(self):
    self.cluster_num = 2
    self.labels_ = []
  def fit(self, X,cluster_num):
    self.cluster_num = int(cluster_num)
    x_t = X
```

```
        gmm = GMM(n_components=self.cluster_num).fit(x_t)
        self.labels_ = gmm.predict(x_t)
cluster = GMMClusterer()
cluster.fit(X)
label = cluster.labels_
```

5.2.2.2 Latent Dirichlet Allocation

LDA is a generative probabilistic model based on the Dirichlet distribution. LDA assumes that each feature of a sample is jointly generated by the sample's cluster distribution θ and the cluster's feature distribution ϕ. Based on the Dirichlet distribution, LDA generates the cluster distribution for the samples and the feature distribution for each cluster. Then, using Gibbs sampling or variational inference methods, LDA iteratively adjusts these distributions to ensure that the generated feature distribution best fits the actual data distribution. Finally, LDA generates the cluster distribution θ for each sample, representing the probability of each sample belonging to different clusters.

The code of this algorithm is as follows:

```
from sklearn.decomposition import LatentDirichletAllocation as LDA
class LDAClusterer:
    def __init__(self):
        self.cluster_num = 2
        self.labels_=[]
    def fit(self, X,cluster_num):
        self.cluster_num = int(cluster_num)
        self.x_data = X
        Mx=(self.x_data*10).astype(int)
        lda = LDA(n_components = self.cluster_num, random_state = None)
        lda.fit(Mx)
        doc_topic_dist = lda.transform(Mx)
        self.labels_ = doc_topic_dist.argmax(axis=1)
cluster = LDAClusterer()
cluster.fit(X)
label = cluster.labels_
```

5.2.3 Matrix-based

5.2.3.1 Nonnegative Matrix Factorization

The NMF model is a dimensionality reduction technique that seeks a low-rank decomposition of a data matrix X with non-negativity constraints based on the basis matrix W and coefficient matrix H. As a data-driven dual-clustering approach ($X^{m\times n} = W^{m\times k} \cdot H^{k\times n}$), NMF allows the simultaneous identification of clusters of features and clusters of individuals in the database. Each row in W corresponds to a sample and indicates the sample's membership across different clusters (which correspond to the latent features in H).

Based on the values in W, samples can be assigned to the cluster in which they have the highest weight, thereby revealing the underlying structure of the data and grouping samples with similar features together.

The code of this algorithm is as follows:

```
import rpy2.robjects as robjects
from rpy2.robjects.packages import importr
from rpy2.robjects import numpy2ri
class NMFClusterer:
  def __init__(self):
    self.cluster_num = 2
    self.labels_ = []
    self.H = []
    self.W = []
  def fit(self, X, cluster_num):
    self.x_data = X
    self.cluster_num = int(cluster_num)
    numpy2ri.activate()
    importr('NMF')
    r = robjects.r
    r.assign("x", self.x_data)
    r.assign("k_num", self.cluster_num)
    r('model <- nmf(x, rank = k_num, nrun = 1)')
    r('W <- basis(model)')
    r('H <- coef(model)')
    r('labels <- apply(W, 1, which.max)')
    self.labels_ = np.array(r('labels'))-1
    self.W = r('W')
    self.H = r('H')
    numpy2ri.deactivate()
cluster = NMFClusterer()
cluster.fit(X)
label = cluster.labels_
```

5.2.4 Decision Trees

5.2.4.1 Unsupervised Random Forest

The URF algorithm uses a random forest predictor to distinguish between 'observed' data and 'synthetic' data. The observed data consists of the original, unlabeled data, while the synthetic data is drawn from a reference distribution. Typically, the Addcl1 method is used to construct the synthetic dataset. In this method, a value is randomly selected from the range of observed values in the first variable of the dataset. The same process is repeated for all variables, generating a synthetic observation with independence across variables. Then, by analysing the distribution of the original samples across the leaf nodes of the decision trees in the random forest, a similarity measure between data points is generated. This similarity is further transformed into RF dissimilarity, which can be used as input for other clustering algorithms to obtain cluster labels for the original data. The procedure of the URF algorithm is as follows:

1. Train the Random Forest Model: Train the random forest model using the original data (observed data) and synthetic data.
2. Obtain Leaf Node Indices: Represent the final position of the samples in each tree.
3. Calculate Similarity Between Samples: Compute the similarity measure between samples by calculating the frequency of shared leaf nodes across all trees.
4. Convert to Dissimilarity Matrix: Typically, dissimilarity can be defined as 1 minus the similarity score, or other distance measures can be used to represent dissimilarity.
5. Use the Dissimilarity Measure as Input for Other Clustering Algorithms.

The code of this algorithm is as follows:

```
from sklearn import cluster, manifold
import random
import URF.forest_cluster.RandomForestEmbedding as rfc
class URFClusterer:
  def __init__(self):
    self.cluster_num = 2
    self.labels_ = []
    def fit(self, X,cluster_num):
    self.cluster_num = int(cluster_num)
    rf = rfc(n_estimators=100, random_state=None, sparse_output=False,n_jobs=4)
    leaves = rf.fit_transform(self.x_data)
    projector = manifold.TSNE(n_components=2, random_state=1234, metric='hamming')
    embedding = projector.fit_transform(leaves)
    clusterer = cluster.KMeans(n_clusters=self.cluster_num, n_init=20)
    clusterer.fit(embedding)
    self.labels_ = clusterer.labels_
cluster = URFClusterer()
cluster.fit(X)
label = cluster.labels_
```

5.2.5 Graph-based

5.2.5.1 Spectral

Another class of methods considered in our analyses is spectral clustering. These methods emerged as an alternative to traditional clustering approaches that were unable to define non-linear discriminative hypersurfaces. The main advantage of spectral methods lies in the definition of an adjacency structure from the original dataset, which avoids imposing a predefined shape for the clusters.

The first step in spectral clustering is to construct an affinity matrix A, where the value in the j-th row and k-th column indicates the similarity between points j and k. This matrix can be regarded as a weighted graph representation of the data. The eigenvalues and eigenvectors of the matrix are then used to partition the data according to a given criterion. Many different types of similarity matrices can be used, with the Laplacian matrix being a popular choice. The steps of un-normalised spectral clustering are as follows:

1. Construct a similarity matrix and a similarity graph.
2. Compute the un-normalised Laplacian L.
3. Compute the first k eigenvectors of L, and let $U \in R^{n \times k}$ be the matrix containing the eigenvectors as columns.
4. For $i = 1, \ldots, n$, let $y_i \in R^k$ be the vector corresponding to the i-th row of U.
5. Cluster the points y_i with the K-means algorithm into clusters.

The code of this algorithm is as follows:

```
from sklearn import decomposition
from sklearn.cluster
import SpectralClustering as Spectral
  def __init__(self):
    self.alpha = 2
    self.labels_ = []
  def fit(self, X,alpha):
    self.alpha = alpha
    self.labels_ = Spectral(n_clusters=self.alpha,eigen_solver='lobpcg').fit_predict(X)
cluster = URFClusterer()
cluster.fit(X)
label = cluster.labels_
```

5.2.6 Complex Method

5.2.6.1 CHIMERA

Before using this algorithm, we need the datasets containing normal control (NC) $X = \{x_1, \ldots, x_m\}$ and patients $Y = \{y_1, \ldots, y_N\}$. The samples are described by two sets of features: $x_m = (x_m^v, x_m^c), y_n = (y_n^v, y_n^c)$. $(x_m^v, y_n^v) \in \mathbb{R}^{D_1}$ are imaging features. $(x_m^c, y_n^c) \in \mathbb{R}^{D_2}$ are covariate features (such as age, gender). This algorithm models the pathological transition between the two groups as transformations T from NC to patient distribution. Conversely, the transformed NC points cover the entire set of patients. Due to the disease heterogeneity, the pathological transitions T might take several directions. Therefore, T consists of multiple possible transformations, each of them (T_k) representing a pathological direction of imaging change. Let us assume that the transformed NC samples are denoted as $X' = \{x_1{}', \ldots, x_m{}'\}$, where the imaging features $x_m^v{}'$ are transformed from x_m^v and the covariate features $x_m^c{}'$ remains the same, $x_m' = T(x_m) = (T(x_m^v), x_m^c)$. For T_k,linear transformations were chosen for the imaging features. Each was described by a pair of parameters (A_k, b_k):

$$T(x_m^v) = \sum_{k=1}^{K} \xi_{km}(A_k x_m^v + b_k) \tag{5.3}$$

where $\sum_k \xi_{km} = 1$ and $\xi_{km} > 0$ for all m. ξ_{km} is the probability of an NC x_m being converted into a patient through pathological transition T_k.

The algorithm assumes that any normal person may be converted into a patient with equal probability $P(x'_m) = \frac{1}{M}$. Covariates are introduced into the distribution matching criteria by combining imaging and covariate-specific distances in a multi-kernel way. The distance between two points is measured by RBF kernels. Therefore, the likelihood of the distribution X' and Y can be described as follows:

$$\begin{aligned} P(X', Y) &= \prod_{n=1}^{N}\sum_{m=1}^{M} P(x'_m)P(y_n|x'_m) \\ &= \prod_{n=1}^{N}\sum_{m=1}^{M} \frac{1}{M}\frac{r^{\frac{D_2}{2}}}{(\sqrt{2\pi}\sigma)^{D_1-D_2}} \\ &\cdot \exp\left\{\frac{\|y_n^v - T(x_m^v)\|^2 + r\|y_n^c - x_m^c\|^2}{-2\sigma^2}\right\} \end{aligned} \tag{5.4}$$

The goal of this algorithm is to maximise the likelihood function of the patient distribution Y and the transformation distribution X', by minimising the following energy function:

$$\begin{aligned} E(X', Y, \Theta) &= -\mathscr{L}(X', Y, \Theta) + \mathscr{R}(\Theta) \\ &= -\log(P(X', Y)) + \frac{\lambda_1}{2\sigma^2}\sum_k \|b_k\|_2^2 + \frac{\lambda_2}{2\sigma^2}\sum_k \|A_k - \mathrm{I}\|_F^2 \end{aligned} \tag{5.5}$$

Here, $\Theta = (A, b, \xi, \sigma^2)$ denotes the parameters of CHIMERA. The log-likelihood $\mathscr{L}$ represents the distribution of X' and Y given the parameters. The regularisation term $\mathscr{R}$ improves the generalisation and robustness of CHIMERA. An EM algorithm is used to optimise the parameters Θ.

The likelihood of a patient y_n to have been generated by the transformation T_k can be estimated by $P_k(y_n) = \sum_m P(y_n|x_m)\xi_{km}$. By choosing the largest likelihood for each patient, we can get the clustering label.

The code of this algorithm is as follows:

```
from CHIMERA.chimera.algorithm_test import clustering_main
from CHIMERA.chimera import algorithm_test
from mlni.base import RB_Input
class ChimeraClusterer:
  def __init__(self):
    self.labels_ = []
    self.name="Chimera"
  def fit(self,input,k,weight_covariate = -1.0, weight_site = 10, lambda_b = 10.0,
    lambda_A = 100.0, transformation_type = 'duo', tol = 0.001, max_iteration =
    20,save_model=True,verbose = False):
    ## go into the core function of chimera
```

```
        label = clustering_main(input, config_arg)
        self.labels_ = label
input_data = RB_Input(feature_tsv=feature_tsv, covariate_tsv=covariate_tsv)
cluster = ChimeraCluster()
cluster.fit(input_data,k)
label = cluster.labels_
```

5.2.6.2 HYDRA

The algorithm also needs two kinds of datasets, including NC and patients ($\mathcal{D} = (x_i, y_i)_{i=1}^n$, $X_i \in \mathbb{R}^d$ and $y_i \in \{-1,1\}$). First, it aims to estimate k hyperplanes that form a convex polytope, which separates the NC and patients with a large margin. Second, it aims to assign each patient to the hyperplane that best separates it from the NCs. Patients who belong to different subtypes will be assigned to different hyperplanes, each of which reflects a respective pathological process. The algorithm assumes that all hyperplanes can correctly divide positive samples (NC), and there is at least one hyperplane that can divide negative samples (patients). By introducing an assignment matrix S^-, samples that have been affected by the same pathological process are assigned to the same hyperplane. We firstly consider hard clustering solution. The assignment matrix is a binary matrix that describes the assignment of the i-th negative class sample to the j-th face of the polytope, $S^- = \left[s_{i,j}\right] \in \{0,1\}^{n^- \times K}$. Thus, the search space $\mathcal{F}_K$ is defined as:

$$\mathcal{F}_K \triangleq \left\{w_j, b_j\right\}_{j=1}^K$$
$$st \begin{cases} \forall j : w_j^T x_i + b_j \geq 1 \text{ if } y_i = +1 \\ w_j^T x_i + b_j \leq -1 \text{ if } y_i = -1 \text{ and } s_{i,j} = 1 \end{cases} \tag{5.6}$$

Analogous to the support vector machine (SVM) formulation, the margin for the j-th face of the polytope is $\frac{2}{\left\|w_j\right\|_2}$. However, due to the piecewise nature of the convex polytope, there are multiple notions of margin for the surface of the polytope. In this work, aiming to keep the problem tractable, we maximise the average margin across all the faces of the polytope. The goal of HYDRA is to maximise the average margin across all the faces of the polytope. After adding the slack terms $\xi_{i,j}$, the objective function is:

$$\underset{w_j, b_j, \xi_j}{\text{Minimise}} \frac{\left\|w_j\right\|_2^2}{2} + C\sum_{i=1}^{n} \xi_{i,j} \text{ subject to} \tag{5.7}$$

$$\begin{cases} w_j^T x_i + b_j \geq 1 - \xi_{i,j} & \text{if } y_i = +1 \\ w_j^T x_i + b_j \leq -1 + \xi_{i,j} & \text{if } y_i = -1 \text{ and } s_{i,j} = 1 \\ \xi_{i,j} \geq 0 & \text{for } j = 1, \ldots, n \end{cases}$$

where C is a penalty parameter on the training error.

The optimisation for the binary assignment S^- is a non-convex problem. Therefore, to make the problem tractable, HYDRA relaxes the hard clustering solution to a soft assignment $S^- = \left[s_{i,j}\right] \in \left[0,1\right], \sum_{j=1}^{K} s_{i,j} = 1, \forall i)$. In order to synthesise the results of multiple runs, the algorithm constructs a consensus matrix. The elements are composed of the frequencies that different samples are classified into the same class. By using spectral clustering on the consensus matrix, we finally get the final clustering partition results. The procedure of HYDRA is composed of the following steps:

1. Initialisation: Initialise S^- by applying determinantal point processes (DPP).
2. Loop: Repeat until convergence (or reach a fixed number of iterations).
   - Fix S^- to update W, b by weighted LIBSVM.
   - Fix W, b to update S^-.

The code of this algorithm is as follows:

```
from Hydra.mlni.clustering import RB_DualSVM_Subtype
from Hydra.mlni.utils import make_cv_partition
class HydraClusterer:
    def __init__(self):
        self.labels_ = []
    def fit(self,input_data, k, cv_repetition=5, cv_strategy='hold_out',
save_models=False,cluster_predefined_c=0.25, class_weight_balanced=True,
weight_initialization_type='DPP',num_iteration=50,num_consensus=30, tol=1e-8, n_threads=16,
verbose=False):
        ## data split
        if cv_strategy == "hold_out":
            split_index = make_cv_partition(input_data.get_y(), cv_strategy, cv_repetition)
        elif cv_strategy == "k_fold":
            split_index = make_cv_partition(input_data.get_y(), cv_strategy, cv_repetition)
        ## Here, semi-supervised clustering
        wf_clustering = RB_DualSVM_Subtype(input_data, split_index, cv_repetition, k,
balanced=class_weight_balanced,num_consensus=num_consensus
num_iteration=num_iteration,tol=tol, predefined_c=cluster_predefined_c,
weight_initialization_type=weight_initialization_type,
n_threads=n_threads, save_models=save_models, verbose=verbose)
        label = wf_clustering.run()
        self.labels_ = label
input_data = RB_Input(feature_tsv=feature_tsv, covariate_tsv=covariate_tsv)
cluster = HydraCluster()
cluster.fit(input_data,k)
label = cluster.labels_
```

References

Chand, G.B., Dwyer, D.B., Erus, G. et al. (2020). Two distinct neuroanatomical subtypes of schizophrenia revealed using machine learning. *Brain* 143 (3): 1027–1038.

Dong, A., Toledo, J.B., Honnorat, N. et al. (2017). Heterogeneity of neuroanatomical patterns in prodromal Alzheimer's disease: links to cognition, progression and biomarkers. *Brain* 140 (3): 735–747.

Drummond, E., Nayak, S., Faustin, A. et al. (2017). Proteomic differences in amyloid plaques in rapidly progressive and sporadic Alzheimer's disease. *Acta Neuropathologica* 133 (6): 933–954.

Drysdale, A.T., Grosenick, L., Downar, J. et al. (2017). Resting-state connectivity biomarkers define neurophysiological subtypes of depression. *Nature Medicine* 23 (1): 28–38.

Dubois, B., Feldman, H.H., Jacova, C. et al. (2007). Research criteria for the diagnosis of Alzheimer's disease: revising the NINCDS–ADRDA criteria. *The Lancet Neurology* 6 (8): 734–746.

Ezzati, A., Zammit, A.R., Habeck, C. et al. (2020). Detecting biological heterogeneity patterns in ADNI amnestic mild cognitive impairment based on volumetric MRI. *Brain Imaging and Behavior* 14 (5): 1792–1804.

Ferreira, D., Nordberg, A., and Westman, E. (2020). Biological subtypes of Alzheimer disease: a systematic review and meta-analysis. *Neurology* 94 (10): 436–448.

Habes, M., Grothe, M.J., Tunç, B. et al. (2020). Disentangling heterogeneity in Alzheimer's disease and related dementias using data-driven methods. *Biological Psychiatry* 88 (1): 70–82.

Hong, S.J., Vogelstein, J.T., Gozzi, A. et al. (2020). Toward neurosubtypes in autism. *Biological Psychiatry* 88 (1): 111–128.

Lenhart, L., Seiler, S., Pirpamer, L. et al. (2021). Anatomically standardized detection of MRI atrophy patterns in early-stage Alzheimer's disease. *Brain Sciences* 11 (11): 1491.

Mann, K., Roos, C.R., Hoffmann, S. et al. (2018). Precision medicine in alcohol dependence: a controlled trial testing pharmacotherapy response among reward and relief drinking phenotypes. *Neuropsychopharmacology* 43 (4): 891–899.

Poulakis, K., Pereira, J.B., Mecocci, P. et al. (2018). Heterogeneous patterns of brain atrophy in Alzheimer's disease. *Neurobiology of Aging* 65: 98–108.

Salmanpour, M.R., Shamsaei, M., Hajianfar, G. et al. (2022). Longitudinal clustering analysis and prediction of Parkinson's disease progression using radiomics and hybrid machine learning. *Quantitative Imaging in Medicine and Surgery* 12 (2): 906–919.

Ten Kate, M., Dicks, E., Visser, P.J. et al. (2018). Atrophy subtypes in prodromal Alzheimer's disease are associated with cognitive decline. *Brain* 141 (12): 3443–3456.

Varol, E., Sotiras, A., Davatzikos, C. et al. (2017). HYDRA: Revealing heterogeneity of imaging and genetic patterns through a multiple max-margin discriminative analysis framework. *Neuroimage* 145 (Pt B): 346–364.

Vogel, J.W., Young, A.L., Oxtoby, N.P. et al. (2021). Four distinct trajectories of tau deposition identified in Alzheimer's disease. *Nature Medicine* 27 (5): 871–881.

Whitwell, J.L., Graff-Radford, J., Tosakulwong, N. et al. (2018). [18F]AV-1451 clustering of entorhinal and cortical uptake in Alzheimer's disease. *Annals of Neurology* 83 (2): 248–257.

Zhang, B., Lin, L., Wu, S. et al. (2021a). Multiple subtypes of Alzheimer's disease base on brain atrophy pattern. *Brain Science* 11 (2): 278.

Zhang, Y., Wu, W., Toll, R.T. et al. (2021b). Identification of psychiatric disorder subtypes from functional connectivity patterns in resting-state electroencephalography. *Nature Biomedical Engineering* 5 (4): 309–323.

6

Computer-aided Diagnosis Systems in Neuroscience Based on Responsible Artificial Intelligence

Carmen Jimenez-Mesa[1,2], David López[1,3], Juan E Arco[1,3], Francisco J Martínez-Murcia[1,4], Javier Ramírez[1,4], María Ruz[3], and Juan M Górriz[1,4,5]*

[1]*DaSCI Andalusian Institute of Data Science and Computational Intelligence, University of Granada, Granada, Spain*
[2]*Department of Communication Engineering, University of Málaga, Spain*
[3]*Mind, Brain and Behavior Research Center, University of Granada, Granada, Spain*
[4]*Data Science and Computational Intelligence Institute (DaSCII), University of Granada, Granada, Spain*
[5]*Department of Psychiatry, University of Cambridge, Cambridge, UK*

*Corresponding author: gorriz@ugr.es

6.1 Introduction

Artificial intelligence (AI) and machine learning (ML) have increasingly permeated various aspects of daily life, offering new opportunities to improve the well-being of individuals across sectors such as business, healthcare and education (Angra & Ahuja, 2017). The growing potential of AI has prompted important questions about how to incorporate fairness, interpretability, privacy and safety into these systems. While some applications may seem overwhelming or concerning, the responsible use of ML offers clear advantages and significant benefits. This is especially important in healthcare, where ML algorithms, for example, can identify patterns in the human genome for personalised treatments. Therefore, computer-aided diagnosis (CAD) systems can provide quicker and more accurate diagnoses, enabling early detection and better prognoses for various diseases (Jimenez-Mesa et al., 2023a; Jafari et al., 2024). Such applications underscore the need for fairness, interpretability, privacy and safety to be embedded within these systems to ensure equitable and trustworthy outcomes.

This chapter explores the interface between neuroimaging and ML through the design of CAD systems, as illustrated in Figure 6.1. The integration of artificial intelligence with advanced non-invasive neuroimaging techniques, briefly outlined in Section 6.2, has significantly enhanced our understanding of human brain function. These approaches excel in detecting subtle changes in brain signals that traditional methods often overlook. Section 6.3 of this chapter will provide an overview of the multivariate approach to neuroimaging data analysis, emphasising its differences and advantages over classical univariate methods. Furthermore, we will delve into diverse AI algorithms, ranging from fundamental

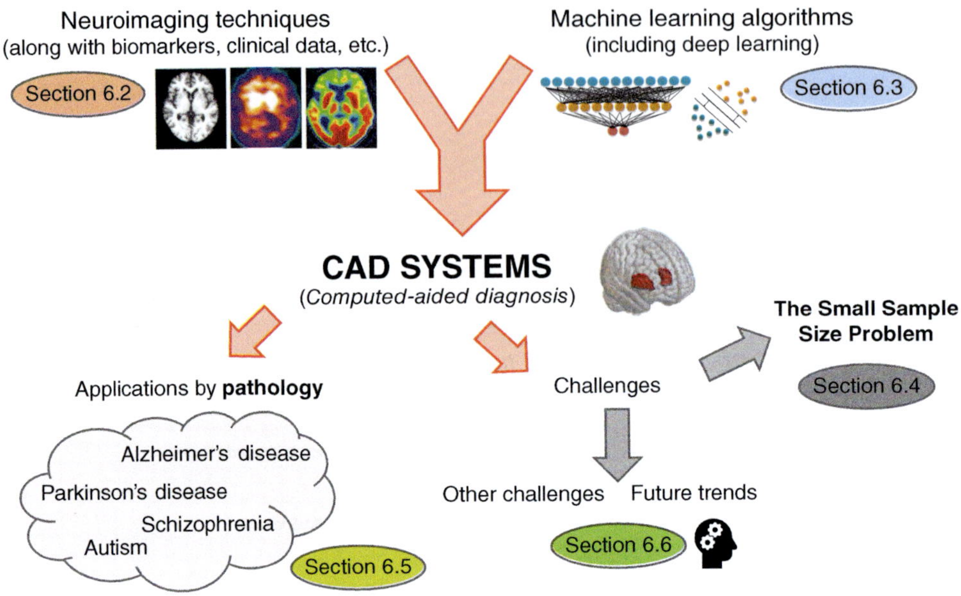

Figure 6.1 General chapter overview.

architectures, such as linear classifiers, to complex deep learning (DL) architectures, including convolutional neural networks (CNNs), autoencoders (AEs) and generative adversarial networks (GANs), among others.

Although ML has become an essential tool across many scientific domains, its application to neuroimaging data introduces a number of specific challenges. In conventional ML settings, algorithms are typically trained on very large datasets, often containing thousands or even millions of samples, with the goal of achieving the highest predictive accuracy in a given task. Neuroimaging studies, however, operate under very different conditions. Datasets usually involve only a few dozen subjects per group, and even in the most favourable cases, the total number of participants rarely exceeds a few hundred. This imbalance often results in having far more features than subjects. Later, in Section 6.4, we will examine the implications of limited sample sizes, particularly in relation to cross-validation (CV) procedures, which are frequently employed to estimate model performance.

In Section 6.5, we explore significant neurodegenerative and neurodevelopmental disorders, including dementia and Alzheimer's disease (AD), Parkinson's disease (PD), schizophrenia (SCZ) and autism. We emphasise their impact on cognitive health and the advancements in neuroimaging and artificial intelligence technologies used for their diagnosis and study. The integration of multidisciplinary approaches underscores ongoing efforts to improve patient outcomes and quality of life across these neurological conditions.

Although the advent of DL has revolutionised neuroimaging analysis, its widespread adoption in clinical settings hinges on interpretability. DL models often function as black boxes, offering accurate predictions but lacking transparency in decision-making processes critical for medical diagnostics. In Section 6.6, we delve into the promising future of neuroimaging in clinical and research scenarios, driven by advancements in Explainable

AI (XAI), transfer learning, data augmentation or multimodal fusion models. In conclusion, while challenges persist in integrating diverse neuroimaging modalities and ensuring AI interpretability, ongoing advancements hold promise for transforming the neuroimaging field. The shift towards models that are not only accurate but also interpretable and robust highlights their transformative role in advancing our knowledge of brain function. By prioritising transparency, generalisability and reliability, these approaches pave the way for more trustworthy insights into the complex dynamics of the human brain.

6.2 Neuroimaging Techniques

Neuroimaging techniques are the imaging modalities on which neuroscience CAD systems are based. Common methods include structural magnetic resonance imaging (sMRI) for detailed brain anatomy, functional MRI (fMRI) for brain activity through blood flow changes, positron emission tomography (PET) and single photon emission computed tomography (SPECT) for brain metabolism and function, and electroencephalography (EEG) and magnetoencephalography (MEG) for real-time electrical and magnetic activity.

6.2.1 Magnetic Resonance Imaging

MRI is among the most widely utilised imaging modalities, both in neuroimaging and across a diverse range of clinical applications. Its non-invasive character and the fact that it does not expose patients to ionising radiation make it advantageous compared to techniques such as computed tomography (CT) (Weishaupt et al., 2006). MRI enables the acquisition of highly detailed anatomical information on soft tissues, organs and vascular structures, thereby supporting accurate diagnostic assessments. Consequently, it has become an essential tool in the detection and monitoring of a wide variety of medical conditions, including neurological disorders, neoplasms, cerebrovascular accidents, musculoskeletal injuries and oncological diseases.

The underlying principle of MRI is based on the interaction between a strong static magnetic field, B_0, and radiofrequency (RF) pulses, which together allow for the reconstruction of high-resolution images of internal structures. This technique exploits the magnetic properties of hydrogen nuclei, which are particularly suitable for imaging because hydrogen is the most abundant element in the human body, accounting for roughly 70% of its mass. This prevalence results in a favourable signal-to-noise ratio (SNR), contributing to the superior image quality achieved by MRI.

After outlining the basic principles of MRI, it is useful to summarise the image formation process. Within the scanner, the main magnetic field B_0 aligns hydrogen nuclei, creating a net magnetic moment. RF pulses tuned to the Larmor frequency then perturb this alignment, generating measurable signals. The Larmor frequency is given by:

$$f_{\text{Larmor}} = \frac{\gamma}{2\pi} B_0, \tag{6.1}$$

where γ is the gyromagnetic ratio for the nucleus ($\gamma_H = 42.6$ MHz/T for hydrogen). When the RF pulse matches this frequency, it disturbs the alignment of the nuclei, temporarily displacing their magnetic moments, a process known as *excitation*. As soon as the RF pulse is turned off, the nuclei gradually return to equilibrium with the static field, a process referred to as relaxation. During relaxation, the protons emit electromagnetic signals that are captured by the receiver coils of the scanner. Two main parameters characterise this process: T1, the time required for spins to realign along the longitudinal axis, and T2, the time associated with the loss of phase coherence in the transverse plane. The differences in these relaxation times across tissues provide the fundamental contrast that enables MRI to generate detailed anatomical and functional images.

The differences in relaxation properties across tissues form the basis for the wide range of MRI modalities. Among the most commonly employed are T1-weighted, T2-weighted and proton density (PD) imaging. T1-weighted acquisitions, which are sensitive to short T1 relaxation times, provide excellent anatomical detail and clear contrast between grey matter (GM) and white matter (WM), thereby facilitating the identification of structural abnormalities. By contrast, T2-weighted images enhance the visibility of tissues with short T2 relaxation times and are particularly effective for detecting pathological processes such as inflammation, oedema or lesions involving cerebrospinal fluid (CSF). PD imaging, on the other hand, primarily reflects the relative abundance of hydrogen nuclei, offering complementary tissue contrast. Figure 6.2a shows an example of a T1-weighted brain scan.

Taken together, these modalities constitute the foundation of neuroimaging research and clinical practice. Their ability to capture different aspects of tissue composition and pathology makes them indispensable not only for diagnosis but also as input data for computational methods. In the following sections, we will discuss how such MRI contrasts are leveraged within CAD systems and how ML techniques can extract meaningful patterns that go beyond visual inspection.

6.2.2 Functional Magnetic Resonance Imaging

Beyond structural imaging, another pivotal modality is fMRI, which enables the assessment of brain activity through haemodynamically mediated signal changes. Unlike purely anatomical contrasts, fMRI captures fluctuations in regional cerebral blood flow that accompany neural activation. An illustrative example is shown in Figure 6.2b. The technique relies on the interaction between T2-weighted effects and local magnetic field inhomogeneities generated by variations in blood oxygenation. Specifically, it exploits the differential magnetic properties of oxyhaemoglobin, which is diamagnetic, and deoxyhaemoglobin, which is paramagnetic.

The accumulation of deoxyhaemoglobin in venous blood introduces microscopic distortions in the surrounding magnetic field. These field variations alter the resonance frequency and phase of nearby hydrogen nuclei, ultimately reducing the T2 relaxation time. As the proportion of deoxyhaemoglobin increases, the measured MR signal weakens, giving rise to what is known as T2*-weighted imaging. Importantly, the concentration of deoxyhaemoglobin ranges from approximately 2% in arterial blood to around 40% in venous blood, providing a physiologically meaningful source of contrast.

When a particular brain region becomes active, the vascular response oversupplies oxygenated blood relative to local metabolic demand. This overcompensation decreases the concentration of deoxyhaemoglobin in the area, thereby reducing local magnetic field

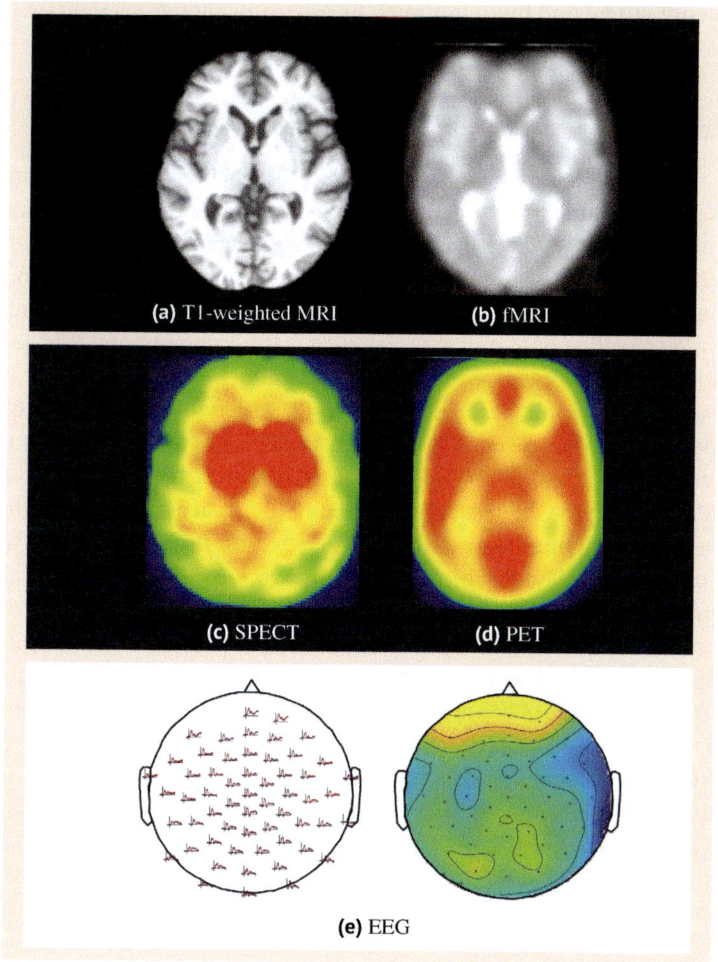

Figure 6.2 Examples of several neuroimaging techniques: (a) T1-weighted MRI; (b) fMRI scan; (c) SPECT-HMPAO scan; (d) PET scan; (e) layout of EEG signal distribution and its topographical frequency.

disturbances and producing a relative increase in signal intensity. This mechanism underlies the Blood Oxygenation Level Dependent (BOLD) contrast, the principal signal exploited in fMRI. Through this effect, fMRI enables researchers and clinicians to non-invasively map functional changes in the human brain with remarkable spatial resolution.

Functional MRI is utilised in both research and clinical settings. In research, fMRI is extensively employed to study the neural mechanisms underlying cognitive functions, sensory processing and neuroplasticity, providing insights into the brain's functional organisation and connectivity. Clinically, fMRI is instrumental in presurgical planning, particularly for identifying critical functional areas such as those responsible for speech, motor skills and sensory functions, thus aiding in the preservation of these areas during neurosurgery. Additionally, fMRI is used in the diagnosis and monitoring of neurological disorders, including epilepsy, stroke and brain tumours, by assessing alterations in brain activity and connectivity associated with these conditions.

6.2.3 Single Photon Emission Computed Tomography

SPECT is an imaging method used in nuclear medicine designed to assess physiological processes such as perfusion and organ function. Its clinical relevance is particularly pronounced in neurology and cardiology, where it contributes to the diagnosis, monitoring and management of disorders, including PD, epilepsy, cerebral tumours and myocardial ischaemia.

The procedure involves the intravenous administration of a radiopharmaceutical, which consists of a biologically active molecule labelled with a gamma-emitting radionuclide. The choice of radioligand, as well as the timing of image acquisition, is critical to optimally visualise the targeted physiological process. A wide variety of radiopharmaceuticals are available, depending on the clinical or research application. An example is provided in Figure 6.2c, which illustrates a ^{99m}Tc-HMPAO SPECT brain scan.

Once introduced into the bloodstream, the radiopharmaceutical distributes throughout the body and accumulates in specific organs or tissues. Its radioactive decay results in the emission of gamma photons, which are captured by a specialised gamma camera. The measured signal intensity reflects the local concentration of the tracer. During data acquisition, the detector rotates around the patient, obtaining multiple two-dimensional projections of tracer distribution. These projections are then reconstructed computationally to generate a three-dimensional (3D) map of radiopharmaceutical uptake, thereby providing insight into tissue function and regional activity.

Despite its lower spatial resolution relative to CT or MRI, SPECT remains a crucial tool when functional information is necessary for therapeutic planning or diagnosis. Furthermore, advances in multimodal imaging (e.g., SPECT-CT) are increasingly enabling the integration of anatomical and functional data, thereby enhancing diagnostic accuracy and overcoming the limitations of each individual technique (Zhang et al., 2020).

6.2.4 Positron Emission Tomography

PET shares conceptual similarities with SPECT, but it relies on a distinct physical principle: the detection of coincident gamma photons generated when a positron emitted by a radiotracer annihilates with an electron. In a typical procedure, a small amount of a radioactive compound, typically fluorodeoxyglucose (FDG), is injected into the bloodstream. As positron–electron annihilation occurs, pairs of photons are released in opposite directions. Their simultaneous detection by the scanner enables accurate localisation of the emission event. The PET detection system is composed of scintillators, which convert incident gamma photons into light, coupled with photodetectors such as photomultiplier tubes or silicon avalanche photodiodes. As in SPECT, multiple projections are acquired around the patient and subsequently reconstructed to form a 3D image of tracer distribution. An illustrative example of a PET-FDG scan is provided in Figure 6.2d. Although the reconstruction principles are conceptually similar to those in SPECT, PET requires sophisticated hardware and processing pipelines, making the technology more expensive and technically demanding (Carlson, 2012).

Clinically, PET is invaluable for characterising tissue metabolism and physiological processes. Its ability to visualise glucose uptake and other biochemical pathways has established it as a central tool for the detection and staging of cancers, the assessment of cardiovascular disease and the evaluation of neurological disorders. Within the context of brain research, PET has been particularly influential in studying neurodegenerative diseases and mapping neurotransmitter systems, thereby complementing structural and functional information derived from MRI and fMRI.

6.2.5 Electroencephalography

EEG is a non-invasive electrophysiological method that measures the brain's electrical activity through electrodes placed on the scalp, as illustrated in Figure 6.2e. Owing to its excellent temporal resolution, EEG has emerged in recent decades among the most widely used techniques for studying the dynamics of brain function. In the clinical domain, it is routinely employed to assess sleep quality, diagnose sleep disorders and detect epilepsy or other seizure-related conditions. Additional applications include assessing anaesthetic depth during surgery and supporting the identification of specific psychiatric disorders, such as attention deficit hyperactivity disorder (ADHD).

Beyond clinical practice, EEG has played a pivotal role in basic and applied neuroscience research. A prominent example is the study of event-related potentials (ERPs), which are averaged EEG responses time-locked to sensory or cognitive stimuli. ERPs have been extensively used to investigate neural processes underlying perception, attention, memory, language and emotion. Another influential application is the development of brain–computer interfaces, which translate brain activity into external commands to control devices such as computers, wheelchairs or robotic arms. These systems have opened promising avenues for restoring or augmenting lost functions in patients with severe motor impairments.

The versatility, relatively low cost and high temporal precision of EEG make it a cornerstone of modern neuroimaging. While it lacks the spatial resolution of modalities such as MRI or PET, its ability to capture rapid neural dynamics has made it indispensable for both clinical practice and neuroscience research. In the context of CAD, EEG offers a rich source of data that, when combined with advanced AI methods, can provide powerful insights into brain function and dysfunction.

6.2.6 Imaging Preprocessing

Preprocessing constitutes a fundamental stage in CAD systems, as it mitigates many of the inherent challenges of medical imaging data. Issues such as noise, variability in acquisition protocols, the presence of artefacts and anatomical misalignments must be addressed before further analysis. By systematically tackling these limitations, preprocessing enhances both the accuracy and the reliability of computational models.

A particularly critical component of this stage is the normalisation of images, both spatially and in terms of intensity. These procedures standardise data across subjects and sessions, ensuring comparability and enabling ML algorithms to operate under

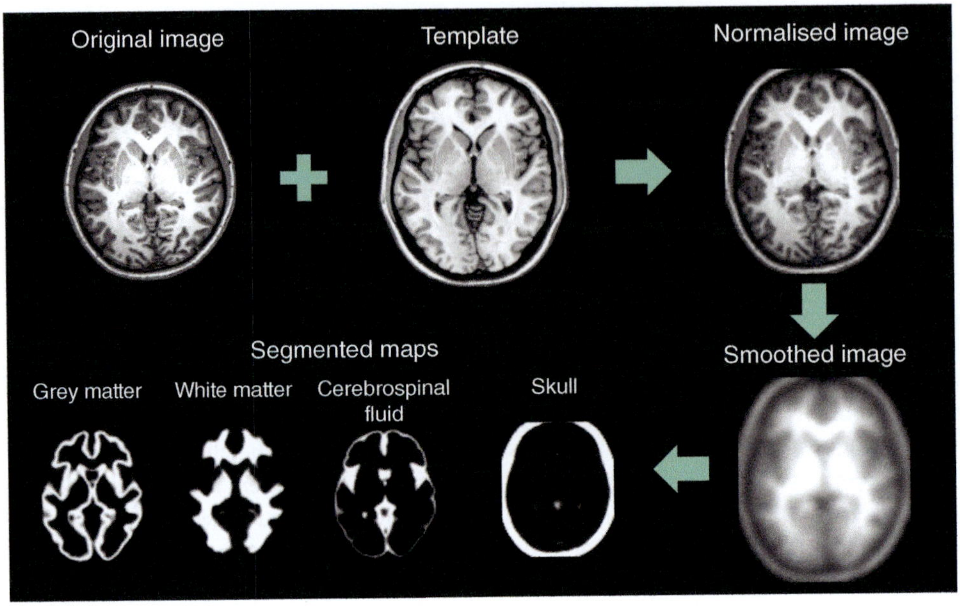

Figure 6.3 An example of typical MRI spatial preprocessing steps. *Source*: Jimenez Mesa, Carmen/CC BY-NC-ND 4.0.

consistent conditions. Normalisation is therefore essential for producing reproducible results and maximising the performance of CAD systems. As an illustrative example, the typical pipeline for spatial preprocessing of MRI scans is presented in Figure 6.3.

Intensity normalisation adjusts voxel intensities in images to a standard scale, allowing for accurate comparison across different subjects or time points. This process addresses variations caused by differences in acquisition settings, equipment or patient characteristics. Intensity normalisation is especially important in functional imaging like SPECT or PET, where intensity values are directly linked to biomarkers and can affect analysis and interpretation. In contrast, structural images like MRI are less affected by intensity variations since these images are typically considered unitless.

Several strategies have been proposed for intensity normalisation, each aiming to reduce interscan variability while preserving relevant information. A widely adopted approach consists of rescaling image intensities through linear transformations, expressed as:

$$I = \frac{I_0}{I_p} \tag{6.2}$$

Here, I denotes the normalised intensity, I_p is a scan-specific reference and I_0 is the original voxel intensity. One common choice for I_p is the mean of the top k% highest intensity values in the scan (with k typically set to 5%), an approach known as normalisation to the maximum (Salas-Gonzalez et al., 2009).

A more general formulation of linear rescaling is given by:

$$I = aI_0 + b, \tag{6.3}$$

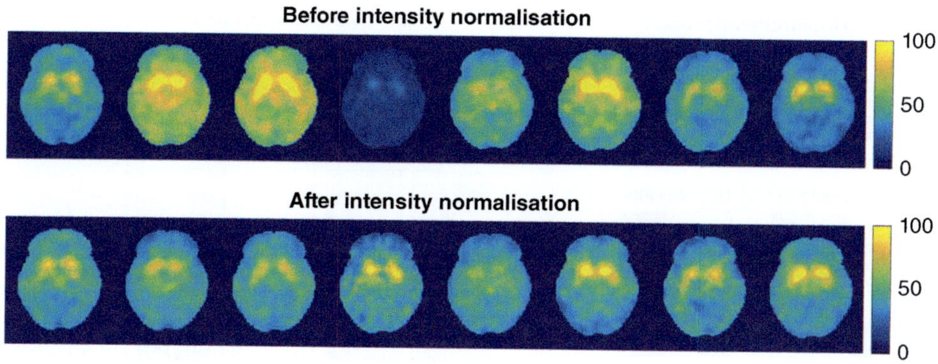

Figure 6.4 Examples of differences between ^{123}I-FP-CIT SPECT scans before and after intensity normalisation using α-stable distributions. *Source*: Jimenez Mesa, Carmen/CC BY-NC-ND 4.0.

where a and b are, respectively, a multiplicative scaling factor and an additive offset. This representation provides additional flexibility for adjusting image intensities. Within this category, methods based on histogram standardisation (Arndt et al., 1996) or probabilistic modelling, such as the α-stable distribution, have been developed. The Gaussian distribution, frequently used in practice, can be regarded as a particular case of the α-stable family (Salas-Gonzalez et al., 2013).

For α-stable normalisation, the transformation can be written as:

$$I = \frac{\gamma^*}{\gamma} I_0 + \left(\mu^* - \frac{\gamma^*}{\gamma} \mu \right), \tag{6.4}$$

where γ^* represents the mean dispersion parameter across all input scans and μ^* corresponds to the average location parameter. By aligning the statistical properties of different images, this procedure mitigates variability introduced by extrinsic factors such as differences in tracer dosage, patient absorption rates or scanner calibration. An illustrative example of this procedure is shown in Figure 6.4.

6.3 Machine Learning in Neuroimaging

In neuroimaging research, traditional analyses have typically considered each voxel, electrode or feature in isolation. With the increasing integration of AI methods, however, multivariate pattern analysis (MVPA) has gained prominence as a complementary approach. Rather than focusing on single responses, MVPA examines distributed patterns of activity and interactions across multiple brain regions, voxels, electrodes or features, enabling the decoding and classification of complex neural information (Haxby et al., 2014). This has two main implications: first, it is more important how information is distributed instead of the power of the average signal across voxels; second, this allows the study of informative patterns that are not locally grouped. Thus, the sensitivity of MVPA is superior to the classical univariate approaches, identifying subtleties that are often ignored (Coutanche et al., 2011) (Figure 6.5).

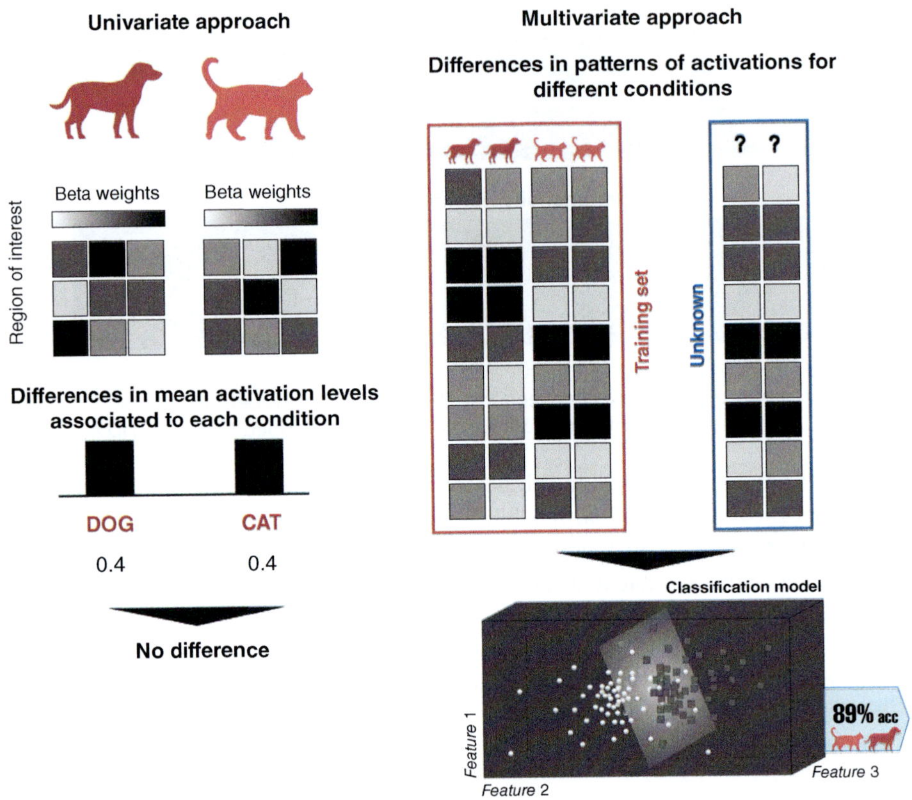

Figure 6.5 How multivariate analysis works? Univariate vs. multivariate approaches for analysing the same neuroimaging dataset. *Source*: Adapted from López García et al. (2023).

To support clinicians and researchers from different disciplines and provide broader access to multivariate approaches, a variety of software tools have been developed for diverse neuroimaging modalities. Among those specifically designed for fMRI analysis are Statistical Parametric Mapping (SPM) (Penny et al., 2011), the Decoding Toolbox (TDT) (Hebart et al., 2015) and the Pattern Recognition for Neuroimaging Toolbox (PRoNTo) (Schrouff et al., 2013). Other widely used MATLAB-based toolboxes include MVPAlab (López-García et al., 2022b), the Amsterdam Decoding and Modeling Toolbox (ADAM) (Fahrenfort et al., 2018) and CoSMoMVPA (Oosterhof et al., 2016). In addition, Python-based alternatives such as MNE-Python (Gramfort et al., 2013) and PyMVPA (Hanke et al., 2009) offer flexible frameworks for multivariate analysis across different neuroimaging data types.

The basics behind these methods are simple: there is information in brain imaging that can be related to the clinical state of the patient, so that it is possible to guess the actual diagnosis of a person by looking at his/her brain. In the context of cognitive neuroscience, the idea is very similar, but the aim is usually to find the relationship between neural patterns and an experimental condition rather than the diagnosis of a certain pathology. For example, the orientation of different visual stimuli can be decoded from voxels in the visual cortex, showing the presence of distributed patterns of activity in this brain region (Haynes & Rees, 2005).

In recent years, the application of ML in neuroimaging has shifted from traditional models, such as linear classifiers and decision trees (see Figure 6.6a), towards more flexible strategies, including nonlinear algorithms and ensemble-based methods (Bishop & Nasrabadi, 2006). Among these, support vector machines (SVMs) have become one of the most widely adopted approaches. By incorporating different kernel functions, SVMs can capture complex and nonlinear relationships within high-dimensional data (Schölkopf et al., 2002). A simple case using a linear kernel is illustrated in Figure 6.6b. Their ability to handle large feature spaces with high accuracy and robustness has made SVMs particularly valuable for medical imaging tasks (Rojas et al., 2013; Gao et al., 2015). Ensemble approaches, such as bagging (Breiman, 1996), boosting (Freund & Schapire, 1996) and random forests (Breiman, 2001), have also gained traction, as they combine multiple weak learners to improve predictive performance and highlight relevant features in complex datasets (Castillo-Barnes et al., 2018).

Although classical ML algorithms have achieved solid results, the advent of DL has represented a major turning point in the field. The simplest form of DL architecture is the multilayer perceptron (MLP), which consists of an input layer, one or more hidden layers, and an output layer (see Figure 6.6c). In particular, CNNs have profoundly transformed this area (Mohapatra et al., 2021). CNNs are characterised by a distinctive architecture that incorporates convolutional layers to identify local patterns in images through the use of filters, together with pooling layers that reduce dimensionality. This design substantially decreases the number of parameters compared to fully connected neural networks. An illustration of a CNN is provided in Figure 6.6d. Owing to this hierarchical structure, deep neural networks (DNNs) have shown remarkable ability to capture complex patterns within the data, achieving significant improvements in both accuracy and generalisation across a wide range of image processing tasks (Gu et al., 2018; Martinez-Murcia et al., 2018).

Building upon this architectural foundation, more specialised DNN designs have been introduced. Among them, Siamese networks represent a prominent example. They are composed of two identical neural branches that simultaneously process separate input objects, as illustrated in Figure 6.6e (Koch et al., 2015). The central goal of this architecture is to assess the degree of similarity or dissimilarity between the inputs. In medical imaging applications, and particularly in the domain of molecular imaging, Siamese networks have shown considerable promise in facilitating the comparison and alignment of scans while improving quantitative reliability (Arco et al., 2022, 2023b).

AEs represent another prevalent DL architecture (Charte et al., 2018). They operate in two phases: the encoder maps the input data into a compact, lower-dimensional latent space, and the decoder subsequently reconstructs the data from this representation, aiming to reproduce the original input as closely as possible. Convolutional layers are frequently incorporated to extract features within the latent representation, enhancing the model's capacity to capture relevant patterns. Due to this design, AEs are particularly valuable for a range of tasks in CAD systems. They enable dimensionality reduction, thereby lowering computational demands (Tuan et al., 2021); image denoising, which improves visual quality (Shao et al., 2021); segmentation of regions of interest (ROIs) (Wu et al., 2020); and even data generation through variational autoencoders, which introduce a stochastic component to generate new samples in the latent space (Sanaat et al., 2021a).

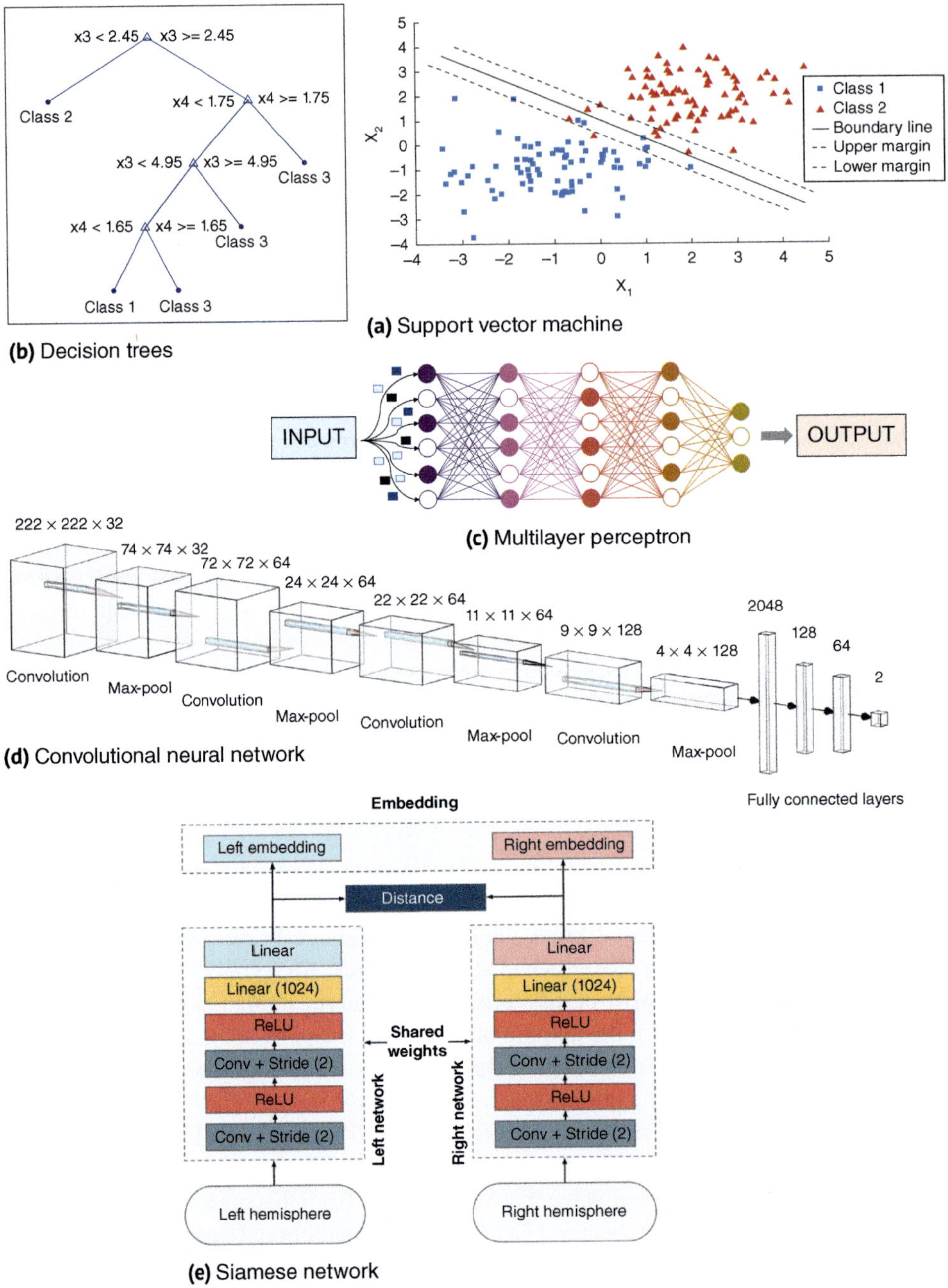

Figure 6.6 Examples of various ML algorithms, illustrating their unique structures: (a) decision trees, which split data based on feature values; (b) SVM, designed to find the optimal hyperplane for classification; (c) multilayer perceptron, a type of neural network with multiple layers; (d) CNNs are usually composed of a convolutional block (convolutional layers and max-pooling) for feature extraction and fully connected layers for classification; and (e) Siamese networks, which compare two inputs to measure their similarity. *Source*: Subfigures (a) and (b) Adapted from Jiménez Mesa (2023)/Universidad De Granada/CC BY-NC-ND 4.0; subfigure (e) Adapted from Arco et al. (2022).

Generative adversarial networks (GANs) have become a prominent class of DL models for producing realistic data samples. In this framework, a generator network synthesises data, while a discriminator network evaluates whether the samples are real or artificially generated. This adversarial training paradigm renders GANs particularly effective for applications in neuroimaging, including image synthesis, image correction and data augmentation (Goodfellow et al., 2014; Wang et al., 2023). Although they require large datasets and careful tuning, GANs show great potential for supporting progress in neuroimaging research while enhancing patient-specific treatment approaches.

Graph convolutional networks (GCNs) offer a different approach by focusing on relationships between data rather than direct feature extraction. Designed for structured data in graph form, GCNs employ node-level information propagation via graph convolution operations, updating the encoding of individual nodes by aggregating information from its neighbouring nodes. This approach has gained increasing attention within medical imaging, especially in the context of the analysis of brain connectivity and the prediction of neurological disorders, with widespread applications in both structural and functional MRI (Zhou et al., 2020).

The development of DL models continues to evolve, with an emerging trend towards attention-based models gaining popularity (Guo et al., 2022). These models focus on processing the most relevant parts of the data to enhance predictive accuracy. Among the attention-based techniques, transformers stand out (Vaswani et al., 2017). Transformers have proven highly effective for sequential data by employing stackable encoder and decoder layers that leverage an attention mechanism, allowing the model to consider all elements of a sequence simultaneously and capture long-range dependencies. The introduction of this architecture marked a turning point in natural language processing (NLP), leading to the development of prominent models, including BERT (Devlin et al., 2018) and GPT (Radford et al., 2018). Given their success in NLP, transformers have been adapted for computer vision tasks (Khan et al., 2022). The Vision Transformer (ViT) (Dosovitskiy et al., 2020) processes images by converting image regions into token sequences so that the architecture can understand detailed spatial patterns. This methodology has led to significant improvements in tasks including image segmentation, multimodal fusion, classification, synthesis and object detection within medical imaging (Shamshad et al., 2023; Tang et al., 2022). Although their application in neuroimaging is still emerging, transformers offer considerable potential for advancing the study of brain function and enhancing diagnostic capabilities.

6.4 Small Sample Size in Neuroimaging

Neuroimaging contexts usually rely on heterogeneous data, with a high number of predictors and a limited sample size, known as the "curse of dimensionality." This affects the training and validation capacity of ML models, in addition to their generalisation ability (Neyshabur et al., 2017a). Feature reduction techniques are commonly employed in ML to alleviate this situation (Jimenez-Mesa et al., 2020).

An additional problem is referred to the use of CV techniques to evaluate the performance of ML models. CV iteratively separates the input dataset into different folds, using all but one subset for training the model and the remaining one for testing its performance. The final results are calculated as the mean of the outcomes from each fold. Although this method is widely employed in literature, there are some limitations that are not usually taken into account. There is a prediction error and a confidence interval whose size depends on the characteristics of the dataset: size, homoscedasticity, shape of the distribution, etc. In studies involving small sample sizes and heterogeneous datasets, CV often substantially underestimates the true generalisation error, reflecting a potential violation of the ergodicity assumption. Previous research has highlighted the considerable variability observed across CV folds in numerous neuroimaging experiments, which has important consequences for predictive inference (Eklund et al., 2016; Jollans et al., 2019). Therefore, recent studies suggest the use of resubstitution when sample size is limited (Braga-Neto et al., 2004; Jimenez-Mesa et al., 2023c), i.e., use the whole input dataset simultaneously. To mitigate the overoptimistic bias inherent in resubstitution, the true error can be approximated using an upper bound (Vapnik, 1982). This bound quantifies the maximum difference between the empirical error and the actual error for a given trained learning algorithm, expressed as $\mu \geq |E_{\text{act}} - E_{\text{emp}}|$. Accordingly, the worst-case estimate of the true error with probability $1 - \eta$ can be defined as:

$$\widehat{E}_{\text{act}} = \widehat{E}_{\text{emp}} + \mu \tag{6.5}$$

It can be viewed as a theoretical maximum for classification performance, allowing all available data to be utilised in the estimation of relevant metrics. Furthermore, measures such as accuracy, sensitivity and specificity can be constrained by this bound, since the errors associated with these metrics represent partial contributions to the total classification error.

6.4.1 Upper Bound's Selection

The state of the art comprises a large number of upper bounds. The most well-known is the one proposed by V. Vapnik, which is based on the VC dimension (Vapnik et al., 1994). This bound is defined as:

$$\mu_{\text{VC}} \leq \sqrt{\frac{h\left(\ln\left(\frac{2n}{h}\right) + 1\right) - \ln\left(\frac{\eta}{4}\right)}{n}} \tag{6.6}$$

where η denotes the significance level, n represents the size of the training set, d is the feature dimensionality and h corresponds to the VC dimension (Vapnik, 1982). For linear models, such as SVMs, the VC dimension is given by $d + 1$. In the case of DL architectures, however, the precise value of h remains a subject of ongoing research (Bartlett et al., 2019).

Górriz et al. (2019) propose an alternative upper bound on the true risk that does not depend on the specific model. This bound is derived from concentration inequalities, assuming that the training samples are positioned in general position (i.g.p.) within $\mathbb{R}^d$. It is primarily applicable to linear classifiers, including linear-kernel SVMs. The corresponding mathematical formulation is expressed as follows:

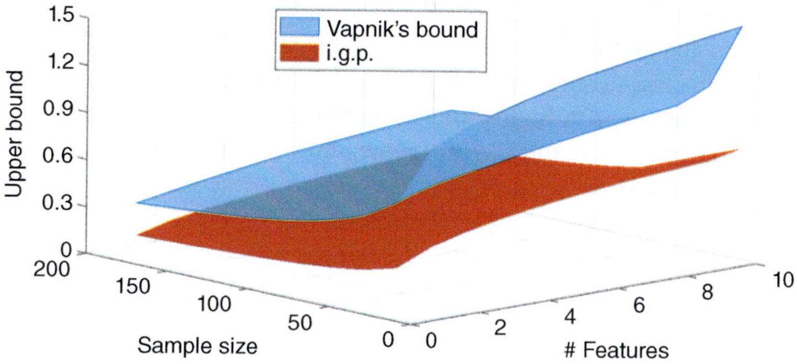

Figure 6.7 Upper bound values are displayed depending on the sample size and feature dimensionality. The blue surface represents Vapnik's bound for linear classifiers, while the red surface corresponds to the i.g.p. bound. *Source*: Figure extracted from Jiménez Mesa (2023) [7]/Universidad De Granada/CC BY-NC-ND 4.0.

$$\mu_{\text{i.g.p}} \leq \sqrt{\frac{1}{2n} \ln \frac{2\sum_{k=0}^{d-1} \binom{n-1}{k}}{\eta}} \tag{6.7}$$

where n denotes the size of the training set, and d represents the dimensionality of the features. Compared with Vapnik's bound, this alternative bound is less restrictive for linear models, as illustrated in Figure 6.7. Both bounds, along with CV, are compared in the work presented by Jimenez-Mesa et al. (2023c). This study introduces a non-parametric framework to evaluate the statistical significance of outcomes generated by CAD systems. The methodology is based on random-effects inference implemented through a label permutation test. Statistical significance is evaluated in terms of both statistical power and the control of Type I error. Several scenarios are analysed in the study, covering both balanced and imbalanced datasets, as well as binary and multi-class problems, all of which share the common denominator of being limited sample sets related to AD. In these scenarios, the CAD systems employ AEs for feature extraction, followed by SVMs for the classification stage. The findings suggest that resubstitution combined with an upper bound correction slightly outperforms K-fold CV in scenarios with small sample sizes, particularly when the datasets exhibit high heterogeneity. In addition, the results obtained clearly reflect a more conservative nature of Vapnik's bound.

The use of PAC-Bayesian bounds provides an additional rigorous approach for assessing model generalisation. For instance, McAllester introduced a dropout-based bound (McAllester, 2013), which incorporates a dropout rate, $\alpha \in [0, 1]$, to reduce the complexity cost of the function. The effect of this bound becomes more pronounced as the dropout rate approaches 1. Its mathematical formulation is expressed as follows:

$$\mu_{\text{PAC}} = \min_{1 \leq i \leq k} \left(\frac{1}{1 - \frac{1}{2\lambda_i}} - 1 \right) \hat{E} + \frac{1}{1 - \frac{1}{2\lambda_i}} \left(\frac{\lambda_i E_{\max}}{n} \left(\frac{1-\alpha}{2} \| \Theta \|^2 + \ln \frac{k}{\eta} \right) \right) \tag{6.8}$$

where k distinct values of the parameter λ (ranging from 1/2 to 10) are evaluated to identify the value that minimises the bound. The quantity being bounded, namely the estimated classifier's loss, is denoted as $\hat{E}$. The highest possible value, E_{max}, must belong to the set of real numbers; in classification tasks, it is typically set to 1. Lastly, Θ represents the set of the classifier's parameters. A highly interesting comparison is made between this upper bound, the i.g.p. upper bound and CV in Jimenez-Mesa et al. (2024). This study analyses one-dimensional features extracted from brain sulci to assess SCZ. Owing to the limited sample size and the heterogeneity of the features, CV was observed to produce inconsistent results when the number of samples varied. In comparison, the PAC-Bayes bound demonstrated a modest improvement over the i.g.p. bound by better adapting to the characteristics of the classification algorithm, in this case, the SVM.

Due to the complexity and vast number of parameters in DL models, applying upper bounds directly in their classification process poses significant computational challenges (Neyshabur et al., 2017a; Zhang et al., 2021). Several upper bounds have been proposed (Valle-Pérez & Louis, 2020; Neyshabur et al., 2017b), yet none have been implemented in real-world scenarios.

6.4.2 Upper Bound's Applications

The application of these bounds, and ML in general, is not limited to classification scenarios. Another significant utility lies within brain mapping. Traditionally, methods based on classical statistics, such as the general linear model (GLM), have been used in this field. However, it is important to note that classical statistical methods assume certain properties that are often not met, particularly in small sample size scenarios. This can result in a significant occurrence of false positives caused by multiple testing (Eklund et al., 2016; Brett et al., 2004). This situation is partially alleviated with alternatives such as Bonferroni correction (Bonferroni, 1936). Nevertheless, works such as that of Eklund et al. (2016) warn of the need to validate the statistical methods used in neuroimaging, especially in clusterwise inference.

To move beyond hypothesis-driven strategies in brain mapping, Gorriz et al. (2021) introduced a data-driven tool for deriving statistical maps from brain images, referred to as *Statistical Agnostic Mapping* (SAM). As indicated by its name, this method avoids parametric techniques in its process, differing from the gold standard framework in this field, the SPM software. Although they are fundamentally different approaches (SPM is essentially a univariate method, while SAM is a multivariate method), the main goal of both is the same: to obtain statistical brain maps. The main differences between SPM and SAM procedures are shown in Figure 6.8. While SPM relies on the GLM to conduct the statistical inference for each voxel of the brain image, SAM is an atlas-based approach where brain images (1) are parcelled into regions, e.g., using AAL116 atlas (Tzourio-Mazoyer et al., 2002); (2) each region is processed and classified individually according to the target labels and (3) performance metrics are used to conduct the statistical inference. Specifically, the data-driven method consists of a feature extraction phase using partial least squares (PLS), reducing thousands of voxels to a few features per region, and a classification stage using

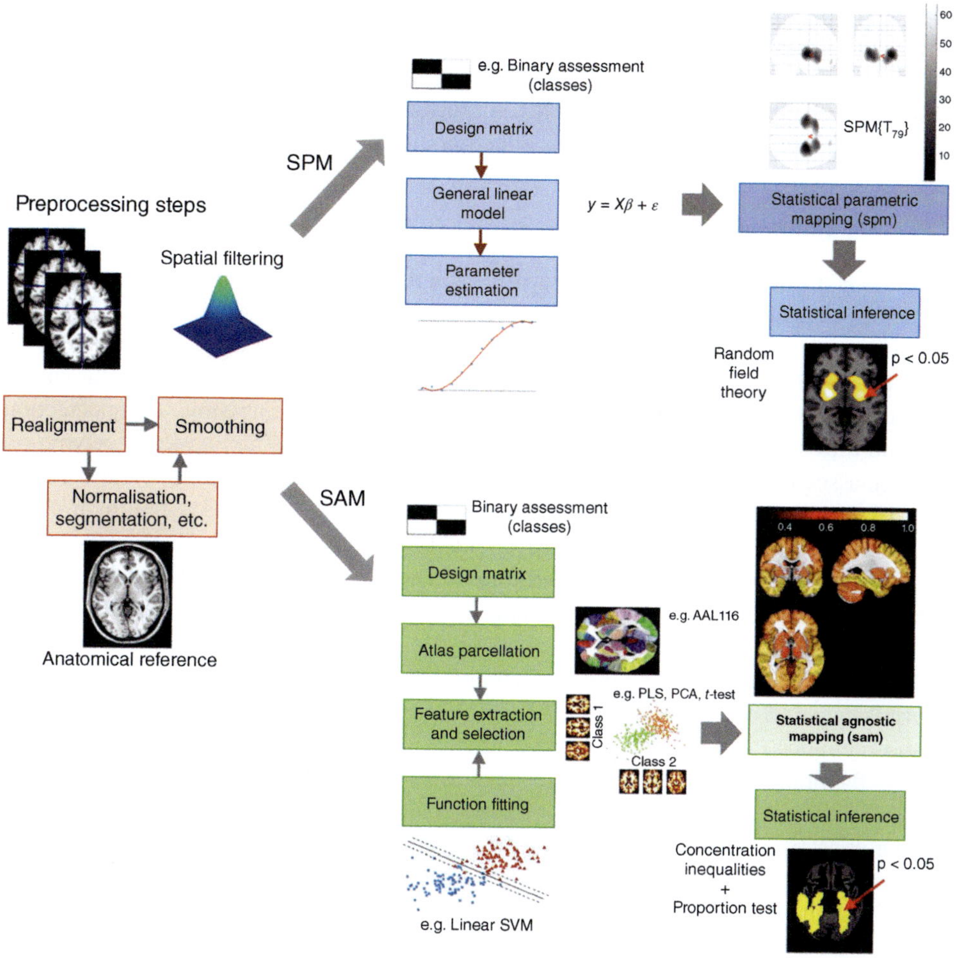

Figure 6.8 Comparison of brain mapping methods SPM (top) and SAM (bottom). SPM is a statistical parametric approach based on the GLM, while SAM is a statistical non-parametric approach based on ML.

a linear SVM. Resubstitution, corrected by an upper bound, is employed to estimate the accuracy. It is important to note that the ROIs analysed in SAM depend on the atlas used, meaning ROIs can be as detailed as required.

This methodology has been successfully applied in various imaging techniques, such as fMRI, fMRI, PET or SPECT (Gorriz et al., 2021; Jimenez-Mesa et al., 2022; Castillo-Barnes et al., 2022), across different conditions (AD, PD, SCZ, etc.) and effect sizes ranging from large to trivial. It has been shown to effectively control false positives and produce consistent results across different sample sizes. Therefore, it can be concluded that SAM is a highly competitive and complementary alternative to SPM, with a higher robustness to varying sample size.

6.5 Pathologies and Case Studies

In this section, several examples of ML applications in neurological conditions are explored. These examples will demonstrate how ML is utilised to enhance diagnosis, prognosis and treatment strategies for these conditions affecting the nervous system.

6.5.1 Dementia and Alzheimer's Disease

Dementia is a progressive condition marked by a decline in cognitive functions, including memory, reasoning and the ability to perform daily activities. According to the World Health Organization, over 55 million people worldwide are affected by dementia, with approximately 10 million new cases reported each year. Beyond the individual, this condition has substantial effects on families, caregivers and society as a whole. Early detection is essential to slow disease progression and enhance quality of life, typically relying on cognitive assessments such as the Mini-Mental State Examination (MMSE), the Clinical Dementia Rating (CDR) and the Clock Drawing Test (CDT) (Mainland & Shulman, 2017; Morris, 1997; Palsetia et al., 2018), complemented by neuroimaging. In this context, CAD systems represent powerful tools in clinical practice (So et al., 2017; Amini et al., 2021), facilitating the work of medical professionals in the accurate identification and monitoring of cognitive impairments.

AD is the leading cause of dementia, comprising 60–80% of cases. In 2023, about 6.7 million Americans aged 65 and older had AD (Alzheimer's Association, 2023). This neurodegenerative disorder affects memory and cognitive functions, marked by abnormal protein accumulation in the brain, leading to neurodegeneration (Modrego, 2006). Early-stage AD involves subtle memory loss and progresses to confusion, mood changes, language difficulties and the need for assistance with daily activities. Early diagnosis is crucial, especially to distinguish AD from related disorders, including its prodromal stage mild cognitive impairment (MCI) (Davatzikos, 2019). The causes of AD are complex, involving genetic, environmental and lifestyle factors. Advanced age is the primary risk factor, with most cases occurring in those aged 65 and older (LOAD). Rare forms, such as dominantly inherited Alzheimer's disease (DIAD), occur earlier due to specific genetic mutations (Moulder et al., 2013).

Currently, there is no cure for AD. However, ongoing improvements in clinical resources for diagnosis, treatment and monitoring are notable. Advances in medical technology and neuroimaging techniques have enabled earlier detection and deeper insights into disease progression. Current research continues to investigate the genetic determinants, cellular mechanisms and the functions of specific brain proteins involved in AD. MRI is extensively used for studying AD, detecting brain atrophy in both WM and GM, particularly in areas like the hippocampus and parahippocampal gyrus, which are crucial in disease development (de Vos et al., 2016). Moreover, nuclear imaging modalities such as PET and SPECT highlight decreased activity in areas including the precuneus, lateral parietal and posterior temporal cortices, providing additional support for diagnostic evaluation (Claus et al., 1994).

The application of ML models to the study of AD is based on two main ideas: (1) the use of neuroimaging associated with the first stages of the disease as a predictor of the outcome of each individual patient and (2) the relationship between the images and other clinical variables to enhance the diagnostic power of the tool. Arco et al. (2023a) developed a DL framework to evaluate differences between hemispheres at several stages of the neurodegeneration caused by AD, demonstrating that an initial asymmetry in the brain regions affected is a clear early predictor of the pathology. Jimenez-Mesa et al. (2023b) proposed a CAD system to automate the diagnosis of cognitive impairment based on the classic version of the CDT. Thus, it could ease the physicians' workload, especially in those cases where the system clearly detects whether the patient is cognitively impaired or not. Other studies combined MRI with additional biological indicators, including CSF analyses or genetic profiling such as APOE status, to find the optimum combination of variables that allows a better understanding of the disease (Hinrichs et al., 2011). The combination and fusion of different types of neuroimaging (mainly MRI, PET and SPECT) within the same model has also emerged as a promising alternative for the study of this disease (Ortiz et al., 2014). This is primarily because structural alterations in certain brain regions do not necessarily impair the functions typically attributed to those areas, highlighting the critical role of neural plasticity in AD research.

The availability of open-access datasets and the development of large-scale repositories have facilitated challenges such as CADDementia (Bron et al., 2015) and TADPOLE (https://tadpole.grand-challenge.org/), which aim to improve CAD systems for Alzheimer's diagnosis by addressing challenges, standardising approaches and enhancing reproducibility in neuroimaging (Button et al., 2013). One notable example is a Kaggle challenge predicting MCI from MRI data (four-class classification task) (Sarica et al., 2018), in which the most successful approach employed a random forest ensemble combined with feature extraction, achieving 61% accuracy for a four-class classification (Dimitriadis et al., 2018). This was later surpassed by a system optimising feature extraction with PLS and classification with SVM, reaching 67% accuracy (Jimenez-Mesa et al., 2020).

6.5.2 Parkinson's Disease

PD constitutes another important cause of dementia, accounting for roughly 3.6% of all dementia cases, with up to 24% of individuals with PD developing dementia as the disease progresses (Alzheimer's Association, 2023). This progressive neurodegenerative disorder arises from the gradual degeneration of dopaminergic neurons within the nigrostriatal pathway, influenced by both genetic and environmental factors (Poewe et al., 2017).

Since dopamine plays a key role in regulating movement and coordination, PD primarily impairs motor functions. Typical clinical manifestations include tremors (involuntary shaking), rigidity (muscle stiffness), bradykinesia (slowed movements), and postural instability (impaired balance) (National Institute of Neurological Disorders and Stroke (US), 2004). Beyond motor symptoms, patients may also present with cognitive alterations, mood alterations, disruptions in sleep patterns and impairments of the autonomic nervous system.

While PD typically manifests in older adults, it can also onset in younger individuals. Similar to AD, there is currently no cure for PD, underscoring the importance of early diagnosis for optimising symptom management. Functional imaging techniques, such as SPECT-DaTSCAN, are instrumental in assessing dopamine transporter levels.

This imaging modality uses a radiopharmaceutical that binds to dopamine transporters in the striatum, revealing distinctive patterns indicative of PD pathology.

Martorell-Marugan et al. (2023) provide a recent comprehensive review of the use of ML in PD research, emphasising its growing importance in this field. Salvatore et al. (2014) and Vanegas et al. (2018) both demonstrate the feasibility of supervised ML algorithms in differentiating between PD and other related conditions, using brain MRI data and EEG-based biomarkers, respectively. Khachnaoui et al. (2020) and Challa et al. (2016) further extend this work by discussing the application of ML and DL in clinical data and non-motor symptoms prediction, achieving impressive accuracy rates.

Numerous works have employed the striatum as the ROI to analyse in the study of this disorder. Kim et al. (2017) proposed an accurate linear regression model to predict clinical scores from features derived from neuroimaging and genetics information. Similarly, Martinez-Murcia et al. (2024) developed a self-supervised DL model to extract latent variables from SPECT images and relate them with the ApoE genetic information of different patients. Results showed a very similar distribution of the 2D projection of the latent space and the aforementioned genotypes.

6.5.3 Schizophrenia

SCZ is a chronic psychiatric disorder marked by disturbances in the perception of reality (psychosis), affecting cognition, emotion and behaviour. Clinical manifestations include hallucinations, delusions, disorganised thinking, social withdrawal and cognitive deficits, which vary in severity and are typically classified as positive symptoms (e.g., hallucinations) and negative symptoms (e.g., diminished emotional expression). Affecting around 1% of the population (Uno & Coyle, 2019), SCZ's exact cause is complex and likely involves genetic, environmental and neurochemical factors, including disruptions in neurotransmitters such as dopamine and glutamate (Insel, 2010). Typically emerging in early adulthood, timely detection and appropriate treatment are critical for managing symptoms and improving overall well-being.

Early detection of SCZ through MRI is actively researched in neuroimaging. Structural abnormalities in the brain are observed in early stages of the disease, including reduced cortical folding, particularly in temporal-parietal-occipital regions (Sallet et al., 2003; Yan et al., 2019). Reductions in the superior temporal gyrus are associated with positive symptoms, while prefrontal area reductions correlate with negative symptoms (Walton et al., 2017, 2018).

A range of studies have demonstrated the potential of ML in the diagnosis and understanding of SCZ. For example, Greenstein et al. (2012) applied ML to identify cognitive deficits and classify childhood-onset SCZ, with promising results. Recent studies have also focused on the analysis of brain sulci patterns from MRI to improve the understanding of the aetiology of SCZ (Jimenez-Mesa et al., 2024; Rollins et al., 2020).

6.5.4 Autism

Autism spectrum disorder (ASD) is a multifaceted neurodevelopmental condition marked by impairments in social interaction, communication and the presence of repetitive behaviours. The designation "spectrum" underscores the considerable variability in both the type and severity of symptoms across individuals. Autism affects individuals of all genders, races, ethnicities and economic backgrounds (Nedungadi et al., 2024). Diagnosis is typically based on behavioural assessments and developmental screenings rather than biological markers. Although the precise aetiology of autism remains incompletely understood, current research suggests that both genetic and environmental factors contribute to its development. Known risk factors include having a sibling with ASD, older parental age, certain genetic conditions and very low birth weight. Recent studies have employed ML-based solutions for exploring the neural basis of this disorder from an anatomical perspective to a function-guided alternative. The idea was to identify the presence of autism and improve our knowledge about the role of certain brain regions in this disorder. Katuwal et al. (2015) explored the use of ML algorithms to analyse structural MRI data for diagnosing autism. The study employed SVM and CNN to identify significant differences in brain structures between individuals with autism and neurotypical controls. Anatomical abnormalities in the temporal pole and cerebellum seem to be extremely relevant for the identification of this disorder, especially in structural MRI (Heinsfeld et al., 2018). Moreover, information extracted from functional MRI also suggests that a differential connectivity between networks can be potentially used as a biomarker for early diagnosis (Plitt et al., 2015).

Górriz et al. (2019) proposed a solution based on PLS and linear classification for distinguishing between individuals with autism and neurotypical controls based on brain structure. Specifically, significant differences were found in regional brain structures, aligning with the Extreme Male Brain (EMB) theory, which posits that autism is an exaggeration of typical male cognitive patterns. The research highlights the importance of considering sex differences in autism studies and suggests that both biological and environmental factors contribute to the observed neuroanatomical patterns. In a similar study, Leming et al. (2020) developed a deep architecture to analyse the connectivity matrices associated with different fMRI tasks. Results demonstrate the important role of several brain regions (corpus callosum, frontal lobes, parietal areas) in brain activity within some cognitive functions, and the differences between male and female in the execution of some routine tasks. Besides, the class activation maps (CAMs) and results based on activation maximisation identified the most relevant patterns in the study of this disorder, but emphasised that the distinction between ASDs and typically developing patients is difficult due to the common variability and complexity of the data.

6.6 Challenges and Trends in Neuroimaging

6.6.1 Multimodal Fusion Models

Multimodal fusion models have become essential tools in ML for neuroscience, allowing the integration of diverse data types, e.g., MRI, PET and genetic information, to achieve a more comprehensive understanding of neurological conditions. In this section, we will explore the fusion of spatiotemporal data, imaging modalities and the integration of imaging with other clinical data, including omics, cognitive assessments and demographic information.

Despite the intrinsic potential of current non-invasive neuroimaging modalities and modern analysis techniques presented in this chapter, the human brain remains an extraordinarily intricate system. Cognitive functions emerge from complex patterns of neural activity that unfold across specific spatial and temporal scales, including oscillatory dynamics, transient signals and rapid information exchange between brain regions. To achieve a comprehensive understanding of these processes, it is necessary to capture neural activity with both temporal and spatial precision. Yet, the inherent limitations of current non-invasive neuroimaging techniques often force researchers to prioritise either temporal or spatial resolution. For instance, M/EEG recordings offer exceptional temporal detail but limited spatial specificity, whereas fMRI provides precise spatial localisation of brain activity at millimetre resolution but with poor temporal fidelity. Addressing this trade-off remains a central challenge in cognitive neuroscience, and recent approaches increasingly explore multimodal data integration as a promising avenue (Cichy & Oliva, 2020; López-García et al., 2022a).

However, linking activity patterns across different brain measurement modalities is not a trivial task, as these techniques capture fundamentally different signals (e.g., haemodynamic versus electrical activity) that do not correspond directly. To tackle this correspondence problem, it is essential to employ a framework that moves beyond raw activity patterns. Representational similarity analysis (RSA), introduced by Kriegeskorte et al. (2008), expanded the scope of MVPA by providing such an abstraction. The central principle of RSA is straightforward: neural activity patterns elicited by similar stimuli or tasks should resemble one another. This approach generalises the notion of voxel- or electrode-specific activation to more abstract similarity structures. To implement this, pairwise comparisons of the original activity patterns are converted into dissimilarity matrices, which can then be compared across brain regions, time points, individuals, neuroimaging modalities or even species (Figure 6.9). Multimodal fusion strategies based on RSA have been applied, among other studies, to trace the spatiotemporal cascade of visual object processing (Cichy et al., 2014) and to examine the dynamic neural patterns underlying face perception (Muukkonen et al., 2020).

Another illustrative example of multimodal fusion involves combining different imaging techniques, such as PET and MRI, to achieve a more complete and informative representation of the condition under study, effectively mitigating the limitations inherent to each individual modality. This strategy leverages both functional data (e.g., SPECT or PET) and structural information (e.g., MRI or CT), offering multiple advantages: higher image quality with improved spatial resolution, enhanced diagnostic accuracy through richer patient-specific data, earlier detection of pathological changes, more efficient treatment planning, and the opportunity to perform more thorough and detailed clinical investigations across a range of diseases and medical conditions.

Numerous studies have demonstrated that combining multiple imaging modalities can enhance diagnostic performance. For instance, Polikar et al. (2010) performed a binary classification analysis to distinguish Alzheimer's patients from healthy controls by integrating MRI, PET and EEG data. In this study, an ensemble of MLPs with majority voting yielded an accuracy of 85.55%. Another example is provided by Song et al. (2021), who fused GM regions from MRI scans with FDG-PET images through registration and mask coding. Subsequently, a 3D CNN and a 3D multi-scale CNN were employed to carry out

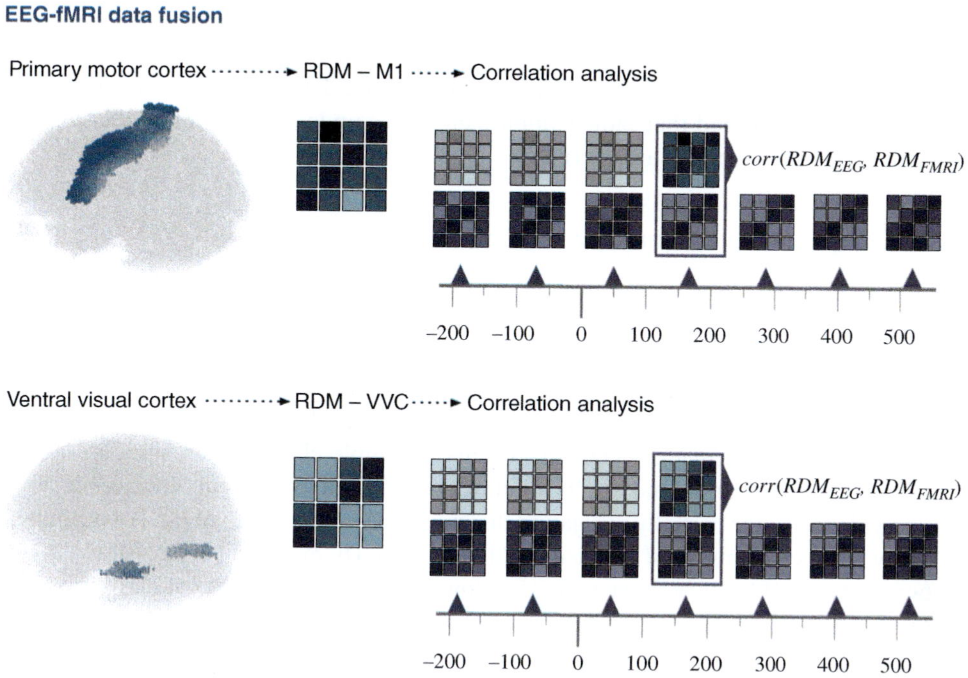

Figure 6.9 Schematic of EEG-fMRI data fusion using representational dissimilarity matrices (RDMs). RDMs are first computed for each EEG time point, while fMRI RDMs are derived from selected brain ROIs. Correlating the EEG and fMRI matrices produces time series of correlation values, capturing the spatiotemporal dynamics underlying the cognitive process under investigation. *Source*: López García et al. (2023)/Universidad De Granada.

both binary and multi-class classification, reaching accuracies of 94.11% for the AD versus healthy control task and 71.52% for the AD versus MCI versus healthy control task. Additionally, Arco et al. (2023b) employed a Siamese neural network to integrate MRI and PET information, allowing the quantification of similarities between the two modalities and highlighting the contributions of specific brain regions to AD progression based on structural and functional changes.

As evident from the examples mentioned earlier, the integration of multimodal imaging represents a promising avenue for future research, with the potential to reveal novel insights and enhance diagnostic accuracy. Nevertheless, combining multiple medical imaging modalities also entails several challenges. These challenges include the technical complexity and substantial costs of installing and managing specialised equipment. In addition, the fusion process can be time-consuming, potentially limiting its use in time-sensitive clinical situations.

Lastly, the combination of imaging and clinical information for improved diagnostics and prognostics also has great significance. Studies have demonstrated the superiority of ML models that combine clinical and imaging data over traditional methods in predicting cardiac events (Betancur et al., 2018) and in developing personalised biomarkers for cancer diagnosis and treatment (Boehm et al., 2022). Special attention is given to the integration of imaging biomarkers and omic data. One key area of study is the correlation between genotype and imaging phenotype (radiogenomics), which could lead to the discovery of non-invasive biomarkers, enabling faster and more cost-effective diagnostics. For example, Guo et al. (2024) demonstrated that radiogenomics enhances understanding of glioblastoma's molecular and phenotypical characteristics.

6.6.2 Interpretability of Neuroimaging Data

One important aspect for the use of CAD systems in neuroimaging is that they should inform about what guides their decision. The emergence of DL models has led to extremely accurate solutions, but with an evident lack of interpretability. In fact, DL alternatives are usually seen as black boxes since only inputs and outputs of the systems are known, whereas the intermediate stages remain completely inaccessible. This is very problematic in the medical or neuroscience context, whereas a simple classification solution is not valid, but the identification and the location of the information. Recent advances in XAI have provided tools for partially mitigating this problem (Górriz et al., 2023). For example, the use of CAMs has improved the visualisation of these DL models (Adebayo et al., 2018; Simonyan et al., 2013; Selvaraju et al., 2017). However, the particularities of neuroimaging (images with three dimensions, information present in multiple clusters that can be far away from the others, complex shapes of the brain, etc.) make it difficult to detect what happens inside the model based only on gradient maps.

Another important aspect in the development of these kinds of tools is that they should be accessible to scientists of a wide range of disciplines. Since the aim of CAD systems is to help clinicians or neuroscientists in the study of different disorders, the tools to be developed must include a visual explanation of the results (Tomaz Da Silva et al., 2021). In fact, the design and validation of the intelligent models must be carried out in collaboration with a multidisciplinary team to adapt each specific stage of the framework to the needs of the professionals who are going to use those tools (Schrouff et al., 2013). Besides, it is important to improve the CAD tools following a similar approach to that used by clinicians in their daily routines, i.e., to analyse the information from different sources (usually results of diagnostic tests) and make a decision based on all the information available (Anand et al., 2023). In this case, the crucial aspect is to assign to each independent source a weight proportional to their estimated importance.

6.6.3 Future Prospects

Transfer learning and data augmentation represent key emerging strategies to improve the reliability and performance of CAD systems in neuroimaging (Garcea et al., 2023). By leveraging previously trained models on extensive datasets from similar domains, ML algorithms can achieve better performance even when neuroimaging data are limited, enhancing both diagnostic accuracy and robustness. Data augmentation artificially increases dataset size and diversity, supporting more comprehensive model training. Nevertheless, while synthetic data can expand variability, they do not always faithfully reflect the characteristics of the clinical population. This discrepancy may affect the reliability and clinical relevance of model predictions, as artificial patterns could overshadow important features present in real patient data. Ongoing research focuses on improving the fidelity and applicability of augmented datasets to enhance system reliability.

Generative AI methods, including GANs and transformer-based models, offer further opportunities for creating synthetic images that complement patient data (Aghakhan Olia et al., 2022; Sanaat et al., 2021b). These approaches are already capable of producing standard-quality scans from low-dose acquisitions and generating information similar to multimodal integration, thus reducing both patient exposure and costs (Watanabe et al., 2021). Transformers, in particular, are expected to become increasingly influential, although strategies such as domain adaptation remain necessary to address limited data scenarios. The ultimate goal is the development of CAD systems that are more comprehensive, interpretable and capable of integrating diverse sources of information.

Such advancements have been facilitated by significant improvements in hardware and computational infrastructure (Huang et al., 2019; Berggren et al., 2020). The rapid growth in computational power has allowed for the efficient training and deployment of increasingly complex models. The adoption of graphics processing units and distributed computing systems has accelerated algorithm execution through massive parallelisation. Continued progress in ML methodologies, alongside corresponding hardware and software innovations, will support the creation of CAD systems capable of real-time decision support, thereby facilitating timely diagnosis and intervention strategies and improving overall clinical efficiency.

References

Adebayo, J., Gilmer, J., Muelly, M. et al. (2018). Sanity checks for saliency maps. *Advances in Neural Information Processing Systems* 31.

Aghakhan Olia, N., Kamali-Asl, A., Hariri Tabrizi, S. et al. (2022). Deep learning–based denoising of low-dose SPECT myocardial perfusion images: quantitative assessment and clinical performance. *European Journal of Nuclear Medicine and Molecular Imaging* 49: 1508–1522.

Alzheimer's Association (2023). Alzheimer's disease facts and figures. *Alzheimer's & Dementia* 19 (4): 1598–1695. https://doi.org/10.1002/alz.13016.

Amini, S., Zhang, L., Hao, B. et al. (2021). An artificial intelligence-assisted method for dementia detection using images from the clock drawing test. *Journal of Alzheimer's Disease* 83 (2): 581–589. https://doi.org/10.3233/JAD-210299.

Anand, V., Gupta, S., Gupta, D. et al. (2023). Weighted average ensemble deep learning model for stratification of brain tumor in MRI images. *Diagnostics* 13 (7): 1320.

Angra, S. and Ahuja, S. (2017). Machine learning and its applications: a review. In: *2017 International Conference on Big Data Analytics and Computational Intelligence (ICBDAC)*, 57–60. IEEE.

Arco, J.E., Ortiz, A., Castillo-Barnes, D. et al. (2022). Quantifying inter-hemispheric differences in Parkinson's disease using Siamese networks. In: *Artificial Intelligence in Neuroscience: Affective Analysis and Health Applications: 9th International Work-Conference on the Interplay Between Natural and Artificial Computation, IWINAC 2022, Puerto de la Cruz,* Tenerife, Spain, May 31–June 3, 2022, *Proceedings, Part I*, 156–165. Springer.

Arco, J.E., Ortiz, A., Castillo-Barnes, D. et al. (2023a). Ensembling shallow Siamese architectures to assess functional asymmetry in Alzheimer's disease progression. *Applied Soft Computing* 134: 109991.

Arco, J.E., Ortiz, A., Gallego-Molina, N.J. et al. (2023b). Enhancing multimodal patterns in neuroimaging by Siamese neural networks with self-attention mechanism. *International Journal of Neural Systems* 33 (04): 2350019.

Arndt, S., Cizadlo, T., O'Leary, D. et al. (1996). Normalizing counts and cerebral blood flow intensity in functional imaging studies of the human brain. *NeuroImage* 3 (3): 175–184.

Bartlett, P.L., Harvey, N., Liaw, C. et al. (2019). Nearly-tight VC-dimension and pseudodimension bounds for piecewise linear neural networks. *The Journal of Machine Learning Research* 20 (1): 2285–2301.

Berggren, K., Xia, Q., Likharev, K.K. et al. (2020). Roadmap on emerging hardware and technology for machine learning. *Nanotechnology* 32 (1): 012002.

Betancur, J., Otaki, Y., Motwani, M. et al. (2018). Prognostic value of combined clinical and myocardial perfusion imaging data using machine learning. *JACC: Cardiovascular Imaging* 11 (7): 1000–1009.

Bishop, C.M. and Nasrabadi, N.M. (2006). *Pattern Recognition and Machine Learning*, volume 4. Springer.

Boehm, K.M., Khosravi, P., Vanguri, R. et al. (2022). Harnessing multimodal data integration to advance precision oncology. *Nature Reviews Cancer* 22 (2): 114–126.

Bonferroni, C. (1936). Teoria statistica delle classi e calcolo delle probabilita. *Pubblicazioni del R Istituto Superiore di Scienze Economiche e Commericiali di Firenze* 8: 3–62.

Braga-Neto, U., Hashimoto, R., Dougherty, E.R. et al. (2004). Is cross-validation better than resubstitution for ranking genes? *Bioinformatics* 20 (2): 253–258. https://doi.org/10.1093/bioinformatics/btg399.

Breiman, L. (1996). Bagging predictors. *Machine Learning* 24 (2): 123–140. https://doi.org/10.1007/BF00058655.

Breiman, L. (2001). Random forests. *Machine Learning* 45 (1): 5–32. https://doi.org/10.1023/A:1010933404324.

Brett, M., Penny, W., and Kiebel, S. (2004). Chapter 44 - introduction to random field theory. In: *Human Brain Function (Second Edition)* (ed. S.J. Richard, Frackowiak, K.J. Friston et al.), 867–879. Burlington: Academic Press.

Bron, E.E., Smits, M., van der Flier, W.M. et al. (2015). Standardized evaluation of algorithms for computer-aided diagnosis of dementia based on structural MRI: the CADDementia challenge. *NeuroImage* 111: 562–579. https://doi.org/10.1016/j.neuroimage.2015.01.048.

Button, K.S., Ioannidis, J.P., Mokrysz, C. et al. (2013). Power failure: why small sample size undermines the reliability of neuroscience. *Nature Reviews Neuroscience* 14 (5): 365–376. https://doi.org/10.1038/nrn3475.

Carlson, N.R. (2012). *Physiology of Behavior*. Pearson Higher Ed.

Castillo-Barnes, D., Jimenez-Mesa, C., Martinez-Murcia, F.J. et al. (2022). Quantifying differences between affine and nonlinear spatial normalization of FP-CIT SPECT images. *International Journal of Neural Systems* 32 (05): 2250019. https://doi.org/10.1142/s0129065722500198.

Castillo-Barnes, D., Ramírez, J., Segovia, F. et al. (2018). Robust ensemble classification methodology for I123-Ioflupane SPECT images and multiple heterogeneous biomarkers in the diagnosis of Parkinson's disease. *Frontiers in Neuroinformatics* 12: 53.

Challa, K.N.R., Pagolu, V.S., Panda, G. et al. (2016). An improved approach for prediction of Parkinson's disease using machine learning techniques. In: *2016 International Conference on Signal Processing, Communication, Power and Embedded System (SCOPES)*, 1446–1451. IEEE.

Charte, D., Charte, F., García, S. et al. (2018). A practical tutorial on autoencoders for nonlinear feature fusion: taxonomy, models, software and guidelines. *Information Fusion* 44: 78–96. https://doi.org/10.1016/j.inffus.2017.12.007.

Cichy, R.M. and Oliva, A. (2020). AM/EEG-fMRI fusion primer: resolving human brain responses in space and time. *Neuron* 107 (5): 772–781.

Cichy, R.M., Pantazis, D., and Oliva, A. et al. (2014). Resolving human object recognition in space and time. *Nature Neuroscience* 17 (3): 455–462.

Claus, J.J., Harskamp, F.V., Breteler, M.M.B. et al. (1994). The diagnostic value of SPECT with Tc 99m HMPAO in Alzheimer's disease: a population-based study. *Neurology* 44 (3 Part1): 454–454.

Coutanche, M.N., Thompson-Schill, S.L., and Schultz, R.T. (2011). Multi-voxel pattern analysis of fMRI data predicts clinical symptom severity. *NeuroImage* 57 (1): 113–123. https://doi.org/10.1016/j.neuroimage.2011.04.016.

Davatzikos, D. (2019). Machine learning in neuroimaging: progress and challenges. *NeuroImage* 197: 652–656. https://doi.org/10.1016/j.neuroimage.2018.10.003.

de Vos, F., Schouten, T.M., Hafkemeijer, A. et al. (2016). Combining multiple anatomical MRI measures improves Alzheimer's disease classification. *Human Brain Mapping* 37 (5): 1920–1929. https://doi.org/10.1002/hbm.23147.

Devlin, J., Chang, M.W., Lee, K. et al. (2018). Bert: pre-training of deep bidirectional transformers for language understanding. *arXiv preprint arXiv:1810.04805*.

Dimitriadis, S.I., Liparas, D., and Tsolaki, M.N. (2018). Random forest feature selection, fusion and ensemble strategy: combining multiple morphological MRI measures to discriminate among healthy elderly, MCI, cMCI and Alzheimer's disease patients: from the Alzheimer's disease neuroimaging initiative (ADNI) database. *Journal of Neuroscience Methods* 302: 14–23. https://doi.org/10.1016/j.jneumeth.2017.12.010.

Dosovitskiy, A., Beyer, L., Kolesnikov, A. et al. (2020). An image is worth 16×16 words: transformers for image recognition at scale. *arXiv*. https://arxiv.org/abs/2010.11929

Eklund, A., Nichols, T.E., and Knutsson, H. (2016). Cluster failure: why fMRI inferences for spatial extent have inflated false-positive rates. *Proceedings of the National Academy of Sciences of the United States of America* 113 (28): 7900–7905.

Fahrenfort, J.J. Van Driel, J., Van Gaal, S. et al. (2018). From ERPs to MVPA using the Amsterdam decoding and modeling toolbox (ADAM). *Frontiers in Neuroscience* 12: 368.

Freund, Y. and Schapire, R.E. (1996). Experiments with a new boosting algorithm. ICML'96: Proceedings of the Thirteenth International Conference on International Conference on Machine Learning.

Gao, X., Chu, C., Li, Y. et al. (2015). The method and efficacy of support vector machine classifiers based on texture features and multi-resolution histogram from 18F-FDG PET-CT images for the evaluation of mediastinal lymph nodes in patients with lung cancer. *European Journal of Radiology* 84 (2): 312–317.

Garcea, F., Serra, A., Lamberti, F. et al. (2023). Data augmentation for medical imaging: a systematic literature review. *Computers in Biology and Medicine* 152: 106391.

Goodfellow, I., Pouget-Abadie, J., Mirza, M. et al. (2014). Generative adversarial nets. *Advances in Neural Information Processing Systems* 27. https://papers.nips.cc/paper_files/paper/2014/hash/f033ed80deb0234979a61f95710dbe25-Abstract.html

Górriz, J.M., Álvarez-Illán, I., Álvarez-Marquina, A. et al. (2023). Computational approaches to explainable artificial intelligence: advances in theory, applications and trends. *Information Fusion* 100: 101945.

Gorriz, J.M., Jimenez-Mesa, C., Romero-Garcia, R. et al. (2021). Statistical agnostic mapping: a framework in neuroimaging based on concentration inequalities. *Information Fusion* 66: 198–212. https://doi.org/10.1016/j.inffus.2020.09.008.

Górriz, J.M., Ramirez, J., and Suckling, J. (2019). On the computation of distribution-free performance bounds: application to small sample sizes in neuroimaging. *Pattern Recognition* 93: 1–13. https://doi.org/10.1016/j.patcog.2019.03.032.

Górriz, J.M., Ramírez, J., Segovia, F. et al. (2019). A machine learning approach to reveal the neurophenotypes of autisms. *International Journal of Neural Systems* 29 (07): 1850058.

Gramfort, A., Luessi, M., Larson, E. et al. (2013). MEG and EEG data analysis with MNE-Python. *Frontiers in Neuroinformatics* 7: 267.

Greenstein, D., Malley, J.D., Weisinger, B. et al. (2012). Using multivariate machine learning methods and structural MRI to classify childhood onset schizophrenia and healthy controls. *Frontiers in Psychiatry* 3: 53.

Gu, J., Wang, Z., Kuen, J. et al. (2018). Recent advances in convolutional neural networks. *Pattern Recognition* 77: 354–377. https://doi.org/10.1016/j.patcog.2017.10.013.

Guo, J., Fathi Kazerooni, A., Toorens, E. et al. (2024). Integrating imaging and genomic data for the discovery of distinct glioblastoma subtypes: a joint learning approach. *Scientific Reports* 14 (1): 4922.

Guo, M.H., Xu, T.X., Liu, J.J. et al. (2022). Attention mechanisms in computer vision: a survey. *Computational Visual Media* 8 (3): 331–368.

Hanke, M., Halchenko, Y.O., Sederberg, P.B. et al. (2009). PyMVPA: a unifying approach to the analysis of neuroscientific data. *Frontiers in Neuroinformatics* 3: 337.

Haxby, J.V., Connolly, A.C., Guntupalli, J.S. et al. (2014). Decoding neural representational spaces using multivariate pattern analysis. *Annual Review of Neuroscience* 37 (1): 435–456.

Haynes, J.D. and Rees, G. (2005). Predicting the orientation of invisible stimuli from activity in human primary visual cortex. *Nature Neuroscience* 8: 686–691

Hebart, M.N., Görgen, K., and Haynes, J.D. (2015). The decoding toolbox (TDT): a versatile software package for multivariate analyses of functional imaging data. *Frontiers in Neuroinformatics* 8: 88.

Heinsfeld, A.S., Franco, A.R., Craddock, R.C. et al. (2018). Identification of autism spectrum disorder using deep learning and the abide dataset. *NeuroImage: Clinical* 17: 16–23.

Hinrichs, C., Singh, V., Xu, G. et al. (2011). Predictive markers for AD in a multi-modality framework: an analysis of MCI progression in the ADNI population. *NeuroImage* 55 (2): 574–589.

Huang, B., Huan, Y., Xu, L.D. et al. (2019). Automated trading systems statistical and machine learning methods and hardware implementation: a survey. *Enterprise Information Systems* 13 (1): 132–144.

Insel, T.R. (2010). Rethinking schizophrenia. *Nature* 468 (7321): 187–193.

Jafari, M., Sadeghi, D., Shoeibi, A. et al. (2024). Empowering precision medicine: AI-driven schizophrenia diagnosis via EEG signals: a comprehensive review from 2002–2023. *Applied Intelligence* 54 (1): 35–79.

Jiménez Mesa, C. (2023). Enhancing diagnostic accuracy in neuroimaging through machine learning: Advancements in statistical classification and mapping. PhD thesis. Universidad de Granada.

Jimenez-Mesa, C., Arco, J.E., Martinez-Murcia, F.J. et al. (2023a). Applications of machine learning and deep learning in SPECT and PET imaging: general overview, challenges and future prospects. *Pharmacological Research* 197: 106984.

Jimenez-Mesa, C., Arco, J.E., Valenti-Soler, M. et al. (2023b). Using explainable artificial intelligence in the clock drawing test to reveal the cognitive impairment pattern. *International Journal of Neural Systems* 33: 2350015. https://doi.org/10.1142/s0129065723500156.

Jimenez-Mesa, C., Castillo-Barnes, D., Arco, J.E. et al. (2022). Analyzing statistical inference maps using MRI images for Parkinson's disease. In: *Artificial Intelligence in Neuroscience: Affective Analysis and Health Applications*, 166–175. Springer International Publishing. https://doi.org/10.1007/978-3-031-06242-1˙17.

Jimenez-Mesa, C., Illán, I.A., Martin-Martin, A. et al. (2020). Optimized one vs one approach in multiclass classification for early Alzheimer's disease and mild cognitive impairment diagnosis. *IEEE Access* 8: 96981–96993. https://doi.org/10.1109/access.2020.2997736.

Jimenez-Mesa, C., Ramirez, J., Suckling, J. et al. (2023c). A non-parametric statistical inference framework for deep learning in current neuroimaging. *Information Fusion* 91: 598–611.

Jimenez-Mesa, C., Ramirez, J., Yi, Z. et al. (2024). Machine learning in small sample neuroimaging studies: novel measures for schizophrenia analysis. *Human Brain Mapping* 45 (5): e26555.

Jollans, L., Boyle, R., Artiges, E. et al. (2019). Quantifying performance of machine learning methods for neuroimaging data. *NeuroImage* 199: 351–365.

Katuwal, G.J., Cahill, N.D., Baum, S.A. et al. (2015). The predictive power of structural MRI in autism diagnosis. *IEEE Engineering in Medicine and Biology Society* 2015: 4270–4273.

Khachnaoui, H., Mabrouk, R., and Khlifa, N. (2020). Machine learning and deep learning for clinical data and PET/SPECT imaging in Parkinson's disease: a review. *IET Image Processing* 14 (16): 4013–4026.

Khan, S., Naseer, M., Hayat, M. et al. (2022). Transformers in vision: a survey. *ACM Computing Surveys (CSUR)* 54 (10s): 1–41.

Kim, M., Kim, J., Lee, L. et al. (2017). Imaging genetics approach to Parkinson's disease and its correlation with clinical score. *Scientific Reports* 7: 46700. https://doi.org/10.1038/srep46700.

Koch, G., Zemel, R., Salakhutdinov, R. et al. (2015). Siamese neural networks for one-shot image recognition. In: *ICML Deep Learning Workshop*, volume 2. Lille.

Kriegeskorte, N., Mur, M., and Bandettini, P.A. (2008). Representational similarity analysis-connecting the branches of systems neuroscience. *Frontiers in Systems Neuroscience* 2: 249.

Leming, M., Górriz, J.M., and Suckling, J. (2020). Ensemble deep learning on large, mixed-site fMRI datasets in autism and other tasks. *International Journal of Neural Systems* 30 (07): 2050012.

López García, D. (2023). Procesado avanzado de señales de electroencefalografia y resonancia magnética en Neurociencia Cognitiva. PhD thesis. Universidad de Granada.

López-García, D., González-Peñalver, J.M., Górriz, J.M. et al. (2022a). Representational similarity analysis: a preliminary step to fMRI-EEG data fusion in MVPAlab. In: *International Work-Conference on the Interplay Between Natural and Artificial Computation*, 84–94. Springer.

López-García, D., Peñalver, J.M., Górriz, J.M. et al. (2022b). MVPAlab: a machine learning decoding toolbox for multidimensional electroencephalography data. *Computer Methods and Programs in Biomedicine* 214: 106549.

Mainland, B.J. and Shulman, K.I. (2017). Clock drawing test. In: *Cognitive Screening Instruments*, 67–108. Springer International Publishing. https://doi.org/10.1007/978-3-319-44775-9˙5.

Martinez-Murcia, F.J., Arco, J.E., Jimenez-Mesa, C., et al. (2024). Bridging imaging and clinical scores in Parkinson's progression via multimodal self-supervised deep learning. *International Journal of Neural Systems* 34 (08): 2450043.

Martinez-Murcia, F.J., Górriz, J.M., Ramírez, J. et al. (2018). Convolutional neural networks for neuroimaging in Parkinson's disease: is preprocessing needed? *International Journal of Neural Systems* 28 (10): 1850035. https://doi.org/10.1142/s0129065718500351.

Martorell-Marugan, J., Chierici, M., Bandres-Ciga, S. et al. (2023). Machine learning applications in the study of Parkinson's disease: a systematic review. *Current Bioinformatics* 18 (7): 576–586.

McAllester, D. (2013). A PAC-Bayesian tutorial with a dropout bound. *arXiv*. https://arxiv.org/abs/1307.2118

Modrego, P.J. (2006). Predictors of conversion to dementia of probable Alzheimer type in patients with mild cognitive impairment. *Current Alzheimer Research* 3 (2): 161–170.

Mohapatra, S., Swarnkar, T., and Das, J. (2021). Deep convolutional neural network in medical image processing. In: *Handbook of Deep Learning in Biomedical Engineering*, 25–60. Elsevier.

Morris, J.C. (1997). Clinical dementia rating: a reliable and valid diagnostic and staging measure for dementia of the Alzheimer type. *International Psychogeriatrics* 9 (S1): 173–176. https://doi.org/10.1017/s1041610297004870.

Moulder, K.L., Snider, B.J., Mills, S.L. et al. (2013). Dominantly inherited Alzheimer network: facilitating research and clinical trials. *Alzheimer's Research & Therapy* 5: 1–7.

Muukkonen, I., Ölander, K., Numminen, J. et al. (2020). Spatio-temporal dynamics of face perception. *NeuroImage* 209: 116531.

National Institute of Neurological Disorders and Stroke (US) (2004). *Parkinson's Disease: Challenges, Progress, and Promise*. National Institute of Neurological Disorders and Stroke, National Institutes.

Nedungadi, P., Shah, S.M., Stokes, M.A. et al. (2024). Mapping autism's research landscape: trends in autism screening and its alignment with sustainable development goals. *Frontiers in Psychiatry*, 14: 1294254.

Neyshabur, B., Bhojanapalli, S., McAllester, D. et al. (2017a). Exploring generalization in deep learning. *Advances in Neural Information Processing Systems* 30: 5947–5956.

Neyshabur, B., Bhojanapalli, S., and Srebro, N. (2017b). A PAC-Bayesian approach to spectrally-normalized margin bounds for neural networks. *arXiv*. https://arxiv.org/abs/1707.09564

Oosterhof, N.N., Connolly, A.C., and Haxby, J.V. (2016). CoSMoMVPA: multi-modal multivariate pattern analysis of neuroimaging data in Matlab/GNU Octave. *Frontiers in Neuroinformatics* 10: 27.

Ortiz, A., Fajardo, D., Górriz, J.M. et al. (2014). Multimodal image data fusion for Alzheimer's disease diagnosis by sparse representation. *Studies in Health Technology and Informatics* 207: 11–8.

Palsetia, D., Rao, G.P., Tiwari, S.C. et al. (2018). The clock drawing test versus mini-mental status examination as a screening tool for dementia: a clinical comparison. *Indian Journal of Psychological Medicine* 40 (1): 1–10.

Penny, W.D., Friston, K.J., Ashburner, J.T. et al. (2011). *Statistical Parametric Mapping: The Analysis of Functional Brain Images*. Elsevier.

Plitt, M., Barnes, K.A., and Martin, A. (2015). Functional connectivity classification of autism identifies highly predictive brain features but falls short of biomarker standards. *NeuroImage: Clinical* 7: 359–366.

Poewe, W., Seppi, K., Tanner, C.M. et al. (2017). Parkinson's disease. *Nature Reviews Disease Primers* 3 (1): 17013. https://doi.org/10.1038/nrdp.2017.13.

Polikar, R., Tilley, C., Hillis, B. et al. (2010). Multimodal EEG, MRI and PET data fusion for Alzheimer's disease diagnosis. In: *2010 Annual International Conference of the IEEE Engineering in Medicine and Biology*, 6058–6061. IEEE.

Radford, A., Narasimhan, K., Salimans, T. et al. (2018). Improving language understanding by generative pre-training.

Rojas, A., Górriz, J.M., Ramírez, J. et al. (2013). Application of empirical mode decomposition (EMD) on DaTSCAN SPECT images to explore Parkinson disease. *Expert Systems with Applications* 40 (7): 2756–2766. https://doi.org/10.1016/j.eswa.2012.11.017.

Rollins, C.P., Garrison, J.R., Arribas, M. et al. (2020). Evidence in cortical folding patterns for prenatal predispositions to hallucinations in schizophrenia. *Translational Psychiatry* 10 (1): 387. https://doi.org/10.1038/s41398-020-01075-y.

Salas-Gonzalez, D., Górriz, J.M., Ramírez, J. et al. (2009). Analysis of SPECT brain images for the diagnosis of Alzheimer's disease using moments and support vector machines. *Neuroscience Letters* 461 (1): 60–64.

Salas-Gonzalez, D., Górriz, J.M., Ramírez, J. et al. (2013). Linear intensity normalization of FP-CIT SPECT brain images using the α-stable distribution. *NeuroImage* 65: 449–455. https://doi.org/10.1016/j.neuroimage.2012.10.005.

Sallet, P.C., Elkis, H., Alves, T.M. et al. (2003). Reduced cortical folding in schizophrenia: an MRI morphometric study. *American Journal of Psychiatry* 160 (9): 1606–1613.

Salvatore, C., Cerasa, A., Castiglioni, I. et al. (2014). Machine learning on brain MRI data for differential diagnosis of Parkinson's disease and progressive supranuclear palsy. *Journal of Neuroscience Methods* 222: 230–237.

Sanaat, A., Mirsadeghi, E., Razeghi, B. et al. (2021a). Fast dynamic brain PET imaging using stochastic variational prediction for recurrent frame generation. *Medical Physics* 48 (9): 5059–5071.

Sanaat, A., Shiri, I., Arabi, H. et al. (2021b). Deep learning-assisted ultra-fast/low-dose whole-body PET/CT imaging. *European Journal of Nuclear Medicine and Molecular Imaging* 48: 2405–2415.

Sarica, A., Cerasa, A., Quattrone, A. et al. (2018). Editorial on special issue: machine learning on MCI. *Journal of Neuroscience Methods* 302: 1–2. https://doi.org/10.1016/j.jneumeth.2018.03.011.

Schölkopf, B., Smola, A.J., Bach, F. et al. (2002). *Learning With Kernels: Support Vector Machines, Regularization, Optimization, and Beyond*. MIT Press.

Schrouff, J., Rosa, M.J., Rondina, J.M. et al. (2013). PRoNTo: pattern recognition for neuroimaging toolbox. *Neuroinformatics* 11: 319–337.

Selvaraju, R.R., Cogswell, M., Das, A. et al. (2017). Grad-CAM: visual explanations from deep networks via gradient-based localization. In: *2017 IEEE International Conference on Computer Vision (ICCV)*. IEEE. https://doi.org/10.1109/iccv.2017.74.

Shamshad, F., Khan, S., Zamir, S.W. et al. (2023). Transformers in medical imaging: a survey. *Medical Image Analysis* 88: 102802.

Shao, W., Rowe, S.P., and Du, Y. (2021). SPECTnet: a deep learning neural network for SPECT image reconstruction. *Annals of Translational Medicine* 9 (9): 819.

Simonyan, K., Vedaldi, A., and Zisserman, A. (2013). Deep inside convolutional networks: visualising image classification models and saliency maps. *arXiv*. https://arxiv.org/abs/1312.6034

So, A., Hooshyar, D., Park, K. et al. (2017). Early diagnosis of dementia from clinical data by machine learning techniques. *Applied Sciences* 7 (7): 651. https://doi.org/10.3390/app7070651.

Song, J., Zheng, J., Li, P. et al. (2021). An effective multimodal image fusion method using MRI and PET for Alzheimer's disease diagnosis. *Frontiers in Digital Health* 3: 637386.

Tang, W., He, F., Liu, Y. et al. (2022). Matr: multimodal medical image fusion via multiscale adaptive transformer. *IEEE Transactions on Image Processing* 31: 5134–5149.

Tomaz Da Silva, L., Esper, N.B., Ruiz, D.D. et al. (2021). Visual explanation for identification of the brain bases for developmental dyslexia on fMRI data. *Frontiers in Computational Neuroscience* 15: 594659.

Tuan, P.M., Phan, T.L., Adel, M. et al. (2021). Autoencoder-based feature ranking for Alzheimer Disease classification using PET image. *Machine Learning with Applications* 6: 100184.

Tzourio-Mazoyer, N., Landeau, B., Papathanassiou, D. et al. (2002). Automated anatomical labeling of activations in SPM using a macroscopic anatomical parcellation of the MNI MRI single-subject brain. *NeuroImage* 15 (1): 273–289. https://doi.org/10.1006/nimg.2001.0978.

Uno, Y. and Coyle, J.T. (2019). Glutamate hypothesis in schizophrenia. *Psychiatry and Clinical Neurosciences* 73 (5): 204–215.

Valle-Pérez, G. and Louis, A.A. (2020). Generalization bounds for deep learning. *arXiv*. https://arxiv.org/abs/2012.04115

Vanegas, M.I., Ghilardi, M.F., Kelly, S.P. et al. (2018). Machine learning for EEG-based biomarkers in Parkinson's disease. In: *2018 IEEE International Conference on Bioinformatics and Biomedicine (BIBM)*, 2661–2665. IEEE.

Vapnik, V., Levin, E., and Le Cun, Y. (1994). Measuring the VC-dimension of a learning machine. *Neural Computation* 6 (5): 851–876.

Vapnik, V.N. (1982). Estimation of dependencies based on empirical data springer. *Information and Control*. ISBN:978-0-387-30865-4.

Vaswani, A., Shazeer, N., Parmar, N. et al. (2017). Attention is all you need. *Advances in Neural Information Processing Systems* 30. https://papers.nips.cc/paper_files/paper/2017/hash/3f5ee243547dee91fbd053c1c4a845aa-Abstract.html

Walton, E., Hibar, D.P., van Erp, T.G.M. et al. (2017). Positive symptoms associate with cortical thinning in the superior temporal gyrus via the enigma schizophrenia consortium. *Acta Psychiatrica Scandinavica* 135 (5): 439–447.

Walton, E., Hibar, D.P., van Erp, T.G.M. et al. (2018). Prefrontal cortical thinning links to negative symptoms in schizophrenia via the enigma consortium. *Psychological Medicine* 48 (1): 82–94.

Wang, R., Bashyam, V., Yang, Z. et al. (2023). Applications of generative adversarial networks in neuroimaging and clinical neuroscience. *NeuroImage* 269: 119898.

Watanabe, S., Ueno, T., Kimura, Y. et al. (2021). Generative image transformer (GIT): unsupervised continuous image generative and transformable model for [123 I] FP-CIT SPECT images. *Annals of Nuclear Medicine* 35: 1203–1213.

Weishaupt, D., Köchli, V.D., Marincek, B. et al. (2006). *How Does MRI Work?: An Introduction to the Physics and Function of Magnetic Resonance Imaging*, volume 2. Springer.

Wu, X., Bi, L., Fulham, M. et al. (2020). Unsupervised positron emission tomography tumor segmentation via gan based adversarial auto-encoder. In: *2020 16th International Conference on Control, Automation, Robotics and Vision (ICARCV)*, 448–453. IEEE.

Yan, J., Cui, Y., Li, Q. et al. (2019). Cortical thinning and flattening in schizophrenia and their unaffected parents. *Neuropsychiatric Disease and Treatment* 15: 935–946.

Zhang, C., Bengio, S., Hardt, M. et al. (2021). Understanding deep learning (still) requires rethinking generalization. *Communications of the ACM* 64 (3): 107–115.

Zhang, Y.D., Dong, Z., Wang, S-H. et al. (2020). Advances in multimodal data fusion in neuroimaging: overview, challenges, and novel orientation. *Information Fusion* 64: 149–187.

Zhou, J., Cui, G., Hu, S. et al. (2020). Graph neural networks: a review of methods and applications. *AI Open* 1: 57–81.

7

Artificial Intelligence in Neurodegenerative Disorders

Timothy Rittman

Department of Clinical Neurosciences, University of Cambridge, Cambridge, UK
Cambridgeshire University Hospitals NHS Foundation Trust, Cambridge, UK

7.1 Introduction

Cognitive and motor neurodegenerative disorders are a significant global challenge. The prevalence of dementias such as Alzheimer's disease is rising rapidly due to lifestyle changes around the world (Wittenberg et al., 2019; Martin Prince et al., 2015), and Parkinson's disease is the fastest-growing neurodegenerative disorder (Ou et al., 2021), possibly driven by environmental factors (Dorsey & Bloem, 2024). The scope and scale of the disease burden caused by neurodegenerative disease is hard to comprehend. In financial terms, the cost to society of caring for people with dementia stood at US$1.3 trillion in 2019 (Wimo et al., 2023). The societal cost of these disorders is huge, not only for the person affected but also for their families, communities and society. However, the standard of diagnosis and the identification of treatable complications remain poor and patchy at best. This is true in developed countries such as the United Kingdom, where the inequality between memory clinics has grown in recent years (Cook et al., 2019; Corrado et al., 2022; Nicholls et al., 2024), and in developing countries, where financial pressures, awareness and attitudes may play a role (Evans-Lacko et al., 2024).

Against this backdrop of diagnostic inequality and financial and societal factors are the emerging therapies for neurodegenerative diseases, most notably Alzheimer's disease. These treatments require an accurate and biologically based diagnosis prior to commencing treatment in those countries where they are licenced (van Dyck et al., 2022). It is therefore imperative to develop new ways of assessing and managing neurodegenerative disorders.

This chapter discusses how big data and artificial intelligence (AI) are beginning to provide tools to address the challenge of neurodegeneration. The focus will be on AI for direct patient benefit, including efforts to improve diagnosis and prognosis and the development of models that support drug discovery. The chapter will also consider the barriers to translating AI tools into clinics for direct benefit to patients.

7.1.1 The Purpose of AI for Neurodegenerative Disorders

Before discussing trends and specific algorithms, it is worth pausing to consider the purpose of AI in this context. Diagnosis is undoubtedly an important part of the assessment of a person with a neurodegenerative disease. Without an accurate diagnosis, it is not possible to recommend specific treatments, and the diagnosis gives people with neurodegenerative disease and their relatives a degree of certainty and clarity over the symptoms that they have been experiencing.

However, a diagnosis on its own is not a treatment. It needs to be coupled with evidence-based medications and management strategies. We can, and should, therefore ask more of digital biomarkers in neurodegenerative disease, going beyond simple diagnostic classification (Landers et al., 2021; Rittman, 2020).

A good biomarker could, for example, predict an individual's likely response to new treatments, heading towards the realm of personalised medicine (Hamburg & Collins, 2010). Even in the absence of specific treatments, a prognostic biomarker can be valuable for patients and clinicians (Fruijtier et al., 2019). Prognosis in neurodegenerative disorders has particular relevance for people planning their lives; for example, making financial decisions such as how much money is needed to support carers or pay for a nursing home, and social decisions including how long relatives might need to put their own lives on hold to become carers. However, health professionals need additional tools and support to improve prognosis in neurodegenerative disorders (Hendriksen et al., 2023). AI tools could therefore provide benefits by assisting in prognosis. Proposals such as the 'Digitized Memory Clinic' have been put forward to implement new technologies, including AI methodologies, to facilitate early diagnosis and ongoing support and monitoring (Gramkow et al., 2024).

From a research perspective, efforts to develop disease-modifying treatments for Alzheimer's disease have shown some early promise (van Dyck et al., 2022; Budd Haeberlein et al., 2022). Despite a statistically significant effect in research studies, these drugs are not yet widely used because of their relatively low efficacy and high burden of side effects. As such, the efforts towards disease-modifying treatments are looking to recruit trial participants earlier, when treatments might be more effective at stopping progression, potentially at a time before symptoms are evident or have caused a loss of day-to-day function. Identifying people who are at high risk for future motor or cognitive dysfunction is not straightforward. Here again, AI may be helpful in identifying high-risk groups prior to the onset of symptoms.

In addition to identifying people appropriate for clinical trials, AI can support drug development. There has been a significant rise in the use of AI in the pharmaceutical industry, with estimates that one-third of new compounds under consideration have been developed using AI (Jayatunga et al., 2022).

To summarise, AI algorithms can be used to support diagnosis in neurodegeneration, but may also support prognosis, selection for clinical trials and development of novel therapies.

7.1.2 Trends in AI for Neurodegeneration

The challenge of synthesising this diffuse and rapidly changing field is reflected in the fact that there are over 200 review articles on the topic of AI in neurodegenerative disorders.

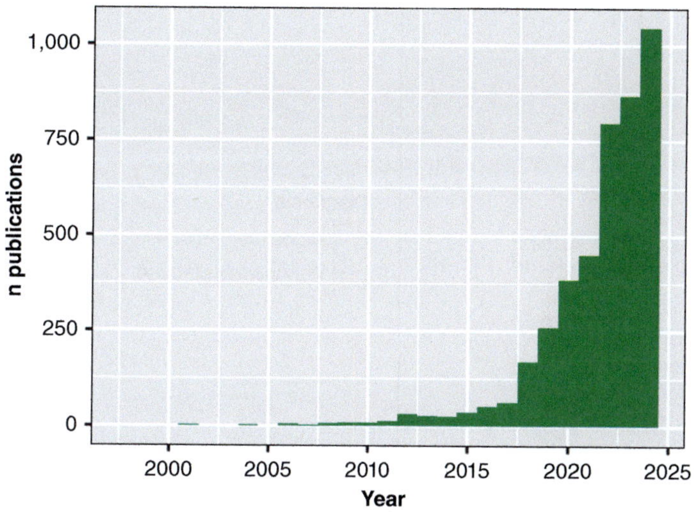

Figure 7.1 An exponential increase in the number of primary publications applying AI to neurodegenerative disease. This plot shows the number of publications between 1997 and December 2024.

There has been an exponential rise in the volume of primary research publications. Between 1997 and 2024, there have been 4,262 such publications, with more than half of those published since 2022 (see Figure 7.1). It is likely that between the time of writing these words and you reading them, there will be thousands more publications.

To make sense of such a rapidly expanding field is challenging. However, there are clear trends within the literature. These trends have developed over time in terms of the type and focus of AI algorithms and the topics that are covered. In this chapter, the trends in diagnosing various neurodegenerative disorders, the use of AI for prognosis and drug development are examined. Given the overwhelming number of papers published in this area, I will confine the references to a few select publications that I hope give a flavor of the field. Inevitably, there will be notable and high-quality publications that I omit, and I encourage you to search for them.

7.1.3 Diagnosis and Datasets

There are two diagnoses that dominate the literature: Alzheimer's disease and Parkinson's disease. Alzheimer's disease is mentioned in 56.8% of the available literature and Parkinson's disease in 29.1% (Figure 7.2).

In the use of AI biomarkers, data has driven innovation. There is no doubt that the availability of the Alzheimer's Disease Neuroimaging Initiative (ADNI) dataset (Jack et al., 2008, 2024a) has led to an explosion of deep learning applications in the field of dementia, and the Parkinson's Progression Markers Initiative (PPMI) has had a similar effect in Parkinson's disease (Marek et al., 2011, 2018).

Given the nature of the ADNI dataset, the majority of these studies are focused on magnetic resonance imaging (MRI) and some on positron emission tomography (PET)

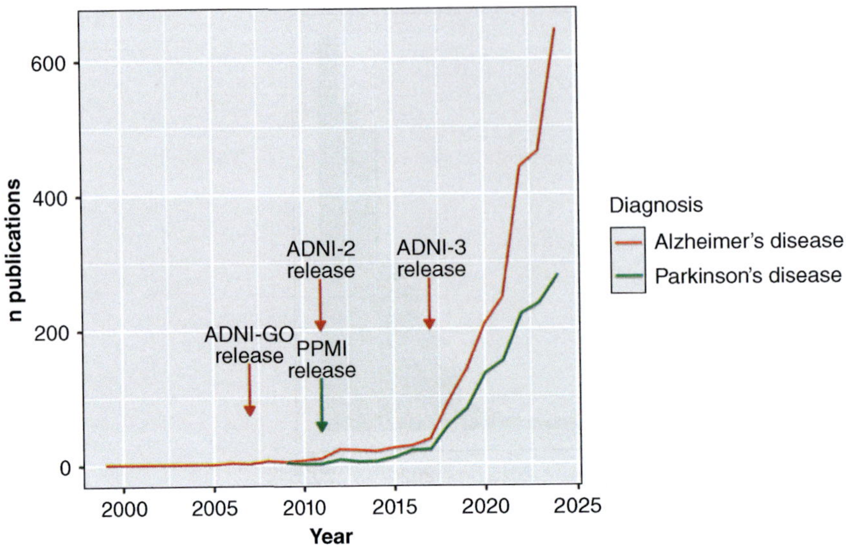

Figure 7.2 The two dominant neurodegenerative disorders in the literature are Alzheimer's disease and Parkinson's disease. The release of large datasets is closely related to large increases in the number of publications for both disorders.

neuroimaging. They are almost exclusively in Alzheimer's disease to the exclusion of people with other types of dementia or people attending memory clinics who have non-dementia memory symptoms (Mendelson et al., 2017). There are efforts to address these limitations. Large datasets of rare neurodegenerative disorders are emerging, a notable example being the Genetic Frontotemporal Dementia Initiative (GenFI) with over 1000 participants with genetic frontotemporal dementia and their relatives (Rohrer et al., 2015). However, there are few datasets for validation once a model has been trained on either ADNI or PPMI datasets. There is also a lack of datasets with multiple diagnoses, which would be required to assess algorithms that attempt multiple diagnoses simultaneously, or in AI terminology, perform 'multi-class classification'.

To overcome the potential biases and lack of heterogeneity in research data, real-world memory clinic datasets are emerging in the United States with the National Alzheimer's Coordinating Center (NACC) (Besser et al., 2018) and in the United Kingdom with the Quantitative MRI in NHS Memory Clinics Study (Lee et al., 2024). The bias towards MRI and Western populations in the dementia datasets is being addressed by the Global Brain Consortium with plans to establish a large EEG dataset, given the wide availability and low cost of EEG (Valdes-Sosa et al., 2022).

Other less common neurodegenerative disorders have started to be explored in more recent years (see Figure 7.3).

Many smaller datasets exist, which can be brought together to provide a more powerful combined dataset for neurodegenerative diseases. There are a number of platforms that have aggregated datasets, including CBRAIN and Dementia Platform UK (DPUK). These platforms provide a 'trusted research environment' where data can be analysed on the cloud, providing a powerful platform and approach to making use of all possible available data for neurodegenerative disorders.

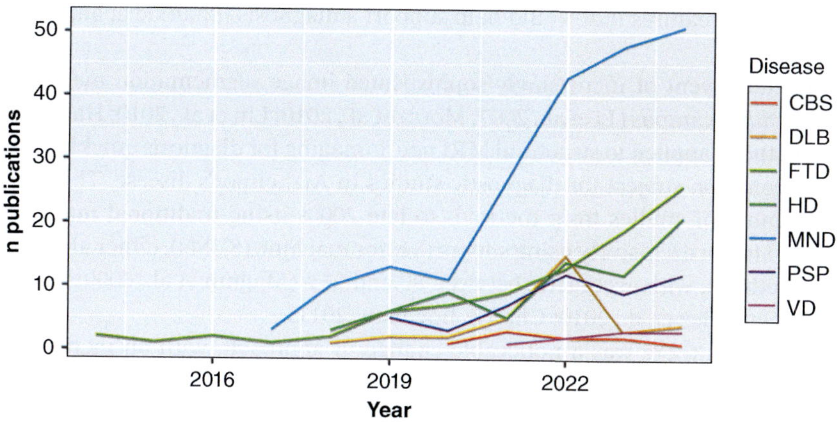

Figure 7.3 The past few years have seen some increase in the use of AI for less common neurodegenerative disorders, mostly notably motor neuron disease, frontotemporal dementia and Huntington's disease. However, the frequency of these publications is an order of magnitude lower than Alzheimer's and Parkinson's disease publications. CBS, corticobasal syndrome; DLB, dementia with Lewy bodies; FTD, frontotemporal dementia; HD, Huntington's disease; MND, motor neuron disease; PSP, progressive supranuclear palsy; VD, vascular dementia.

The UK Biobank is an increasingly valuable resource for studying neurodegenerative disorders. The size of the data alone is staggering, with half a million people having been assessed thoroughly with demographic, functional and cognitive assessments. In addition, 100,000 participants have undergone neuroimaging with MRI (Sudlow et al., 2015; Littlejohns et al., 2020). People recruited to the UK Biobank are between the ages of 45 and 65 years, among whom a number have already gone on to develop dementia. Big data approaches have identified subtle early changes in cognition and function across a range of neurodegenerative disorders, 5–9 years prior to diagnosis (Swaddiwudhipong et al., 2022). This highlights the potential for the UK Biobank and similar ageing studies to be a fertile source of early biomarkers for neurodegenerative disease. The feasibility of this has been demonstrated by one study that developed a deep learning algorithm for identifying Alzheimer's disease in the ADNI dataset, validated it in the NACC dataset and then went on to identify a high-risk group in the UK Biobank (Azevedo et al., 2023). While it was not possible to be certain that the people identified would go on to develop Alzheimer's disease, the high-risk population had the subtle cognitive and functional changes identified in the non-imaging dataset of people who went on to get dementia.

Other similar biobanks are being established, including the 'All of Us' study in the United States (The All of Us Research Program Investigators 2019), the Australian Health Biobank (Light et al., 2021) and the Longitudinal Aging Study in India Diagnostic Assessment of Dementia (LASI-DAD) (Lee et al., 2019).

7.1.4 Alzheimer's Disease

By far the most common application of AI to neurodegenerative disease is in the diagnosis of Alzheimer's disease. The earliest studies examined aspects of cognitive and clinical assessment, identifying features that would help support a diagnosis (Shankle et al., 1998; Mundt et al., 2001).

However, with the advent of increasingly sophisticated image segmentation methods, particularly of the hippocampus (Li et al., 2007; Morra et al., 2010; Liu et al., 2010; Hazarika et al., 2022), AI methods applied to structural MRI neuroimaging for diagnosis quickly took over as the most common subject for diagnostic studies in Alzheimer's disease. This was the focus of a number of studies from the mid- to late 2000s using traditional machine learning methods, almost exclusively using support vector machine (SVMs), either alone or alongside other methods, with accuracies between 80 and 95% (Magnin et al., 2009; Morra et al., 2010; Salas-Gonzalez et al., 2010; Chen & Bowman, 2011).

From the early 2010s, there was a noticeable change in approach, with an increasing range of data types being assessed, including genetic factors (Wei et al., 2011; Khanna et al., 2018; Yang et al., 2020), and a wider array of clinical questions being addressed, such as subtypes of dementia (Ridgway et al., 2012; Alexander et al., 2023). While many studies continued to use structural MRI, combinations of structural MRI sequences with other markers became increasingly common, such as structural and functional neuroimaging (Wee et al., 2012), and combining MRI with FDG-PET, CSF and clinical measures (Zhang et al., 2022; Thung et al., 2014). In addition, the size of the ADNI dataset enabled the training of neural network models for the diagnosis of Alzheimer's disease using structural MRI (Cárdenas-Peña et al., 2017; Jin et al., 2024; Azevedo et al., 2023). These studies all report high accuracy in ADNI data (typically >90%), though Azevedo et al. found a reduced accuracy when the model was validated in the NACC dataset.

Functional MRI methods have also been studied, though to a far smaller extent. These studies have appeared since 2015 (Challis et al., 2015) and have used a mixture of ADNI or local datasets (Hojjati et al., 2019; Zhou et al., 2024; Nguyen et al., 2019), including large datasets from regions in China. These papers use a wide range of approaches in terms of the aims of classification, the features extracted from the functional MRI data, and whether structural MRI was included in the classifier. However, since 2021, deep learning approaches have been applied with classification accuracy between 77 and 94% (Amini et al., 2021; Chelladurai et al., 2023; Sadeghi et al., 2024; Zhou et al., 2024). Although none of these studies were validated in independent datasets, they highlight the potential of functional MRI in Alzheimer's disease diagnosis. Whether functional MRI outperforms structural MRI is not yet clear.

There has been an explosion in EEG studies since 2020. Earlier studies found modest accuracy (~80%) in identifying Alzheimer's disease using relatively simple machine learning models applied to EEG data (Fan et al., 2018; Farina et al., 2020; Miltiadous et al., 2021) or EEG combined with fNIRS (Cicalese et al., 2020). However, there is some evidence that EEG added to structural MRI data can improve classifications in such models (Jesus et al., 2021). In recent years, neural network models have achieved much higher classification accuracy, mainly above 90% (Rodrigues et al., 2021; Alessandrini et al., 2022; Perez-Valero et al., 2022; AlSharabi et al., 2023; Yan et al., 2024). In addition, EEG may be predictive of underlying Alzheimer pathology (Kim et al., 2023; Zhang et al., 2022; Yan et al., 2024).

Other modalities have been applied to Alzheimer's disease such as SPECT (Chaves et al., 2012; López et al., 2009), psychological assessments (Mundt et al., 2001), handwriting (Kang et al., 2024), speech using machine learning (Lee et al., 2024) and a transformer model (Noto et al., 2024). The numbers included in these initial studies were small, and the methods were largely unvalidated or validated in equally small populations.

A significant methodological advance in AI over recent years has been the widespread application of transformer and large language models. Vision transformers in particular are suitable for neuroimaging data and have been applied for the diagnosis of Alzheimer's disease, with one study demonstrating a 4.1–4.7% gain in accuracy to identify Alzheimer's disease using structural MRI over machine learning and convolutional neural network models (Shaffi et al., 2024). Given the advantages of transformer models to combine multiple data types, a number of studies have used multimodal approaches in Alzheimer's disease diagnosis, combining clinical MRI and genetic data (Machado Reyes et al., 2024), functional and structural MRI data (Bi et al., 2024), or MRI and PET data (Gao et al., 2023). All these transformer models achieve a high classification accuracy (>95%) to detect Alzheimer's disease, highlighting the potential of transformers. However, the results require independent validation. One study has used an explainable AI method to provide transparency to an ensemble of vision transformers applied to structural MRI in Alzheimer's disease and mild cognitive impairment (MCI) (Saoud & AlMarzouqi, 2024), although the classification accuracy achieved was more modest than other deep learning or transformer approaches.

Compared to diagnosis, far fewer studies have assessed other potential properties of a biomarker in terms of prognostication or response to treatments. One early notable exception is the use of symptom reporting to identify those at risk of psychological distress in dementia (Zhou et al., 2006). Most studies purporting to assess prognosis in Alzheimer's disease actually focus on conversion from MCI to Alzheimer's disease, or classification of MCI vs Alzheimer's disease. Studies have attempted to identify a population at risk of Alzheimer's disease in healthy cohorts using machine learning (Buegler et al., 2020) or a neural network model (Azevedo et al., 2023).

The literature best focused on prognosis includes studies assessing distinct trajectories using structural MRI (Vogel et al., 2021), functional MRI (Chen et al., 2023), or multimodal MRI, cognitive and biomarker data (Zhang et al., 2024a). These use statistical models rather than AI models specifically, and their application to individual people is not yet tested. However, assessing individual disease trajectories would be helpful to better identify people at risk of rapid decline and to provide a more nuanced interpretation of clinical trials. In this respect, the prediction of decline on a measure widely used as a primary outcome in clinical trials, the Clinical Dementia Rating Scale, has been assessed using deep learning applied to multimodal imaging, clinical and biomarker data (Wang et al., 2024a). A separate study used EEG to predict response to treatment with transcranial direct current stimulation using a random forest algorithm (Andrade et al., 2023), demonstrating the potential for EEG in patient stratification.

Overall, there is a large and rapidly growing literature on AI tools in Alzheimer's disease for diagnosis and prognosis. The application of deep learning models has increased model accuracy, but most of these tools require validation in independent datasets given the risk of overfitting and the lack of diversity within research cohort populations.

7.1.5 Mild Cognitive Impairment

MCI describes a stage of 'pre-dementia' where people have mild cognitive symptoms, usually memory symptoms, but insufficient to cause an impact on daily life. The term was coined by Ronald Petersen in a series of papers outlining a clinical entity of people with memory symptoms, but insufficient to be given a diagnosis of dementia (Petersen et al., 1997; Petersen et al., 2001; Petersen, 2004; Gauthier et al., 2006; Petersen, 2007; Petersen et al., 2009). MCI equivalents have been proposed for Parkinson's disease (Yarnall et al., 2014), mild vascular cognitive impairment (Consoli et al., 2012) and mild behavioural impairment (Ismail et al., 2017; Pan et al., 2021). Identifying people with MCI is potentially of great interest for disease-modifying trials, with the tantalising possibility of catching and treating people with dementia prior to severe symptoms.

Distinguishing MCI from Alzheimer's disease and healthy controls has been a focus for a number of studies using AI (Morra et al., 2008; Salvatore et al., 2015; Ebadi et al., 2017; Lombardi et al., 2020; Wu et al., 2024). However, MCI reflects a continuum between normal healthy ageing and Alzheimer's disease. Therefore, trying to classify and differentiate MCI from Alzheimer's disease does not reflect distinct underlying pathologies, but rather symptomatic stages superimposed on a single underlying disease process. Hence, the focus shifted swiftly to predicting conversion from MCI to Alzheimer's disease (Moradi et al., 2015; Alahmadi et al., 2016; Ardekani et al., 2017).

The ADNI dataset's decision to include people with MCI has driven significant interest in MCI. This includes an international AI challenge to identify MCI from structural neuroimaging alone (Amoroso et al., 2018a). Predicting progression from MCI to Alzheimer's disease has used a wide array of imaging data, including fluorodeoxyglucose PET (Cabral et al., 2015), functional MRI (De Marco et al., 2017), structural MRI (Giorgio et al., 2020), structural and functional MRI (Gullett et al., 2021), MRI and PET (Li et al., 2015), demographic and MRI data (Velazquez et al., 2021), structural MRI combined with Apo-E4 genetic status (Shu et al., 2021) and diffusion tensor MRI (Dyrba et al., 2015; Ebadi et al., 2017; Wu et al., 2024c). One study using structural MRI and cognition found good replication of results in a clinical cohort (Lee et al., 2024) using a Generalized Matrix Learning Vector Quantization model for trajectory modelling.

Despite its popularity, the concept of MCI does not naturally fit with classification tools typical for AI models. The differentiation between MCI and Alzheimer's disease is subjective. The main distinguishing difference is in the level of impairment on day-to-day functions, which may differ between people. A person who has an active life and is engaged in a wide range of activities may have an impairment in their day-to-day ability to function from a relatively minor cognitive disturbance and be labelled with dementia, whereas a sedentary person with limited day-to-day activities may tolerate a much greater level of cognitive impairment without any impact on their daily function and be given a label of MCI.

Secondly, the application of the diagnosis of MCI varies significantly between different clinicians, different clinics and different cultures and backgrounds. From an anecdotal perspective, the diagnosis of MCI in our tertiary referral neurology-led memory clinic in the United Kingdom is relatively uncommon, with only 6% of patients carrying an MCI label, and a large group of patients with attentional and working memory symptoms. Datasets from other parts of the world set in similar tertiary neurology centres typically report a much larger group of patients with MCI, but few (if any) patients with attentional or working memory symptoms. This is likely to reflect local diagnostic emphasis and potentially other factors such as stigma around psychological explanation for memory symptoms. The uncertainty over the definition of MCI is recognised in an attempt to produce consensus criteria that concluded 'MCI should be considered a clinical syndrome with heterogeneous underlying pathologies, and not a diagnosis in its own right' (Dunne et al., 2021).

With the drive to establish biomarker-supported diagnosis of Alzheimer's disease, it is likely that the concept of MCI will be refined or changed. Although the International Working Group (Dubois et al., 2021) and the Alzheimer's Association Workgroup (Jack et al., 2024b) differ in terms of the emphasis and reliance on biomarkers to support a diagnosis, both move towards the use of biological evidence to identify a group of people with evidence of Alzheimer's pathology and at risk of future memory decline. It is therefore likely that future studies applying AI to predict future decline at the earliest stages will move away from the clinical definition of MCI towards application in biomarker-supported asymptomatic or oligosymptomatic groups.

In this context, increasingly complex and multimodal models are being applied to people with MCI, for example by combining multiple modelling approaches into a single framework (Yang et al., 2024) or using complex models such as long short-term memory with multiple data types for future predictions of cognitive decline (Zhang et al., 2024b). A multimodal transformer model has combined speech, language and vision assessment to identify MCI with reasonable performance (Poor et al., 2024). There is also a move towards earlier predictive models of cognitive decline in people without cognitive impairment (Hou et al., 2023; Yao et al., 2024; Woods et al., 2024).

7.1.6 Application of AI to Parkinson's Disease

Similar to Alzheimer's disease, studies in Parkinson's disease have used AI neuroimaging approaches to support diagnosis. Given the relative paucity of structural change in the brain in Parkinson's disease, some MRI studies have used structural MRI with small samples and modest accuracy of ~80–90% (Duchesne et al., 2008; Chakraborty et al., 2020; Fang et al., 2020; Shi et al., 2021; Li et al., 2023; Solana-Lavalle et al., 2024; Camacho et al., 2023). Structural MRI has also been used to predict response to treatment with levodopa, achieving accuracies of 0.7–0.83 depending on the algorithm applied (Yan et al., 2024).

Given these relatively modest results, a number of studies have assessed alternative MRI sequences, including diffusion tensor imaging (DTI) (Correia et al., 2020; Chattopadhyay et al., 2023), functional MRI (Amoroso et al., 2018b; Cieri et al., 2024) and neuromelanin-sensitive image sequences (Shinde et al., 2019; Ben Bashat et al., 2022; Welton et al., 2024). Of these, the neuromelanin-specific sequences arguably best reflect the disease process by detecting the effect of neuropathological changes in the substantia nigra itself.

Non-MRI neuroimaging approaches include applying AI to DaTSCANs with accuracies between 75 and 97.5% (Tang et al., 2019; Magesh et al., 2020; Zhou & Tagare, 2021; Khachnaoui et al., 2022). EEG has also been used extensively for diagnosis (Chaturvedi et al., 2017; Anjum et al., 2020; Lee et al., 2021; Aljalal et al., 2022; Göker, 2023; Obayya et al., 2023; Zhao et al., 2024) and for identifying complications such as cognitive impairment in deep brain stimulation (Geraedts et al., 2021) and depression (Espinoza et al., 2022).

Non-imaging methods have also been widely used in Parkinson's disease. Prior to the ubiquity of smartphones and smartwatches, customised sensors were applied to monitor motor symptoms in Parkinson's disease, all reporting a high level of concordance with clinical measures of movement, tremor, or gait (Patel et al., 2009; Rigas et al., 2012; Thomas et al., 2018; Haji Ghassemi et al., 2019). As the types of data collected from these sensors expanded, more complex statistical and machine learning approaches were applied – for example, to use wearable sensor data to predict the presence of freezing with ~85% accuracy using machine learning approaches (support vector machines, K-nearest neighbors) (Borzì et al., 2021; Mesin et al., 2022), or using metrics of gait and arm movement in a convolutional neural network to assess treatment response to levodopa (Ileșan et al., 2022). Various body-worn sensor approaches with machine learning or AI approaches have been used to differentiate between Parkinson's disease and essential tremor (Lin et al., 2023) as alternatives to standard clinical measures (Cai et al., 2023; Bremm et al., 2024), identifying time in the 'on' or 'off' state (Wu et al., 2024b), to monitor tremor (Roriguez et al., 2024) and for daily monitoring of freezing (Zampogna et al., 2024). Despite these advances, the application of sensors in the home setting rather than the laboratory setting may be less accurate, with one study finding 12–20% reduction in correlation with clinical measures at home (Crowe et al., 2024), and there can be significant variability in measurements affecting predictions (Khalil et al., 2024). However, conceptually, the long-term and objective monitoring of motor symptoms with wearable data and AI has significant advantages over the subjective report of patients and carers, or the brief snapshot of a clinical assessment.

Speech analysis is an area where AI has been applied to detect changes in Parkinson's disease. Features of speech can be extracted and used to train machine learning classifiers, for example, to monitor disease severity (Bayestehtashk et al., 2015; Tunc et al., 2020) or for diagnosis (El Moudden et al., 2017; Vaiciukynas et al., 2017; Dastjerd et al., 2019; Campi et al., 2023; Warule et al., 2023). One study has shown that speech analysis with machine learning is potentially robust across different languages, being trained in German and Spanish, and then applied in Czech (Rios-Urrego et al., 2024). In line with trends in neuroimaging, more recent studies have used more complex convolutional neural networks or transformer architectures, which have the advantage of not requiring the extraction of specific features (Faragó et al., 2023; Escobar-Grisales et al., 2023; Eguchi et al., 2024). In one such transformer model, the investigators applied explainable AI architectures to identify that their model focused on the muffled frequencies of Parkinson's disease (Jeong et al., 2024), demonstrating that these complex models are not the 'black boxes' they used to be. These recent studies have focused solely on disease classification, and while they have generally performed better than simpler machine learning models, they have not yet been validated in independent datasets or applied in clinical settings.

7.1.7 Other Neurodegenerative Disorders

A range of other neurodegenerative disorders have been studied, although on a smaller scale given the lack of large datasets and lower disease prevalence (see Figure 7.3).

The most studied of these rarer neurodegenerative conditions is motor neuron disease (mainly the commonest amyotrophic lateral sclerosis variant), despite being relatively uncommon, with a prevalence of around 5.7/100,000 (Burchardt et al., 2022). Unlike other disorders, the focus of AI for clinical use has been on using biomarkers for survival prediction and evaluating genetic risk. The earliest application of artificial neural networks to neurodegeneration identified novel genetic risks in MND (Penco et al., 2008). This study only included 54 people with amyotrophic lateral sclerosis and 208 controls, and only one of the genes implicated (hepatic lipase) was subsequently identified in genome-wide association studies, so the results should be treated with caution.

Neuroimaging in motor neuron disease using DTI, changes in brain volume to predict survival using a random forest algorithm (Thome et al., 2022), or a neural network approach (van der Burgh et al., 2017), achieved modest accuracies of 60–65%. A higher accuracy of 80% was achieved by applying an SVM approach to fluorodeoxyglucose PET (Canosa et al., 2023). A proof-of-concept study achieved high classification of motor neuron disease against controls using multiparametric mapping (Kocar et al., 2021). More realistically, for diagnosis, neurophysiological measures (Quintãom et al., 2021) found up to 80% accuracy (though without sensitivity or specificity reported), and muscle ultrasound demonstrated a high specificity of 98% in early amyotrophic lateral sclerosis but lower sensitivity of 78% (Fukushima et al., 2022). Taken together, these suggest that, at present, imaging and neurophysiological changes are mostly insufficient for high diagnostic confidence in motor neuron disease diagnosis or survival prediction.

Various combinations of blood-based biomarkers have been used to predict survival. One of the first approaches used a hackathon/challenge approach to identify distinct trajectories of progression in amyotrophic lateral sclerosis (Kueffner et al., 2019). A metabolomic method showed poor performance in predicting progression (Bjornevik et al., 2019). In contrast, using a battery of clinical measures can assign people with good accuracy to groups predicting speed of decline, using Uniform Manifold Approximation and Projection for Dimension Reduction (Grollemund et al., 2020) or dynamic Bayesian networks (Gromicho et al., 2022). Both of these models allow for visualisation and interpretability and were trained on relatively large longitudinal datasets ($n = 664$ and $n = 4{,}382$). This is a good example of data permitting complex statistical/machine learning models.

Given that there is a definitive genetic test for Huntington's disease, studies have not focused on diagnostic classification, but on developing models of future decline. These have used MRI neuroimaging (Rizk-Jackson et al., 2011; Mason et al., 2018; Polosecki et al., 2020; Ghofrani-Jahromi et al., 2024), blood biomarkers (Vishnuppriya & Tamilpavai, 2024), or a combination of measures (Ko et al., 2022; Ghazaleh et al., 2021; Raschka et al., 2024). Similar to Parkinson's disease, AI has been applied to assess disease severity with wearable sensors with high accuracy (Scheid et al., 2022) and speech with more modest results (Riad et al., 2022; Nunes et al., 2024).

There are increasing AI studies of frontotemporal dementia. This is due to the availability of data through the Neuroimaging in Frontotemporal Dementia (NIFD), Genetic Frontotemporal Dementia Initiative (GeNFI) datasets and more recently the ARTFL-LEFFTDS Longitudinal Frontotemporal Lobar Degeneration (ALLFTD) datasets. The GeNFI and ALLFTD studies have moved towards harmonising their approaches, which is reflected in the recent significant increase in studies (Figure 7.3). More neuroimaging studies have been carried out in frontotemporal dementia compared to motor neuron disease or Huntington's disease. Given that frontotemporal dementia is often accompanied by characteristic structural brain changes of atrophy in the frontal or temporal lobes, a number of studies report high accuracy using structural/volumetric MRI (Kim et al., 2019; Termine et al., 2022; Pérez-Millan et al., 2023) or fluorodeoxyglucose PET (Ford et al., 2021). Investigating heterogeneity in progression has been investigated using structural MRI (Olm et al., 2023), fluorodeoxyglucose PET in genetic disease (Corriveau-Lecavalier et al., 2024) and functional MRI (Sadeghi et al., 2024). These studies have the potential to shed light on disease heterogeneity, in addition to possibly providing individual prognostic information.

Fewer publications have studied the relatively common neurodegenerative disorders of vascular dementia and dementia with Lewy bodies. I have been involved in a recent review of AI and neuroimaging for features of cerebrovascular disease (Lohner et al., 2024). To my surprise, almost two-thirds of the eligible studies ($n = 27$) focused on Alzheimer's disease and only four on vascular dementia. The earlier papers on vascular dementia focused on applying machine learning to neuroimaging to distinguish Alzheimer's disease from vascular dementia with relatively modest success (Zheng et al., 2019; Castellazzi et al., 2020), with more recent efforts using near-infrared spectroscopy (Huang et al., 2024) or SPECT scans (Ni et al., 2024). One study used multimodal MRI and PET imaging markers to identify patterns in Alzheimer's disease, dementia with Lewy bodies and vascular dementia (Wang et al., 2024b). The modest differentiation from Alzheimer's disease is not surprising, given the common co-occurrence of vascular change in Alzheimer's disease and shared risk factors (Polidori et al., 2012; Livingston et al., 2024).

A handful of studies have examined dementia with Lewy bodies, again focusing on differentiating it from Alzheimer's disease using MRI neuroimaging (Wada et al., 2019), EEG (Dauwan et al., 2016; San-Martin et al., 2021), metabolic profiles (Iizuka & Kameyama, 2020), fluorodeoxyglucose PET (Etminani et al., 2022) and speech (Yamada et al., 2022; Kobayashi et al., 2024). The relative paucity of studies reflects the lack of openly available large datasets for vascular dementia and dementia with Lewy bodies. There is an effort to harmonise data between a large range of studies in the European DLB Consortium (https://www.e-dlb.com/), though at present the data is only available to investigators within the consortium.

Less common neurodegenerative disorders have also been the subject of AI approaches, including the atypical parkinsonian disorders of progressive supranuclear palsy, corticobasal syndrome and multiple system atrophy. Of these, progressive supranuclear palsy is the most studied, in particular trying to distinguish it from Parkinson's disease using MRI neuroimaging (Salvatore et al., 2014; Talai et al., 2021; Albrecht et al., 2024), wearables (De Vos et al., 2020), speech (Kang et al., 2023) and blood biomarkers (Bianco et al., 2024). Some studies have looked to distinguish between multiple disorders such as Parkinson's disease, progressive supranuclear palsy and corticobasal syndrome (Correia et al., 2020; Koga et al., 2021b). Given the relative clinical severity of progressive supranuclear palsy and the specific imaging change of midbrain atrophy, these papers have achieved high classification (>90%) using simpler machine learning models. In addition, a number of studies have examined longitudinal trajectories in progressive supranuclear palsy using MRI neuroimaging (Scotton et al., 2022; Scotton et al., 2023) or PET (Hong et al., 2023). One study examined patterns of progressive brain atrophy in both progressive supranuclear palsy and corticobasal syndrome (Saito et al., 2022). Simple machine learning approaches have also been applied to neuropathology in progressive supranuclear palsy, either to quantify neuropathological changes of tau protein accumulation (Pansuwan et al., 2023) or to distinguish progressive supranuclear palsy from corticobasal degeneration (Koga et al., 2021a).

Far fewer studies have focused on multiple system atrophy, but those that have attempted to distinguish it from Parkinson's disease using MRI neuroimaging (Abos et al., 2019, Campabadal et al., 2022; Pang et al., 2022; Chen et al., 2023) or PET (Sun et al., 2024). As with progressive supranuclear palsy, these achieved high accuracy rates with simple machine learning classifiers. Recent efforts in multiple system atrophy have applied AI to metabolomic (Kwon et al., 2022) or neuropathological data (Kim et al., 2019) to understand underlying disease mechanisms.

There are increasing efforts to build large datasets of atypical parkinsonian disorders, such as the Rare Disease Research UK Parkinson Plus Syndrome ExPRESS study (https://rd-research.org.uk/node/express/) in the United Kingdom and the Four Repeat Tauopathy Neuroimaging Initiative in the United States. Given that a number of studies already report high accuracy for MSA and progressive supranuclear palsy, it is likely that these datasets will be useful to understand disease mechanisms and investigate multimodal classification. Corticobasal syndrome is much less studied because the underlying neuropathology is much less certain and may be corticobasal degeneration, progressive supranuclear palsy, or Alzheimer's disease. It is possible that larger datasets will enable the prediction of underlying pathology in corticobasal syndrome.

The accuracy and scope of AI applied to less common or less studied neurodegenerative disorders are likely to grow in line with the more advanced literature on Alzheimer's disease and Parkinson's disease, with increasingly sophisticated and multimodal algorithms and prognostic algorithms. This will need to be supported by larger, open-access datasets such as those emerging for frontotemporal dementia and the atypical parkinsonian disorders.

7.1.8 Barriers to Clinical Use

Few studies have used independent data to validate their findings, and even fewer have validated their AI findings in real-world memory clinic studies (Klöppel et al., 2015; Borchert et al., 2023; Lee et al., 2024). This is a critical first step towards the translation of AI algorithms into routine clinical practice.

In addition to technical barriers and considerations of accuracy, there are a number of logistical and regulatory barriers that must be overcome for AI to be genuinely useful for neurodegenerative disorders. This is a fast-moving field, but to date I am only aware of one AI algorithm that has successfully received FDA approval for use in the diagnosis of dementia (https://brainsee.ai/). Unfortunately, no details are available about the algorithm, so it is hard to make a decision on its accuracy, validity and overall applicability to clinical populations.

Overcoming these logistical and practical algorithms requires a community of clinicians, managers, support staff such as radiographers, computer scientists, IT technicians, nurses and administrative staff. For this effort and infrastructure to be justified, we need good evidence that AI algorithms make a difference in practical, day-to-day matters such as confidence in diagnosis, quality of life and reducing hospital admissions. To do this requires medical device trials, and to my knowledge there are very few such trials ongoing for neurodegenerative disease.

7.1.9 Drug Development

AI has been applied to drug development in neurodegenerative disorders in a number of approaches and across different conditions.

Some of the earliest, and more recent, approaches used models to repurpose existing medications. While a few papers have used simpler machine learning algorithms for Alzheimer's disease (Rodriguez et al., 2021; Gupta et al., 2022), these studies have largely used network models combined with neural network algorithms for vascular dementia (Nam et al., 2019), Parkinson's disease (Haneczok et al., 2023) and Alzheimer's disease (Fang et al., 2022; Wu et al., 2022, 2024a; Hsieh et al., 2023). More recently, a large language model (ChatGPT-4) has been used to identify potential drug targets (Yan et al., 2023).

An extension of the repurposing approach is using machine learning or deep learning approaches to screen potential compounds for targets in neurodegenerative disease. This has included targets such as tau aggregation (Segura et al., 2024), beta-amyloid oligomers (Lee et al., 2021), cholinesterase inhibitors (Baidya et al., 2024) and combinations of targets (Wu et al., 2023) in Alzheimer's disease; parkin (Scott et al., 2020) and other targets (Sohrabi et al., 2021) in Parkinson's disease; and Huntington's disease (Metzger et al., 2022). Some approaches have looked at potential targets across multiple neurodegenerative disorders, such as TDP-43 in Alzheimer's disease and motor neuron disease (König et al., 2024).

An established approach to identifying potential compounds is to use Quantitative Structure–Activity Relationship (QSAR) analysis, which predicts the behaviour of compounds and how they might interact with other molecules. There are variations of the models used for QSAR analysis, but machine learning versions have been applied to identify acetylcholinesterase (AChE) and butyrylcholinesterase (BuChE) inhibitors (Shamsi et al., 2021; Lv et al., 2010) and beta-secretase inhibitors (Ponzoni et al., 2019) in Alzheimer's disease; and alpha-synuclein inhibitor and LRRK2 targets in Parkinson's disease. Combining artificial neural networks with QSAR models has been applied to identify 5-HT6 antagonists in Alzheimer's disease (de Sousa et al., 2024).

Specific targets can be modelled using deep learning to predict molecular docking. This has been applied to identify MAO-B inhibitors in Parkinson's disease (Kashyap et al., 2024) and inhibitors of beta-amyloid oligomerisation in Alzheimer's disease (Segura et al., 2024).

An end-to-end in silico pipeline has been proposed for Alzheimer's disease target identification, compound selection and simulating molecular docking (Alshabrmi et al., 2024).

Of note, the vast majority of published literature on drug development is from 1–2 years of writing this chapter. This is a rapidly moving field, and it is highly likely that it will continue to change rapidly in the coming years. With increasing confidence and accuracy, and relative cost-efficiency compared to cellular or animal models, it is likely that an increasing number of potential new treatments will be developed in silico.

7.1.10 Other Applications of AI

There are a few other noteworthy applications of AI in the field of dementia. One promising approach is home monitoring for diagnosis and disease progression (Vizitiu et al., 2024) or specific symptoms such as sleep (Rigny et al., 2024). AI has been applied to retinal neuroimaging as an early biomarker of dementia in neurodegeneration, though some work remains to achieve high levels of classification (Costanzo et al., 2023). Synthetic neuroimaging data might overcome privacy and security concerns and enable analysis of otherwise sparse and imbalanced datasets (Sánchez Reyna et al., 2024). Finally, AI has been used to demonstrate the value of social media in facilitating public discussion on dementia (Domingo-Espiñeira et al., 2024).

7.1.11 Conclusions and the Future

AI has been widely applied to neurodegenerative disorders, and it has been most commonly used in the diagnosis of both Alzheimer's disease and Parkinson's disease. The last few years have seen an explosion in studies and a broadening of the scope of application of AI. In particular, there has been a much greater application to more clinically relevant questions such as prognosis, and a wider range of non-Alzheimer neurodegenerative conditions have been studied. The promise of AI-driven drug repurposing and development is an exciting one and is starting to yield results. However, it will be important for these methods to demonstrate that they are able to not only validate existing drug targets but also develop truly novel targets not previously identified.

Increasing use of deep learning has demonstrated improved accuracy, but it will be important to validate these findings in independent datasets to guard against overfitting models to their training data. On the near horizon are foundational AI models, which are multipurpose AI models able to be applied to a wide range of contexts, including the diagnosis of various diseases (Zhou et al., 2023; Moor et al., 2023).

The main challenges for the field of AI neurodegeneration are to address key clinical questions and to overcome the barriers that currently exist to implement meaningful application of AI machine learning in clinical practice. Overcoming these barriers has the potential to unleash an exciting new range of tools for the benefit of patients, both in terms of research and clinical utility.

A note on how AI was used to write this chapter: after a literature review identified a large number of papers, with manual screening of titles/abstracts to identify ~4,000 publications, the references were shared with ChatGPT-4 in a BibTex format to help identify themes within the literature. ChatGPT-4 was also asked to identify search terms that would help identify specific themes. However, reading, interpreting and summarising the articles were done by a human, and AI had no role in writing the chapter text.

References

Abos, A., Baggio, H.C., Segura, B. et al. (2019). Differentiation of multiple system atrophy from Parkinson's disease by structural connectivity derived from probabilistic tractography. *Scientific Reports* 9 (1): 16488.

Alahmadi, H.H., Shen, Y., Fouad, S. et al. (2016). Classifying cognitive profiles using machine learning with privileged information in mild cognitive impairment. *Frontiers in Computational Neuroscience* 10: 117.

Albrecht, F., Mueller, K., Ballarini, T. et al. (2024). Structural parameters are superior to eigenvector centrality in detecting progressive supranuclear palsy with machine learning & multimodal MRI. *Heliyon* 10 (15): e34910.

Alessandrini, M., Biagetti, G., Crippa, P. et al. (2022). EEG-based Alzheimer's disease recognition using robust-PCA and LSTM recurrent neural network. *Sensors (Basel)* 22 (10): 3696.

Alexander, T.D., Nataraj, C., and Wu, C. (2023). A machine learning approach to predict quality of life changes in patients with Parkinson's disease. *Annals of Clinical and Translational Neurology* 10 (3): 312–320.

Alshabrmi, F.M., Aba Alkhayl, F.F., Rehman, A. (2024). Novel drug discovery: advancing Alzheimer's therapy through machine learning and network pharmacology. *European Journal of Pharmacology* 976: 176661.

Amini, M., Pedram, M., Moradi, A. et al. (2021). Diagnosis of Alzheimer's disease severity with fMRI Images using robust multitask feature extraction method and convolutional neural network (CNN). *Computational and Mathematical Methods in Medicine* 2021: 5514839.

Amoroso, N., Diacono, D., Fanizzi, A. et al. (2018a). Deep learning reveals Alzheimer's disease onset in MCI subjects: results from an international challenge. *Journal of Neuroscience Methods* 302: 3–9.

Amoroso, N., La Rocca, M., Monaco, A. et al. (2018b). Complex networks reveal early MRI markers of Parkinson's disease. *Medical Image Analysis* 48: 12–24.

Andrade, S.M., da Silva-Sauer, L., de Carvalho, C.D. et al. (2023). Identifying biomarkers for tDCS treatment response in Alzheimer's disease patients: a machine learning approach using resting-state EEG classification. *Frontiers in Human Neuroscience* 17: 1234168.

Anjum, M.F., Dasgupta, S., Mudumbai, R. et al. (2020). Linear predictive coding distinguishes spectral EEG features of Parkinson's disease. *Parkinsonism & Related Disorders* 79: 79–85.

Ardekani, B.A., Bermudez, E., Mubeen, A.M. et al. (2017). Prediction of incipient Alzheimer's disease dementia in patients with mild cognitive impairment. *Journal of Alzheimer's Disease* 55 (1): 269–281.

Azevedo, T., Bethlehem, R.A.I., Whiteside, D.J. et al. (2023). Identifying healthy individuals with Alzheimer's disease neuroimaging phenotypes in the UK Biobank. *Communications Medicine (London)* 3 (1): 100.

Baidya, A.T.K., Goswami, A.K., Das, B. et al. (2024). AI-enabled ultra-large virtual screening identifies potential inhibitors of choline acetyltransferase for theranostic purposes. *ACS Chemical Neuroscience* 15 (22): 4156–4170.

Bayestehtashk, A., Asgari, M., Shafran, I. et al. (2015). Fully automated assessment of the severity of Parkinson's disease from speech. *Computer Speech & Language* 29 (1): 172–185.

Ben Bashat, D., Thaler, A., Lerman Shacham, H. et al. (2022). Neuromelanin and T(2)*-MRI for the assessment of genetically at-risk, prodromal, and symptomatic Parkinson's disease. *NPJ Parkinsons Disease* 8 (1): 139.

Besser, L., Kukull, W., Knopman, D.S. et al. (2018). Version 3 of the national Alzheimer's coordinating center's uniform data set. *Alzheimer Disease & Associated Disorders* 32 (4): 351.

Bi, Y., Abrol, A., Fu, Z. et al. (2024). A multimodal vision transformer for interpretable fusion of functional and structural neuroimaging data. *Human Brain Mapping* 45(17): e26783.

Bianco, M.G., Cristiani, C.M., Scaramuzzino, L. et al. (2024). Combined blood neurofilament light chain and third ventricle width to differentiate progressive supranuclear palsy from Parkinson's disease: a machine learning study. *Parkinsonism Related Disorders* 123: 106978.

Bjornevik, K., Zhang, Z., O'Reilly, É.J. et al. (2019). Prediagnostic plasma metabolomics and the risk of amyotrophic lateral sclerosis. *Neurology* 92 (18): e2089–e2100.

Borchert, R.J., Azevedo, T., Badhwar, A. et al. (2023). Artificial intelligence for diagnostic and prognostic neuroimaging in dementia: a systematic review. *Alzheimer's & Dementia* 19 (12): 1–20.

Borzì, L., Mazzetta, I., Zampogna, A. et al. (2021). Prediction of freezing of gait in Parkinson's disease using wearables and machine learning. *Sensors (Basel)* 21 (2): 614.

Bremm, R.P., Pavelka, L., Garcia, M.M. et al. (2024). Sensor-based quantification of MDS-UPDRS III subitems in Parkinson's disease using machine learning. *Sensors (Basel)* 24 (7): 2195.

Budd Haeberlein, S., Aisen, P.S., Barkhof, F. et al. (2022). Two randomized phase 3 studies of aducanumab in early Alzheimer's disease. *The Journal of Prevention of Alzheimer s Disease* 9(2): 197–210. https://doi.org/10.14283/jpad.2022.30.

Buegler, M., Harms, R., Balasa, M. et al. (2020). Digital biomarker-based individualized prognosis for people at risk of dementia. *Alzheimer's & Dementia: Diagnosis, Assessment & Disease Monitoring* 12 (1): e12073.

Burchardt, J.M., Mei, X.W., Ranger, T. et al. (2022). Analysis of incidence of motor neuron disease in England 1998–2019: use of three linked datasets. *Amyotrophic Lateral Sclerosis & Frontotemporal Degeneration* 23 (5–6): 363.

Cabral, C., Morgado, P.M., Campos Costa, D. et al. (2015). Predicting conversion from MCI to AD with FDG-PET brain images at different prodromal stages. *Computers in Biology and Medicine* 58: 101–109.

Cai, G., Shi, W., Wang, Y. et al. (2023). Specific Distribution of digital gait biomarkers in Parkinson's disease using body-worn sensors and machine learning. *The Journals of Gerontology Series A Biological Sciences and Medical Sciences* 78 (8): 1348–1354.

Camacho, M., Wilms, M., Mouches, P. et al. (2023). Explainable classification of Parkinson's disease using deep learning trained on a large multi-center database of T1-weighted MRI datasets. *Neuroimage Clinical* 38: 103405.

Campabadal, A., Abos, A., Segura, B. et al. (2022). Differentiation of multiple system atrophy subtypes by gray matter atrophy. *Journal of Neuroimaging* 32 (1): 80–89.

Campi, M., Peters, G.W., and Toczydlowska, D. (2023). Ataxic speech disorders and Parkinson's disease diagnostics via stochastic embedding of empirical mode decomposition. *PLoS One* 18 (4): e0284667.

Cárdenas-Peña, D., Collazos-Huertas, D., and Castellanos-Dominguez, G. (2017). Enhanced data representation by kernel metric learning for dementia diagnosis. *Frontiers in Neuroscience* 11: 413.

Castellazzi, G., Cuzzoni, M.G., Cotta Ramusino, M. et al. (2020). A machine learning approach for the differential diagnosis of Alzheimer and vascular dementia fed by MRI selected features. *Frontiers in Neuroinformatics* 14: 25.

Chakraborty, S., Aich, S., and Kim, H.C. (2020). 3D textural, morphological and statistical analysis of voxel of interests in 3T MRI scans for the detection of Parkinson's disease using artificial neural networks. *Healthcare (Basel)* 8 (1): 34.

Challis, E., Hurley, P., Serra, L. et al. (2015). Gaussian process classification of Alzheimer's disease and mild cognitive impairment from resting-state fMRI. *Neuroimage* 112: 232–243.

Chattopadhyay, T., Singh, A., Laltoo, E. et al. (2023). *Comparison of anatomical and diffusion MRI for detecting Parkinson's disease using deep convolutional neural network*. United States.

Chaturvedi, M., Hatz, F., Gschwandtner, U. et al. (2017). Quantitative EEG (QEEG) measures differentiate Parkinson's disease (PD) patients from healthy controls (HC). *Frontiers in Aging Neuroscience* 9: 3.

Chaves, R., Ramírez, J., Górriz, J.M. et al. (2012). Effective diagnosis of Alzheimer's disease by means of large margin-based methodology. *BMC Medical Informatics and Decision Making* 12: 79.

Chelladurai, A., Narayan, D.L., Divakarachari, P.B. et al. (2023). fMRI-Based Alzheimer's disease detection using the SAS method with multi-layer perceptron network. *Brain Sciences* 13 (6): 893.

Chen, P., Yao, H., Tijms, B.M. et al. (2023). Four distinct subtypes of Alzheimer's disease based on resting-state connectivity biomarkers. *Biological Psychiatry* 93 (9): 759–769.

Chen, S. and Bowman, F.D. (2011). A novel support vector classifier for longitudinal high-dimensional data and its application to neuroimaging data. *Statistical Analysis and Data Mining* 4 (6): 604–611.

Cicalese, P.A., Li, R., Ahmadi, M.B. et al. (2020). An EEG-fNIRS hybridization technique in the four-class classification of Alzheimer's disease. *Journal of Neuroscience Methods* 336: 108618.

Cieri, F., Giriprakash, P.P., Nandy, R. et al. (2024). Functional connectivity differences of the olfactory network in Parkinson's disease, mild cognitive impairment and cognitively normal individuals: a resting-state fMRI study. *Neuroscience* 559: 8–16.

Consoli, A., Pasi, M., and Pantoni, L. (2012). Vascular mild cognitive impairment: concept, definition, and directions for future studies. *Aging Clinical and Experimental Research* 24 (2): 113–116.

Cook, L., Souris, H., and Isaacs, J. (2019). *The 2019 National Memory Service Audit, 23.* Dementia Clinical Network, NHS England.

Corrado, O., Essel, R., Fitch-Bunce, C. et al. (2022). *National Audit of Dementia Memory Assessment Services Spotlight Audit* 2021. London: Royal College of Psychiatrists.

Correia, M.M., Rittman, T., Barnes, C.L. et al. (2020). Towards accurate and unbiased imaging-based differentiation of Parkinson's disease, progressive supranuclear palsy and corticobasal syndrome. *Brain Communications* 2 (1): fcaa051.

Corriveau-Lecavalier, N., Barnard, L.R., Przybelski, S.A. et al. (2024). Assessing network degeneration and phenotypic heterogeneity in genetic frontotemporal lobar degeneration by decoding FDG-PET. *Neuroimage Clinical* 41: 103559.

Costanzo, E., Lengyel, I., Parravano, M. et al. (2023). Ocular biomarkers for Alzheimer disease dementia: an umbrella review of systematic reviews and meta-analyses. *Jama Ophthalmology* 141 (1): 84–91.

Crowe, C., Sica, M., Kenny, L. et al. (2024). Wearable-enabled algorithms for the estimation of Parkinson's symptoms evaluated in a continuous home monitoring setting using inertial sensors. *IEEE Transactions on Neural Systems and Rehabilitation Engineering*. 9.

Dastjerd, N.K., Sert, O.C., Ozyer, T. et al. (2019). Fuzzy classification methods based diagnosis of Parkinson's disease from speech test cases. *Current Aging Science* 12 (2): 100–120.

Dauwan, M., van der Zande, J.J., van Dellen, E. et al. (2016). Random forest to differentiate dementia with Lewy bodies from Alzheimer's disease. *Alzheimer's & Dementia: Diagnosis, Assessment & Disease Monitoring* 4: 99–106.

De Marco, M., Beltrachini, L., Biancardi, A. et al. (2017). Machine-learning support to individual diagnosis of mild cognitive impairment using multimodal MRI and cognitive assessments. *Alzheimer Disease & Associated Disorders* 31 (4): 278–86.

de Sousa, D.S., da Silva, A.P., Chiari, L.P.A. et al. (2024). Predicting biological activity and design of 5-HT(6) antagonists through assessment of ANN-QSAR models in the context of Alzheimer's disease. *Journal of Molecular Modeling* 30 (10): 350.

De Vos, M., Prince, J., Buchanan, T. et al. (2020). Discriminating progressive supranuclear palsy from Parkinson's disease using wearable technology and machine learning. *Gait Posture* 77: 257–263.

Domingo-Espiñeira, J., Fraile-Martínez, Ó., García Montero, C. et al. (2024). Analyzing public discourse of dementia from Spanish and English tweets: a comparative analysis with other neurological disorders. *Frontiers in Neurology* 15: 1459578.

Dorsey, E.R. and Bloem, B.R. (2024). Parkinson's disease is predominantly an environmental disease. *Journal of Parkinson's Disease* 14 (3): 451–465.

Dubois, B., Villain, N., Frisoni, G.B. et al. (2021). Clinical diagnosis of Alzheimer's disease: recommendations of the international working group. *The Lancet Neurology* 20 (6): 484–496.

Duchesne, S., Caroli, A., Geroldi, C. et al. (2008). MRI-based automated computer classification of probable AD versus normal controls. *IEEE Transactions on Medical Imaging* 27 (4): 509–520.

Dunne, R.A., Aarsland, D., O'Brien, J.T. et al. (2021). Mild cognitive impairment: the Manchester consensus. *Age and Ageing* 50 (1): 72–80.

Dyrba, M., Barkhof, F., Fellgiebel, A. et al. (2015). Predicting prodromal Alzheimer's disease in subjects with mild cognitive impairment using machine learning classification of multimodal multicenter diffusion-tensor and magnetic resonance imaging data. *Journal of Neuroimaging* 25 (5): 738–747.

Ebadi, A., Dalboni da Rocha, J.L., Nagaraju, D.B. et al. (2017). Ensemble classification of Alzheimer's disease and mild cognitive impairment based on complex graph measures from diffusion tensor images. *Frontiers in Neuroscience* 11: 56.

Eguchi, K., Yaguchi, H., Kudo, I. et al. (2024). Differentiation of speech in Parkinson's disease and spinocerebellar degeneration using deep neural networks. *Journal of Neurology* 271 (2): 1004–1012.

El Moudden, I., Ouzir, M., and ElBernoussi, S. (2017). Feature selection and extraction for class prediction in dysphonia measures analysis: a case study on Parkinson's disease speech rehabilitation. *Technology and Health Care* 25 (4): 693–708.

Escobar-Grisales, D., Ríos-Urrego, C.D., and Orozco-Arroyave, J.R. (2023). Deep learning and artificial intelligence applied to model speech and language in Parkinson's disease. *Diagnostics (Basel)* 13 (13): 2163.

Espinoza, A.I., May, P., Anjum, M.F. et al. (2022). A pilot study of machine learning of resting-state EEG and depression in Parkinson's disease. *Clinical Parkinsonism & Related Disorders* 7: 100166.

Etminani, K., Soliman, A., Davidsson, A. et al. (2022). A 3D deep learning model to predict the diagnosis of dementia with Lewy bodies, Alzheimer's disease, and mild cognitive impairment using brain 18F-FDG PET. *European Journal of Nuclear Medicine and Molecular Imaging* 49 (2): 563–584.

Evans-Lacko, S., Aguzzoli, E., Comas-Herrera, A. et al. (2024). World Alzheimer Report 2024: Global Changes in Attitudes to Dementia. *Alzheimer's Disease International.* https://www.alzint.org/u/World-Alzheimer-Report-2024.pdf, accessed October 20, 2024.

Fang, E., Ann, C.N., Maréchal, B. et al. (2020). Differentiating Parkinson's disease motor subtypes using automated volume-based morphometry incorporating white matter and deep gray nuclear lesion load. *Journal of Magnetic Resonance Imaging* 51 (3): 748–756.

Fang, J., Zhang, P., Wang, Q. et al. (2022). Artificial intelligence framework identifies candidate targets for drug repurposing in Alzheimer's disease. *Alzheimer's Research & Therapy* 14 (1): 7.

Farагó, P., Ştefănigă, S.A., Cordoş, C.G. et al. (2023). CNN-Based identification of Parkinson's disease from continuous speech in noisy environments. *Bioengineering (Basel)* 10 (5): 531.

Farina, F.R., Emek-Savaş, D.D., Rueda-Delgado, L. et al. (2020). A comparison of resting state EEG and structural MRI for classifying Alzheimer's disease and mild cognitive impairment. *Neuroimage* 215: 116795.

Ford, J.N., Sweeney, E.M., Skafida, M. et al. (2021). Heuristic scoring method utilizing FDG-PET statistical parametric mapping in the evaluation of suspected Alzheimer disease and frontotemporal lobar degeneration. *American Journal of Nuclear Medicine and Molecular Imaging* 11 (4): 313–326.

Fruijtier, A.D., Visser, L.N.C., van Maurik, I.S. et al. (2019). ABIDE Delphi study: topics to discuss in diagnostic consultations in memory clinics. *Alzheimer's Research & Therapy* 11 (1): 77.

Fukushima, K., Takamatsu, N., Yamamoto, Y. et al. (2022). Early diagnosis of amyotrophic lateral sclerosis based on fasciculations in muscle ultrasonography: a machine learning approach. *Clinical Neurophysiology* 140: 136–144.

Gao, F., Lv, X., Dai, L. et al. (2023). A combination model of AD biomarkers revealed by machine learning precisely predicts Alzheimer's dementia: China aging and neurodegenerative initiative (CANDI) study. *Alzheimer's & Dementia* 19 (3): 749–760.

Gauthier, S., Reisberg, B., Zaudig, M. et al. (2006). Mild cognitive impairment. *The Lancet* 367 (9518): 1262–1270.

Geraedts, V.J., Koch, M., Contarino, M.F. et al. (2021). Machine learning for automated EEG-based biomarkers of cognitive impairment during Deep Brain Stimulation screening in patients with Parkinson's disease. *Clinical Neurophysiology* 132 (5): 1041–1048.

Ghazaleh, N., Houghton, R., Palermo, G. et al. (2021). Ranking the predictive power of clinical and biological features associated with disease progression in Huntington's disease. *Frontiers in Neurology* 12: 678484.

Ghofrani-Jahromi, M., Poudel, G.R., Razi, A. et al. (2024). Prognostic enrichment for early-stage Huntington's disease: An explainable machine learning approach for clinical trial. *Neuroimage Clinical* 43: 103650.

Giorgio, J., Landau, S.M., Jagust, W.J. et al. (2020). Modelling prognostic trajectories of cognitive decline due to Alzheimer's disease. *Neuroimage Clinical* 26: 102199.

Göker, H. (2023). Automatic detection of Parkinson's disease from power spectral density of electroencephalography (EEG) signals using deep learning model. *Physical and Engineering Sciences in Medicine* 46 (3): 1163–1174.

Gramkow, M.H., Waldemar, G., and Frederiksen, K.S. (2024). The digitized memory clinic. *Nature Reviews Neurology* 20 (12): 738–746.

Grollemund, V., Chat, G.L., Secchi-Buhour, M.S. et al. (2020). Development and validation of a 1-year survival prognosis estimation model for Amyotrophic Lateral Sclerosis using manifold learning algorithm UMAP. *Scientific Reports* 10 (1): 13378.

Gromicho, M., Leão, T., Oliveira Santos, M. et al. (2022). Dynamic Bayesian networks for stratification of disease progression in amyotrophic lateral sclerosis. *European Journal of Neurology* 29 (8): 2201–2210.

Gullett, J.M., Albizu, A., Fang, R. et al. (2021). Baseline neuroimaging predicts decline to dementia from amnestic mild cognitive impairment. *Frontiers in Aging Neuroscience* 13: 758298.

Gupta, C., Xu, J., Jin, T. et al. (2022). Single-cell network biology characterizes cell type gene regulation for drug repurposing and phenotype prediction in Alzheimer's disease. *PLoS Computational Biology* 18 (7): e1010287.

Haji Ghassemi, N., Hannink, J., Roth, N. et al. (2019). Turning analysis during standardized test using on-shoe wearable sensors in Parkinson's disease. *Sensors (Basel)* 19 (14).

Hamburg, M.A. and Collins, F.S. (2010). The path to personalized medicine. *New England Journal of Medicine* 363 (4): 301–304.

Haneczok, J., Delijewski, M., and Moldzio, R. (2023). AI molecular property prediction for Parkinson's disease reveals potential repurposing drug candidates based on the increase of the expression of PINK1. *Computer Methods and Programs in Biomedicine* 241: 107731.

Hazarika, R.A., Maji, A.K., Syiem, R. et al. (2022). Hippocampus segmentation using U-Net convolutional network from brain magnetic resonance imaging (MRI). *Journal of Digital Imaging* 35 (4): 893–909.

Hendriksen, H.M.A., Van Gils, A.M., Van Harten, A.C. et al. (2023). Communication about diagnosis, prognosis, and prevention in the memory clinic: perspectives of European memory clinic professionals. *Alzheimer's Research & Therapy* 15 (1): 131.

Hojjati, S.H., Ebrahimzadeh, A., and Babajani-Feremi, A. (2019). Identification of the early stage of Alzheimer's disease using structural MRI and resting-state fMRI. *Frontiers in Neurology* 10: 904.

Hong, J., Lu, J., Liu, F. et al. (2023). Uncovering distinct progression patterns of tau deposition in progressive supranuclear palsy using [(18)F]Florzolotau PET imaging and subtype/stage inference algorithm. *EBioMedicine* 97: 104835.

Hou, X.H., Suckling, J., Shen, X.N. et al. (2023). Multipredictor risk models for predicting individual risk of Alzheimer's disease. *Journal of Translational Medicine* 21 (1): 768.

Hsieh, K.L., Plascencia-Villa, G., Lin, K.H. et al. (2023). Synthesize heterogeneous biological knowledge via representation learning for Alzheimer's disease drug repurposing. *iScience* 26 (1): 105678.

Huang, W., Zeng, R., Li, Y. et al. (2024). Identification of Alzheimer's disease and vascular dementia based on a deep forest and near-infrared spectroscopy analysis method. *Spectrochimica Acta Part a Molecular and Biomolecular Spectroscopy* 326: 125209.

Iizuka, T. and Kameyama, M. (2020). Spatial metabolic profiles to discriminate dementia with Lewy bodies from Alzheimer disease. *Journal of Neurology* 267 (7): 1960–1969.

Ileşan, R.R., Cordoş, C.G., Mihăilă, L.I. et al. (2022). Proof of concept in artificial-intelligence-based wearable gait monitoring for Parkinson's disease management optimization. *Biosensors (Basel)* 12 (4): 189.

Ismail, Z., Agüera-Ortiz, L., Brodaty, H. et al. (2017). The mild behavioral impairment checklist (MBI-C): a rating scale for neuropsychiatric symptoms in pre-dementia populations. *Journal of Alzheimer's Disease* 56 (3): 929–938.

Jack Jr., C.R., Andrews, J.S., Beach, T.G. et al. (2024b). Revised criteria for diagnosis and staging of Alzheimer's disease: Alzheimer's association workgroup. *Alzheimer's & Dementia* 20 (8): 5143–5169.

Jack, C.R., Arani, A., Borowski, B.J. et al. (2024a). Overview of ADNI MRI. *Alzheimer's & Dementia*, 2024 Sep 11.

Jack, C.R., Bernstein, M.A., Fox, N.C. et al. (2008). The Alzheimer's disease neuroimaging initiative (ADNI): MRI methods. *Journal of Magnetic Resonance Imaging* 27 (4): 685–691.

Jayatunga, M.K.P., Xie, W., Ruder, L. et al. (2022). AI in small-molecule drug discovery: a coming wave? *Nature Reviews Drug Discovery* 21 (3): 175–176.

Jeong, S.M., Song, Y.D., Seok, C.L. et al. (2024). Machine learning-based classification of Parkinson's disease using acoustic features: Insights from multilingual speech tasks. *Computers in Biology and Medicine* 182: 109078.

Jesus, B.J., Cassani, R., McGeown, W.J. et al. (2021). Multimodal prediction of Alzheimer's disease severity level based on resting-state EEG and structural MRI. *Frontiers in Human Neuroscience* 15: 700627.

Jin, Z., Gong, J., Deng, M. et al. (2024). Deep learning-based diagnosis algorithm for Alzheimer's disease. *Journal of Imaging* 10 (12): 333.

Kang, K., Nunes, A.S., Sharma, M. et al. (2023). Utilizing speech analysis to differentiate progressive supranuclear palsy from Parkinson's disease. *Parkinsonism & Related Disorders* 115: 105835.

Kang, L., Zhang, X., Guan, J. et al. (2024). Early Alzheimer's disease diagnosis via handwriting with self-attention mechanisms. *Journal of Alzheimer's Disease* 102 (1): 173–180.

Khachnaoui, H., Khlifa, N., and Mabrouk, R. (2022). Machine learning for early Parkinson's disease identification within SWEDD group using clinical and DaTSCAN SPECT imaging features. *Journal of Imaging* 8 (4): 97.

Khalil, R.M., Shulman, L.M., Gruber-Baldini, A.L. et al. (2024). Machine learning and statistical analyses of sensor data reveal variability between repeated trials in Parkinson's disease mobility assessments. *Sensors (Basel)* 24 (24): 8096.

Khanna, S., Domingo-Fernández, D., Iyappan, A. et al. (2018). Using multi-scale genetic, neuroimaging and clinical data for predicting Alzheimer's disease and reconstruction of relevant biological mechanisms. *Scientific Reports* 8 (1): 11173.

Kim, J.P., Kim, J., Park, Y.H. et al. (2019). Machine learning based hierarchical classification of frontotemporal dementia and Alzheimer's disease. *Neuroimage Clinical* 23: 101811.

Kim, N.H., Park, U., Yang, D.W. et al. (2023). PET-validated EEG-machine learning algorithm predicts brain amyloid pathology in pre-dementia Alzheimer's disease. *Scientific Reports* 13 (1): 10299.

Klöppel, S., Peter, J., Ludl, A. et al. (2015). Applying automated MR-based diagnostic methods to the memory clinic: a prospective study. *Journal of Alzheimer's Disease* 47 (4): 939–954.

Ko, Y.F., Kuo, P.H., Wang, C.F. et al. (2022). Quantification analysis of sleep based on smartwatch sensors for Parkinson's disease. *Biosensors (Basel)* 12 (2): 74.

Kobayashi, M., Yamada, Y., Shinkawa, K. et al. (2024). Vocal expression of emotions discriminates dementia with Lewy bodies from Alzheimer's disease. *Alzheimer's & Dementia: Diagnosis, Assessment & Disease Monitoring* 16 (2): e12594.

Koga, S., Ghayal, N.B., and Dickson, D.W. (2021a). Deep learning-based image classification in differentiating tufted astrocytes, astrocytic plaques, and neuritic plaques. *Journal of Neuropathology and Experimental Neurology* 80 (4): 306–312.

Koga, S., Zhou, X., and Dickson, D.W. (2021b). Machine learning-based decision tree classifier for the diagnosis of progressive supranuclear palsy and corticobasal degeneration. *Neuropathology and Applied Neurobiology* 47 (7): 931–941.

König, L.E., Rodriguez, S., Hug, C. et al. (2024). *TYK2 as a Novel Therapeutic Target in AlzHeimer's Disease with TDP-43 Inclusions.* United States.

Kueffner, R., Zach, N., Bronfeld, M. et al. (2019). Stratification of amyotrophic lateral sclerosis patients: a crowdsourcing approach. *Scientific Reports* 9 (1): 690.

Kwon, D.H., Hwang, J.S., Kim, S.G. et al. (2022). Cerebrospinal fluid metabolome in Parkinson's disease and multiple system atrophy. *International Journal of Molecular Sciences* 23 (3): 1879.

Landers, M., Saria, S., and Espay, A.J. (2021). Will artificial intelligence replace the movement disorders specialist for diagnosing and managing Parkinson's disease? *Journal of Parkinson's Disease* 11 (s1): S117–S122.

Lee, D., Park, D., Kim, I. et al. (2021). Plasmonic nanoparticle amyloid corona for screening Aβ oligomeric aggregate-degrading drugs. *Nature Communications* 12 (1): 639.

Lee, J., Banerjee, J., Khobragade, P.Y. et al. (2019). LASI-DAD study: a protocol for a prospective cohort study of late-life cognition and dementia in India. *BMJ Open* 9 (7): e030300.

Lee, L.Y., Vaghari, D., Burkhart, M.C. et al. (2024). Robust and interpretable AI-guided marker for early dementia prediction in real-world clinical settings. *eClinicalMedicine* 74: 102725. https://www.thelancet.com/journals/eclinm/article/PIIS2589-5370(24)00304-3/fulltext

Li, C., Hui, D., Wu, F. et al. (2023). Automatic diagnosis of Parkinson's disease using artificial intelligence base on routine T1-weighted MRI. *Frontiers in Medicine (Lausanne)* 10: 1303501.

Li, S., Shi, F., Pu, F. et al. (2007). Hippocampal shape analysis of Alzheimer disease based on machine learning methods. *AJNR American Journal of Neuroradiology* 28 (7):1339–1345.

Li, F., Tran, L., Thung, K.H. et al. (2015). A robust deep model for improved classification of AD/MCI Patients. *IEEE Journal of Biomedical and Health Informatics* 19 (5): 1610–1616.

Light, E., Wiersma, M., Dive, L. et al. (2021). Biobank networking and globalisation: perspectives and practices of Australian biobanks. *Australian Health Review* 45 (2): 214–222.

Littlejohns, T.J., Holliday, J., Gibson, L.M. et al. (2020). The UK Biobank imaging enhancement of 100,000 participants: rationale, data collection, management and future directions. *Nature Communications* 11 (1): 2624.

Liu, Y., Paajanen, T., Zhang, Y. et al. (2010). Analysis of regional MRI volumes and thicknesses as predictors of conversion from mild cognitive impairment to Alzheimer's disease. *Neurobiology of Aging* 31 (8): 1375–1385.

Livingston, G., Huntley, J., Liu, K.Y. et al. (2024). Dementia prevention, intervention, and care: 2024 report of the lancet standing commission. *The Lancet* 404 (10452): 572–628.

Lohner, V., Badhwar, A., Detcheverry, F.E. et al. (2024). Machine learning applications in vascular neuroimaging for the diagnosis and prognosis of cognitive impairment and dementia: a systematic review and meta-analysis. *medRxiv*. p. 2024.12.17.24319166. https://www.medrxiv.org/content/10.1101/2024.12.17.24319166v1

Lombardi, G., Crescioli, G., Cavedo, E. et al. (2020). Structural magnetic resonance imaging for the early diagnosis of dementia due to Alzheimer's disease in people with mild cognitive impairment. Cochrane Dementia and Cognitive Improvement Group, editor. *Cochrane Database of Systematic Reviews*. http://doi.wiley.com/10.1002/14651858.CD009628.pub2

López, M.M., Ramírez, J., Górriz, J.M. et al. (2009). SVM-based CAD system for early detection of the Alzheimer's disease using kernel PCA and LDA. *Neuroscience Letters* 464 (3): 233–238.

Lv, W. and Xue, Y. (2010). Prediction of acetylcholinesterase inhibitors and characterization of correlative molecular descriptors by machine learning methods. *European Journal of Medicinal Chemistry* 45 (3): 1167–1172.

Machado Reyes, D., Chao, H., Hahn, J. et al. (2024). Initiative FTADN. Identifying progression-specific Alzheimer's subtypes using multimodal transformer. *Journal of Personalized Medicine* 14 (4): 421.

Magesh, P.R., Myloth, R.D., and Tom, R.J. (2020). An explainable machine learning model for early detection of Parkinson's disease using LIME on DaTSCAN imagery. *Computers in Biology and Medicine* 126: 104041.

Magnin, B., Mesrob, L., Kinkingnéhun, S. et al. (2009). Support vector machine-based classification of Alzheimer's disease from whole-brain anatomical MRI. *Neuroradiology* 51 (2): 73–83.

Marek, K., Chowdhury, S., Siderowf, A. et al. (2018). The Parkinson's progression markers initiative (PPMI) – establishing a PD biomarker cohort. *Annals of Clinical and Translational Neurology* 5 (12): 1460–1477.

Marek, K., Jennings, D., Lasch, S. et al. (2011). Parkinson Progression Marker Initiative. The Parkinson Progression Marker Initiative (PPMI). *Progress in Neurobiology* 95 (4): 629–635.

Martin Prince, A., Wimo, A., Guerchet, M. et al. (2015). *World Alzheimer Report 2015 The Global Impact of Dementia - An Analysis of Prevalence, Incidence, Cost and Trends*. London: Alzheimer's Disease International.

Mason, S.L., Daws, R.E., Soreq, E. et al. (2018). Predicting clinical diagnosis in Huntington's disease: An imaging polymarker. *Annals of Neurology* 83 (3): 532–543.

Mendelson, A.F., Zuluaga, M.A., Lorenzi, M. et al. (2017). Selection bias in the reported performances of AD classification pipelines. *NeuroImage: Clinical* 14: 400–416.

Mesin, L., Porcu, P., Russu, D. et al. (2022). A multi-modal analysis of the freezing of gait phenomenon in Parkinson's disease. *Sensors (Basel)* 22 (7): 2613.

Metzger, J.J., Pereda, C., Adhikari, A. et al. (2022). Deep-learning analysis of micropattern-based organoids enables high-throughput drug screening of Huntington's disease models. *Cell Reports Methods* 2 (9): 100297.

Miltiadous, A., Tzimourta, K.D., Giannakeas, N. et al. (2021). Alzheimer's disease and frontotemporal dementia: a robust classification method of EEG signals and a comparison of validation methods. *Diagnostics (Basel)* 11 (8): 1437.

Moor, M., Banerjee, O., Abad, Z.S.H. et al. (2023). Foundation models for generalist medical artificial intelligence. *Nature* 616 (7956): 259–265.

Moradi, E., Pepe, A., Gaser, C. et al. (2015). Machine learning framework for early MRI-based Alzheimer's conversion prediction in MCI subjects. *Neuroimage* 104: 398–412.

Morra, J.H., Tu, Z., Apostolova, L.G. et al. (2008). Validation of a fully automated 3D hippocampal segmentation method using subjects with Alzheimer's disease mild cognitive impairment, and elderly controls. *Neuroimage* 43 (1): 59–68.

Morra, J.H., Tu, Z., Apostolova, L.G. et al. (2010). Comparison of AdaBoost and support vector machines for detecting Alzheimer's disease through automated hippocampal segmentation. *IEEE Transactions on Medical Imaging* 29 (1): 30–43.

Mundt, J.C., Ferber, K.L., Rizzo, M. et al. (2001). Computer-automated dementia screening using a touch-tone telephone. *Archives of Internal Medicine* 161 (20): 2481–2487.

Nam, Y., Kim, M., Chang, H.S. et al. (2019). Drug repurposing with network reinforcement. *BMC Bioinformatics* 20 (Suppl 13): 383.

Nguyen, D.T., Ryu, S., Qureshi, M.N.I. et al. (2019). Hybrid multivariate pattern analysis combined with extreme learning machine for Alzheimer's dementia diagnosis using multi-measure rs-fMRI spatial patterns. *PLoS One* 14 (2): e0212582.

Ni, Y.C., Lin, Z.K., Cheng, C.H. et al. (2024). Classification prediction of Alzheimer's disease and vascular dementia using physiological data and ECD SPECT images. *Diagnostics (Basel)* 14 (4): 365.

Nicholls, D., Quirk, D.A., Corrado, D.O. et al. (2024). *Royal college of psychiatrists (2023/24) national audit of dementia spotlight audit in memory assessment services national report.* Royal College of Psychiatrist. https://www.rcpsych.ac.uk/improving-care/ccqi/national-clinical-audits/national-audit-of-dementia

Noto, S., Sekiyama, Y., Nagata, R. et al. (2024). Analysis of speech features in Alzheimer's disease with machine learning: a case-control study. *Healthcare (Basel)* 12 (21): 2194.

Nunes, A.S., Pawlik, M., Mishra, R.K. et al. (2024). Digital assessment of speech in Huntington disease. *Frontiers in Neurology* 15: 1310548.

Obayya, M., Saeed, M.K., Maashi, M. et al. (2023). A novel automated Parkinson's disease identification approach using deep learning and EEG. *PeerJ Computer Science* 9:e1663.

Olm, C.A., Burke, S.E., Peterson, C. et al. (2023). Event-based modeling of T1-weighted MRI is related to pathology in frontotemporal lobar degeneration due to tau and TDP. *Neuroimage Clinical* 37: 103285.

Ou, Z., Pan, J., Tang, S. et al. (2021). Global trends in the incidence, prevalence, and years lived with disability of Parkinson's disease in 204 countries/territories from 1990 to 2019. *Frontiers in Public Health* 9: 776847.

Pan, Y., Shea, Y.F., Li, S. et al. (2021). Prevalence of mild behavioural impairment: a systematic review and meta-analysis. *Psychogeriatrics* 21 (1): 100–111.

Pang, H., Yu, Z., Yu, H. et al. (2022). Multimodal striatal neuromarkers in distinguishing parkinsonian variant of multiple system atrophy from idiopathic Parkinson's disease. *CNS Neuroscience & Therapeutics* 28 (12): 2172–2182.

Pansuwan, T., Quaegebeur, A., Kaalund, S.S. et al. (2023). Accurate digital quantification of tau pathology in progressive supranuclear palsy. *Acta Neuropathologica Communications* 11 (1): 178.

Patel, S., Lorincz, K., Hughes, R. et al. (2009). Monitoring motor fluctuations in patients with Parkinson's disease using wearable sensors. *IEEE Transactions on Information Technology in Biomedicine* 13 (6): 864–873.

Penco, S., Buscema, M., Patrosso, M.C. et al. (2008). New application of intelligent agents in sporadic amyotrophic lateral sclerosis identifies unexpected specific genetic background. *BMC Bioinformatics* 9: 254.

Pérez-Millan, A., Contador, J., Juncà-Parella, J. et al. (2023). Classifying Alzheimer's disease and frontotemporal dementia using machine learning with cross-sectional and longitudinal magnetic resonance imaging data. *Human Brain Mapping* 44 (6): 2234–2244.

Perez-Valero, E., Morillas, C., Lopez-Gordo, M.A. et al. (2022). An automated approach for the detection of Alzheimer's disease from resting state electroencephalography. *Frontiers in Neuroinformatics* 16: 924547.

Petersen, R.C. (2004). Mild cognitive impairment as a diagnostic entity. *Journal of Internal Medicine* 256 (3): 183–194.

Petersen, R.C. (2007). Mild cognitive impairment: current research and clinical implications. *Seminars in Neurology* 27 (01): 022–031.

Petersen, R.C., Doody, R., Kurz, A. et al. (2001). Current concepts in mild cognitive impairment. *Archives of Neurology* 58 (12): 1985–1992.

Petersen, R.C., Roberts, R.O., Knopman, D.S. et al. (2009). Mild cognitive impairment: ten years later. *Archives of Neurology* 66 (12): 1447–1455.

Petersen, R.C., Smith, G.E., Waring, S.C. et al. (1997). Aging, memory, and mild cognitive impairment. *International Psychogeriatrics* 9 (S1): 65–69.

Polidori, M.C., Pientka, L., and Mecocci, P. (2012). A review of the major vascular risk factors related to Alzheimer's disease. *Journal of Alzheimer's Disease* 32 (3): 521–530.

Polosecki, P., Castro, E., Rish, I. et al. (2020). Resting-state connectivity stratifies premanifest Huntington's disease by longitudinal cognitive decline rate. *Scientific Reports* 10 (1): 1252.

Ponzoni, I., Sebastián-Pérez, V., Martínez, M.J. et al. (2019). QSAR classification models for predicting the activity of inhibitors of beta-secretase (BACE1) associated with Alzheimer's disease. *Scientific Reports* 9 (1): 9102.

Poor, F.F., Dodge, H.H., and Mahoor, M.H. (2024). A multimodal cross-transformer-based model to predict mild cognitive impairment using speech, language and vision. *Computers in Biology and Medicine* 182: 109199.

Quintãom, C., Vigário, R., Santos, M.M. et al. (2021). Surface electromyography for testing motor dysfunction in amyotrophic lateral sclerosis. *Neurophysiologie Clinique* 51 (5): 454–465.

Riad, R., Lunven, M., Titeux, H. et al. (2022). Predicting clinical scores in Huntington's disease: a lightweight speech test. *Journal of Neurology* 269 (9): 5008–5021.

Ridgway, G.R., Lehmann, M., Barnes, J. et al. (2012). Early-onset Alzheimer disease clinical variants: multivariate analyses of cortical thickness. *Neurology* 79 (1): 80–84.

Rigas, G., Tzallas, A.T., Tsipouras, M.G. et al. (2012). Assessment of tremor activity in the Parkinson's disease using a set of wearable sensors. *IEEE Transactions on Information Technology in Biomedicine* 16 (3): 478–487.

Rios-Urrego, C.D., Rusz, J., and Orozco-Arroyave, J.R. (2024). Automatic speech-based assessment to discriminate Parkinson's disease from essential tremor with a cross-language approach. *NPJ Digital Medicine* 7 (1): 37.

Rittman, T. (2020). Neurological update: neuroimaging in dementia. *Journal of Neurology* 1–6.

Rizk-Jackson, A., Stoffers, D., Sheldon, S. et al. (2011). Evaluating imaging biomarkers for neurodegeneration in pre-symptomatic Huntington's disease using machine learning techniques. *Neuroimage* 56 (2): 788–796.

Rodriguez, S., Hug, C., Todorov, P. et al. (2021). Machine learning identifies candidates for drug repurposing in Alzheimer's disease. *Nature Communications* 12 (1): 1033.

Rohrer, J.D., Nicholas, J.M., Cash, D.M. et al. (2015). Presymptomatic cognitive and neuroanatomical changes in genetic frontotemporal dementia in the genetic frontotemporal dementia Initiative (GENFI) study: a cross-sectional analysis. *Lancet Neurology* 14: 253–262.

Sadeghi, M.A., Stevens, D., Kundu, S. et al. (2024). Detecting Alzheimer's disease stages and frontotemporal dementia in time courses of resting-state fMRI data using a machine learning approach. *Journal of Imaging Informatics in Medicine* 37 (6): 2768–2783.

Saito, Y., Kamagata, K., Wijeratne, P.A. et al. (2022). Temporal progression patterns of brain atrophy in corticobasal syndrome and progressive supranuclear palsy revealed by subtype and stage inference (SuStaIn). *Frontiers of Neurology* 13: 814768.

Salas-Gonzalez, D., Górriz, J.M., Ramírez, J. et al. (2010). Computer-aided diagnosis of Alzheimer's disease using support vector machines and classification trees. *Physics in Medicine & Biology* 55 (10): 2807–2817.

Salvatore, C., Cerasa, A., Battista, P. et al. (2015). Magnetic resonance imaging biomarkers for the early diagnosis of Alzheimer's disease: a machine learning approach. *Frontiers of Neuroscience* 9: 307.

Salvatore, C., Cerasa, A., Castiglioni, I. et al. (2014). Machine learning on brain MRI data for differential diagnosis of Parkinson's disease and Progressive Supranuclear Palsy. *Journal of Neuroscience Methods* 222: 230–237.

San-Martin, R., Fraga, F.J., Del Percio, C. et al. (2021). Classification of patients with Alzheimer's disease and dementia with Lewy bodies using resting EEG selected features at sensor and source levels: a proof-of-concept study. *Current Alzheimer Research* 18 (12): 956–969.

Saoud, L.S. and AlMarzouqi, H. (2024). Explainable early detection of Alzheimer's disease using ROIs and an ensemble of 138 3D vision transformers. *Scientific Reporters* 14 (1): 27756.

Scheid, B.H., Aradi, S., Pierson, R.M. et al. (2022). Predicting severity of Huntington's disease with wearable sensors. *Frontiers of Digital Health* 4: 874208.

Scott, H.L., Buckner, N., Fernandez-Albert, F. et al. (2020). A dual druggable genome-wide siRNA and compound library screening approach identifies modulators of parkin recruitment to mitochondria. *Journal of Biological Chemistry* 295 (10): 3285–3300.

Scotton, W.J., Bocchetta, M., Todd, E. et al. (2022). A data-driven model of brain volume changes in progressive supranuclear palsy. *Brain Communications* 4 (3): fcac098.

Scotton, W.J., Shand, C., Todd, E. et al. (2023). Uncovering spatiotemporal patterns of atrophy in progressive supranuclear palsy using unsupervised machine learning. *Brain Communications* 5 (2): fcad048.

Segura, L., Santos, N., Flores, R. et al. (2024). Exploring tau fibril-disaggregating and antioxidating molecules binding to membrane-bound amyloid oligomers using machine learning-enhanced docking and molecular dynamics. *Molecules* 29 (12): 2818.

Shaffi, N., Viswan, V., and Mahmud, M. (2024). Ensemble of vision transformer architectures for efficient Alzheimer's disease classification. *Brain Informatics* 11 (1): 25.

Shankle, W.R., Mania, S., Dick, M.B. et al. (1998). Simple models for estimating dementia severity using machine learning. *Studies in Health Technology and Informatics* 52 (1): 472–476.

Shinde, S., Prasad, S., Saboo, Y. et al. (2019). Predictive markers for Parkinson's disease using deep neural nets on neuromelanin sensitive MRI. *Neuroimage Clinical* 22: 101748.

Shu, Z.Y., Mao, D.W., Xu, Y.Y. et al. (2021). Prediction of the progression from mild cognitive impairment to Alzheimer's disease using a radiomics-integrated model. *Therapeutic Advances in Neurological Disorders* 14: 17562864211029551.

Sohrabi, S., Mor, D.E., Kaletsky, R. et al. (2021). High-throughput behavioral screen in C. elegans reveals Parkinson's disease drug candidates. *Communications Biology* 4 (1): 203.

Sudlow, C., Gallacher, J., Allen, N. et al. (2015). UK Biobank: an open access resource for identifying the causes of a wide range of complex diseases of middle and old age. *PLoS Medicine* 12 (3): e1001779.

Swaddiwudhipong, N., Whiteside, D.J., Hezemans, F.H. et al. (2022). Pre-diagnostic cognitive and functional impairment in multiple sporadic neurodegenerative diseases. *Alzheimer's & Dementia* 19 (5): 1752–1763. https://onlinelibrary.wiley.com/doi/abs/10.1002/alz.12802

Tang, J., Yang, B., Adams, M.P. et al. (2019). Artificial neural network-based prediction of outcome in Parkinson's disease patients using DaTscan SPECT imaging features. *Molecular Imaging and Biology* 21 (6): 1165–1173.

Termine, A., Fabrizio, C., Caltagirone, C. et al. (2022). Initiative OBOTFLDN. A reproducible deep-learning-based computer-aided diagnosis tool for frontotemporal dementia using MONAI and clinica frameworks. *Life (Basel)* 12 (7): 947.

Thomas, I., Westin, J., Alam, M. et al. (2018). A treatment-response index from wearable sensors for quantifying Parkinson's disease motor states. *IEEE Journal of Biomedical and Health Informatics* 22 (5): 1341–1349.

Thome, J., Steinbach, R., Grosskreutz, J. et al. (2022). Classification of amyotrophic lateral sclerosis by brain volume, connectivity, and network dynamics. *Human Brain Mapping* 43 (2): 681–699.

Thung, K.H., Wee, C.Y., Yap, P.T. (2014). Neurodegenerative disease diagnosis using incomplete multi-modality data via matrix shrinkage and completion. *Neuroimage* 91: 386–400.

Tunc, H.C., Sakar, C.O., Apaydin, H. et al. (2020). Estimation of Parkinson's disease severity using speech features and extreme gradient boosting. *Medical & Biological Engineering & Computing* 58 (11): 2757–2773.

Vaiciukynas, E., Verikas, A., Gelzinis, A. et al. (2017). Detecting Parkinson's disease from sustained phonation and speech signals. *PLoS One* 12 (10): e0185613.

Valdes-Sosa, P.A., Leon, M.A.B., Lopera, F. et al. (2022). EEG in the Global Brain Consortium, aiming to strengthen linkages between neuroscientists across borders and disciplines to advance equitable solutions to priority health challenges worldwide. *Alzheimer's & Dementia* 18 (S6): e059945.

van der Burgh, H.K., Schmidt, R., Westeneng, H.J. et al. (2017). Deep learning predictions of survival based on MRI in amyotrophic lateral sclerosis. *Neuroimage Clinical* 13: 361–369.

van Dyck, C.H., Swanson, C.J., Aisen, P. et al. (2022). Lecanemab in early Alzheimer's disease. *New England Journal of Medicine* 388 (1): 9–21.

Vishnuppriya, C. and Tamilpavai, G. (2024). Performance comparison between deep neural network and machine learning based classifiers for Huntington disease prediction from human DNA sequence. *IEEE/ACM Transactions on Computational Biology and Bioinformatics* 1–12.

Vogel, J.W., Young, A.L., Oxtoby, N.P. et al. (2021). Four distinct trajectories of tau deposition identified in Alzheimer's disease. *Nature Medicine* 27 (5): 871–881.

Wada, A., Tsuruta, K., Irie, R. et al. (2019). Differentiating Alzheimer's disease from dementia with Lewy bodies using a deep learning technique based on structural brain connectivity. *Magnetic Resonance in Medical Sciences* 18 (3): 219–224.

Wang, C., Tachimori, H., Yamaguchi, H. et al. (2024a). A multimodal deep learning approach for the prediction of cognitive decline and its effectiveness in clinical trials for Alzheimer's disease. *Translational Psychiatry* 14 (1): 105.

Wang, D., Honnorat, N., Toledo, J.B. et al. (2024b). Deep learning reveals pathology-confirmed neuroimaging signatures in Alzheimer's, vascular and Lewy body dementias. *Brain:* awae388.

Warule, P., Mishra, S.P., and Deb, S. (2023). Time-frequency analysis of speech signal using Chirplet transform for automatic diagnosis of Parkinson's disease. *Biomedical Engineering Letters* 13 (4): 613–623.

Wee, C.Y., Yap, P.T., Zhang, D. et al. (2012). Identification of MCI individuals using structural and functional connectivity networks. *Neuroimage* 59 (3): 2045–2056.

Wei, W., Visweswaran, S., and Cooper, G.F. (2011). The application of naive Bayes model averaging to predict Alzheimer's disease from genome-wide data. *Journal of the American Medical Informatics Association* 18 (4): 370–375.

Welton, T., Hartono, S., Lee, W. et al. (2024). Classification of Parkinson's disease by deep learning on midbrain MRI. *Frontiers in Aging Neuroscience* 16: 1425095.

Wimo, A., Seeher, K., Cataldi, R. et al. (2023). The worldwide costs of dementia in 2019. *Alzheimer's & Dementia* 19 (7): 2865–2873.

Wittenberg, R., Hu, B., Barranza-Araiza, L. et al. (2019). *Projections of Older People with Dementia and Costs of Dementia Care in the United Kingdom, 2019–2040*. Care Policy and Evaluation Centre, London School of Economics and Political Science. https://www.alzheimers.org.uk/sites/default/files/2019-11/cpec_report_november_2019.pdf

Wu, K., Liu, Q., Long, K. et al. (2024a). Deciphering the role of lipid metabolism-related genes in Alzheimer's disease: a machine learning approach integrating Traditional Chinese Medicine. *Frontiers in Endocrinology (Lausanne)* 15: 1448119.

Wu, X., Li, Z., Chen, G. et al. (2023). Hybrid neural network approaches to predict drug-target binding affinity for drug repurposing: screening for potential leads for Alzheimer's disease. *Frontiers in Molecular Biosciences* 10: 1227371.

Wu, X., Ma, L., Wei, P. et al. (2024b). Wearable sensor devices can automatically identify the ON-OFF status of patients with Parkinson's disease through an interpretable machine learning model. *Frontiers in Neurology* 15: 1387477.

Wu, Y., Liu, Q., Qiu, Y. et al. (2022). Deep learning prediction of chemical-induced dose-dependent and context-specific multiplex phenotype responses and its application to personalized Alzheimer's disease drug repurposing. *PLoS Computational Biology* 18 (8): e1010367.

Wu, Y., Wang, X., and Fang, Y. (2024c). Predicting mild cognitive impairment in older adults: a machine learning analysis of the Alzheimer's Disease Neuroimaging Initiative. *Geriatrics Gerontology International* 24 (1): 96–101.

Yamada, Y., Shinkawa, K., Nemoto, M. et al. (2022). Speech and language characteristics differentiate Alzheimer's disease and dementia with Lewy bodies. *Alzheimer's & Dementia: Diagnosis, Assessment & Disease Monitoring* 14 (1): e12364.

Yan, C., Grabowska, M.E., Dickson, A.L. et al. (2023). Leveraging generative AI to prioritize drug repurposing candidates: validating identified candidates for Alzheimer's disease in real-world clinical datasets. medRxiv : the preprint server for health sciences. *United States.* 2023.07.07.23292388.

Yan, C., Grabowska, M.E., Dickson, A.L. et al. (2024). Leveraging generative AI to prioritize drug repurposing candidates for Alzheimer's disease with real-world clinical validation. *NPJ Digital Medicine* 7 (1): 46.

Yang, J., Feng, X., Laine, A.F. et al. (2020). Characterizing Alzheimer's disease with image and genetic biomarkers using supervised topic models. *IEEE Journal of Biomedical and Health Informatics* 24 (4):1180–1187.

Yang, H., Mao, J., Ye, Q. et al. (2024). Distance-based novelty detection model for identifying individuals at risk of developing Alzheimer's disease. *Frontiers in Aging Neuroscience* 16: 1285905.

Yao, S., Zhang, Y., Chen, J. et al. (2024). Enhancing identification performance of cognitive impairment high-risk based on a semi-supervised learning method. *Journal of Biomedical Informatics* 157: 104699.

Yarnall, A.J., Breen, D.P., Duncan, G.W. et al. (2014). Characterizing mild cognitive impairment in incident Parkinson disease: The ICICLE-PD study. *Neurology* 82 (4): 308–316.

Zhang, S., Yuan, J., Sun, Y. et al. (2024a). Machine learning on longitudinal multi-modal data enables the understanding and prognosis of Alzheimer's disease progression. *iScience* 27 (7): 110263.

Zhang, X., Fan, H., Guo, C. et al. (2024b). Establishment of a mild cognitive impairment risk model in middle-aged and older adults: a longitudinal study. *Neurological Sciences* 45 (9): 4269–4278.

Zhang, Y., Ghose, U., Buckley, N.J. et al. (2022). Predicting AT(N) pathologies in Alzheimer's disease from blood-based proteomic data using neural networks. *Frontiers in Aging Neuroscience* 14: 1040001.

Zheng, Y., Guo, H., Zhang, L. et al. (2019). Machine learning-based framework for differential diagnosis between vascular dementia and Alzheimer's disease using structural MRI features. *Frontiers in Neurology* 10: 1097.

Zhou, X., Kedia, S., Meng, R. et al. (2024). Deep learning analysis of fMRI data for predicting Alzheimer's Disease: a focus on convolutional neural networks and model interpretability. *PLoS One* 19 (12): e0312848.

Zhou, X., Xu, J., and Zhao, Y. (2006). Machine learning methods for anticipating the psychological distress in patients with Alzheimer's disease. *Australasian Physical & Engineering Sciences in Medicine* 29 (4): 303–309.

Zhou, Y. and Tagare, H.D. (2021). Self-normalized Classification of Parkinson's Disease DaTscan Images. *Proceedings (IEEE Int Conf Bioinformatics Biomed)* 2021: 1205–1212.

Zhou, Y., Chia, M.A., Wagner, S.K. et al. (2023). A foundation model for generalizable disease detection from retinal images. *Nature* 622 (7981): 156–163.

8

Applications of Machine Learning in the Genetics of Brain Disorders

Matthew Bracher-Smith[1,2] *and Valentina Escott-Price*[1,2]

[1]*Division of Psychological Medicine & Clinical Neurosciences, Cardiff University, UK*
[2]*Dementia Research Institute, Cardiff University, UK*

8.1 Introduction

Numerous fields have recently been upturned by advances in machine learning (ML) and artificial intelligence, but the genetics of brain disorders, like many other disciplines in the biosciences, has been only partially affected. Understanding why this is the case, what past efforts have achieved and what promising avenues lie ahead will be the focus of this chapter. The methods themselves and their potential impacts are exciting, but to explore their challenges, we must first appreciate the problem at hand. To this end, we begin by looking at brain disorders in neurology and psychiatry and the current reigning champion of genetic prediction in complex traits – polygenic risk scores (PRSs) – before diving into the most frequently applied ML approaches. We cover common pitfalls and promises in ML applications before touching on the future of federated learning (FL) approaches in genetics. In applying supervised ML to the genetics of brain disorders, we seek to estimate a function that maps from our inputs (genetic data) X to the outcome (disease status) y. Here, we will work from outcome definition to genetic data and then the methods for estimating the mapping between the two.

8.1.1 Prediction in Neurology and Psychiatry

Brain disorders are complex and often highly heritable traits that can be caused by genetics, environmental factors or a combination (Bellenguez et al., 2020; Baselmans et al., 2021; Bellou et al., 2020). Overall, the impact of these disorders on the individual and society is severe, with global prevalence expected to increase threefold by 2050 (Nichols et al., 2022). Their clinical diagnosis is also challenging due to shared symptoms and comorbidities (Scheltens et al., 2021). Dementia and other neurodegenerative disorders, for instance, cause a range of symptoms, including memory loss, depression/anxiety and physical impairments such as incontinence (Bature et al., 2017). The most common form of dementia, Alzheimer's disease (AD), accounts for more than 75% of cases (Duong et al., 2017).

Diagnosing the correct form of dementia has long proven difficult due to different forms sharing phenotypic characteristics (Karantzoulis & Galvin, 2011). Currently, the only method to confirm a diagnosis of a specific type of dementia is through a postmortem brain biopsy (Paraskevaidi et al., 2018). Even in studies with postmortem confirmation, the presence of brain pathology for multiple dementia types, known as mixed dementia, is still reasonably common (Rahimi et al., 2014). By contrast, diagnosis of neurodevelopmental (psychiatric) disorders relies on symptoms; there is no clinical diagnostic test. Assessment is done through psychiatric history and a Mental State Examination (MSE). Many psychiatric and neurodegenerative disorders share symptoms, and the border between the disorders, or even the absence of a disorder, can be ambiguous.

Compounding these challenges, the neurodegenerative disease or disorder status, which is available in large studies, is often determined through clinical symptoms prior to death, which can lead to potential misclassification due to the inability to confirm the diagnosis pathologically postmortem. Meanwhile, genetic data often serve as an undercurrent that influences risk over the course of an individual's life, potentially manifesting as dementia after the study period. This right censoring means we may correctly predict from genetic data that an individual will eventually get dementia, but not observe this outcome during the study period, and thus mark such an instance as an error during model evaluation. Together, these issues highlight a fundamental barrier in applying ML in this area: outcomes are both defined and measured with some inaccuracy, leading to noisy labels, a known challenge in modern deep learning (DL) applications (Song et al., 2020).

Despite the challenges, efforts to improve the prediction of these conditions are essential. While there are currently no specific treatments to prevent brain disorders or reverse their course, notwithstanding the recent success in clinical trials for Lecanemab (Dyck et al., 2023) and Donanemab (Sims et al., 2023) in targeting amyloid plaques, determining an individual's risk of onset at an early stage can enable clinicians to improve quality of life during disease progression. Early prediction can also provide insights for patients and caregivers, enabling them to prepare for the personal implications of these disorders. Implementation of personalised medicine in the fields of neurology, psychiatry and neuroscience can trigger a shift in medical practise towards presymptomatic diagnosis and appropriate interventions. Prevention strategies can be employed, with a strong focus on individualised care, before clinical and neuropsychological symptoms become manifest.

Biomarkers, including those derived from neuroimaging, blood and cerebrospinal fluid, will be essential in identifying subjects with prodromal disease, in stratifying people for treatment trials, and for analysing data from such trials. However, they remain expensive and sometimes invasive. In genomics, the development of non-expensive and comprehensive genome-wide arrays, which paved the way for genome-wide association studies (GWAS), means genotypes can potentially act as genetic predictors in brain disorders.

8.1.2 The Genetic Landscape in Brain Disorders (with Reference to AD)

Brain disorders have complex genetic architectures, characterised by an interplay of both common (often taken as a frequency of >1%) and rare (<1%) genetic variants. These variants include chromosomal abnormalities like trisomy 21, large structural changes known as copy number variations (CNVs), small changes such as insertions/deletions (indels) and single-nucleotide polymorphisms (SNPs). In general, disorders follow the typical rule for complex traits: variants that are common in the population have a small effect on disease risk, while those with a comparatively large impact are rare (Bodmer et al., 2008). AD marks an exception to this, as the apolipoprotein E (*APOE*) ε4 allele is both common and has a substantial effect size, with an odds ratio (OR) of approximately 3.7 for heterozygotes and 14.9 for homozygotes compared to the ε3 allele (Corder et al., 1993). Rare variants in genes such as *PSEN1, PSEN2* and *APP* are well-documented in early-onset familial AD, with these mutations often resulting in near-certain disease onset before age 65 years. This is more congruent with other disorders where a large burden of rare variants, in those disorders in which they make a known contribution to disease risk, typically indicates a more severe presentation or course of the disease or an earlier onset, as has been observed in schizophrenia (SCZ) and autism (Owen Michael, 2014).

How much rare and common variants contribute to risk varies by disorder. For instance, in Parkinson's disease (PD), common variants in genes such as *LRRK2* and *SNCA* have been identified, with the *LRRK2* G2019S mutation being present in around 4% of familial cases and 1% of sporadic cases depending on the population studied (Healy et al., 2008). In contrast, SCZ has a significant burden of rare genetic variants, including CNVs such as the 22q11.2 deletion (Rees and Kirov, 2021). Furthermore, rare coding variants in groups of genes that are loss-of-function (LoF) intolerant also contribute significantly to SCZ risk (Singh et al., 2017). Bipolar disorder (BD), however, has fewer rare variants compared to SCZ and is more influenced by common variants with smaller effect sizes, such as those found in *CACNB2* and *ANK3*, both of which have ORs around 1–1.2, which is more typical of common variants (Hara et al., 2022; Mullins et al., 2021). These features impact modelling choices from ML, as ultra-rare variants are likely to need aggregating, given standard data sizes, to be learnable, and the impact of common variants with weak effect sizes is likely to need to be constrained in the model (through penalisation or knowledge embedding) to limit the effect of noise.

There is also a clear distinction between psychiatric and neurodegenerative disorders with regards to pleiotropy. Psychiatric conditions exhibit substantial genetic correlation between them, particularly for SCZ and BD, characterised by a strong polygenic component from many common variants of small effect (Grotzinger et al., 2022). Correlation between neurodegenerative diseases, however, is relatively low (Brainstorm Consortium, 2018) and is likely to be driven by the effect of *APOE* due to its association with ageing and other comorbid age-related disorders (Foley et al., 2020; Deelen et al., 2019). This impacts considerations of whether to use multivariate analyses that simultaneously model related outcomes, something more likely to be productive in psychiatry than neurology. Prediction in neurodegenerative conditions, however, can still take advantage of outcome correlation through measures such as Mini-Mental State Examination (MMSE) and multi-task learning, for instance. Likelihood of misclassification (label noise) also impacts the choice of loss function in ML. Methods that use an exponential loss, which greatly penalises gross classification errors, like AdaBoost, are unlikely to perform as well as other common loss functions like the negative log-likelihood or hinge loss.

Overall, while common variants like *APOE* ε4 in AD have large effects, the majority of SNPs affect disease through an accumulation of many small effects, while rare coding variants also play critical roles in diseases like SCZ and AD, which are prototypical disorders within the categories of developmental psychiatric and neurodegenerative disorders, respectively. The heterogeneity in genetic contributions across different brain disorders underscores the importance of comprehensive genetic studies to fully elucidate the underlying mechanisms and potential therapeutic targets, and the need for distinct modelling approaches in different disorders.

8.2 Genetic Prediction

8.2.1 Emergence of Polygenic Risk Scores

GWAS unearthed a large number of genetic associations with complex disorders, all of which have a small individual effect. However, there are vast numbers of variants that do not reach genome-wide significance but may still collectively contribute to disease risk. The PRS approach has been widely used across different traits for estimating polygenic risk and disease prediction. PRS combines small additive effects of specific loci across the genome and has shown potential to stratify individuals into risk categories based on their genetic profile for common genetic disorders.

PRS analysis requires two independent datasets. For the first, summary statistics from association studies are sufficient, as this dataset is used to select the SNPs, the risk score alleles and their effects on disease risk. The second dataset is used to test whether the PRS differs in cases and controls and requires the genotypes for each individual.

The simplest PRS method (clumping and thresholding; Wray et al., 2007) allows us to capture SNPs that are the most (even if not significantly) associated with the disease in a linkage disequilibrium (LD) block, defined by the correlation structure of the genome in a particular population. The PRS is calculated from the effect size (β)-weighted sum of associated alleles for each individual in the test data. For M SNPs in a set, PRS for an individual (j) combines SNP genotypes g_i, $i = 1,\ldots, M$ into a single regression predictor using single-SNP effect sizes $\left(\log(OR_i) = \beta_i\right)$ taken from a previous study as coefficients

$$\text{PRS}_j = \sum_{i=1}^{M} \beta_i \, g_i^j$$

Various methodological approaches have been developed to derive PRS scores, including several Bayesian approaches such as PRS-CS (Ge et al., 2019) and quickPRS (Zhang et al., 2021), among others. PRS-CS utilises a high-dimensional Bayesian regression framework by placing a continuous shrinkage (CS) prior on effect sizes, which is robust to varying genetic architectures, and enables multivariate modelling of local LD patterns (Ge et al., 2019). QuickPRS is an approximate version of MegaPRS that implements an LDAK model (Speed et al., 2020) with various summary statistics tools, including BayesR (Lloyd-Jones et al., 2019). It assumes non-equal contribution of SNPs to the heritability of a trait, as compared to the previous models like GCTA and LDSC (Speed et al., 2017). The PRS approach encompasses more of the causal variance, as a genetic risk score is calculated based not solely on genome-wide significant polymorphisms, but on all nominally associated variants at a defined significance threshold (typically resulting in thousands of variants being included). This type of analysis has demonstrated that there is a significant polygenic contribution to other complex genetic diseases (Escott-Price et al., 2015; International Schizophrenia Consortium et al., 2009).

The utility of risk scores in precision medicine is an open question. Originally, PRSs were designed to summarise genome-wide genotype data into a single variable that measures genetic liability to a disorder or a trait. PRS studies often reach sufficiently high statistical significance levels (small *p*-value) to suggest trait polygenicity, but the prediction accuracy is usually not sufficient for clinical utility (Lewis and Vassos, 2017). For example, the prediction accuracy from PRS in SCZ, a highly heritable disorder, has an area under the curve (AUC) of about 73%, whereas in BD, the AUC is lower, at around 65% (Koch et al., 2023). Nevertheless, PRS has been suggested as a useful tool for the selection of individuals for clinical trials in individuals of European ancestry across different traits (Agerbo et al., 2015; Escott-Price et al., 2017b; Lall et al., 2017; Natarajan et al., 2017). The PRS prediction accuracy of risk for AD is relatively high at 70–75% (Harrison et al., 2020), especially if the diagnosis is based on pathological confirmation (AUC up to 84%) rather than clinical assessment (Escott-Price et al., 2017a), though even under ideal scenarios, the main driver of predictive performance is still *APOE* status (Karlsson et al., 2022). In PD, a polygenic basis was confirmed and shown to correlate with age at disease onset (Escott-Price et al., 2015). This method can also be used to identify overlap in genetic determinants between related disorders, e.g., the genetic correlation between SCZ and BD, or depression and anxiety mentioned previously (Demirkan et al., 2011). While the polygenic method undoubtedly introduces noise by including some variants that are not involved in disease susceptibility (i.e., false-positives), this is more than offset by the increased power to identify those at the highest/lowest risk of disease. The use of publicly available effect sizes from large GWAS, and the reduction to a single variable, also means the sample size requirements for adequate power in the test set and the multiple testing burden are relatively modest.

8.2.2 Limitations of PRS and Opportunities for ML

Genetic risk prediction modelling through PRS has become a common approach in research to assess an individual's risk of developing a given disease (Pate et al., 2019). However, in the genetics of complex diseases, it is unknown whether and to what extent non-additive genetic interaction effects contribute to risk (Hardy & Escott-Price, 2019). PRS, as originally implemented, has fundamental limitations to its ability to model the relationships between the genetic variants and disease.

First, they consider only a linear additive combination of variants. This means that when considering the relationship between a single variant, such as a SNP, and the log(odds) of disease status, for example, we cannot account for deviations from linear relationships between the genetic variants (SNPs) and disease status, known as dominance deviations in statistical genetics. In other words, if there is an increase of 0.5 in log(odds) from a single risk allele, then an additional risk allele at the same SNP must also give another 0.5 increase in log(odds). Such dominance deviations are well-documented, and the relationship between a given allele and a trait may exhibit underdominance, overdominance or complete dominance (Falconer, 1996). Between loci, PRS typically only accounts for an

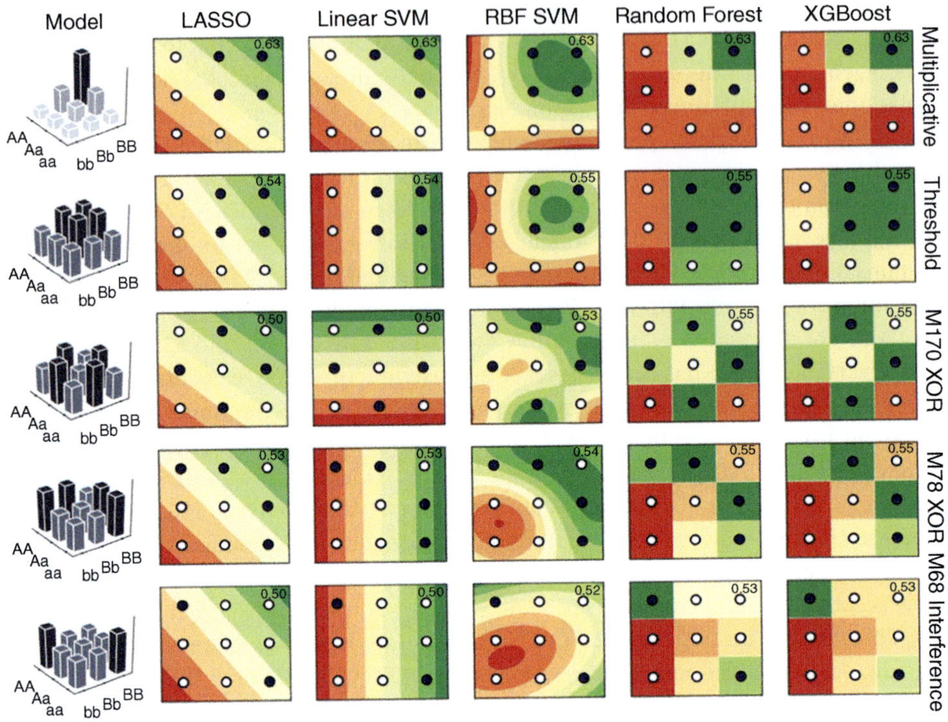

Figure 8.1 Non-linear methods are required for complex two-locus interactions. Subplots show the decision boundary for different classifiers as contour plots from a simulation of two-locus interactions. Points denote different genotype combinations between alleles on the *x* and *y* axes, where dark blue points represent combinations that increase disease risk. Green regions indicate the positive side of the decision boundary, and red regions indicate the negative side. Ideally, dark blue points should be encircled by dark green areas to show they have been correctly assigned as high risk. The cross-validated AUC is given in the upper right of each subplot. The rotated 3D interaction model is shown in the left-hand column on the odds scale; names for interaction models on the right-hand side are taken from Li and Reich (2000). The two-locus simulation methodology was based on that described by Zhang and Liu (2007).

additive relationship, which precludes accounting for epistasis, defined as statistical interactions by Fisher (1919). Complex interactions between two or more genetic loci cannot be learned using only linear additive methods (Figure 8.1).

Second, PRS considers only external weights in forming a risk score. This allows them to make use of summary statistics of large meta-analysed GWAS, but also means effect sizes are not updated to reflect different relationships in the test data. In recent years, novel PRS approaches have enabled different methods for combining variants, such as multiplicatively or for updating the effect sizes in new datasets. However, these are small modifications or wrapper approaches that do not allow for the joint estimation of the effects of many variants and do not allow for combining variables in a flexible manner that can incorporate both within- and between-locus deviations from additivity. The need to address such a situation is particularly strong in AD, where the statistical and biological effects of the *APOE* ε4 allele make *APOE*-dependent effects likely (Fortea et al., 2024).

The limitations to PRS present an opportunity for ML approaches, which address these challenges by allowing for the joint estimation of variant effects under flexible models that can learn non-linear interactions between variants (Figure 8.1). Such an opportunity is not without its challenges, though – PRS has remained popular due to its theoretical and computational simplicity, qualities that ML methods, being complex and computationally intensive, do not exhibit. PRS has also achieved relatively high prediction accuracy in European datasets, while ML methods have been plagued by biased implementations, leaving unanswered the important questions of whether they can achieve similar or better prediction accuracy or identify novel variants that impact the outcome.

8.2.3 The Need for Alternative Predictive Approaches

While the likelihood of complex interactions and dominance deviations occurring along the path from genetics to disease diagnosis motivates the statistical and biological cases for exploring methods beyond linear additive PRS, a separate argument follows from issues around data privacy. Individual genetic data are not easy to share between studies due to the General Data Protection Regulation (GDPR), where genetic data are included in the list of sensitive data. Only secure computational platforms with a legally compliant process of data handling and processing offer an opportunity to combine genetic data from a number of studies. This means that combining data on a central server for analysis is typically precluded, and studies must instead rely on local analyses, using classical linear methods, before meta-analysing summary statistics. Approaches thus far have overwhelmingly been centralised learning (CL) on a single dataset using a handful of out-of-the-box methods. However, recent advances in edge computing and federated ML have enabled new opportunities. There is a need for utilising these approaches, which enable previously siloed data to become available in large-scale collaborations without the need to transfer data, give access to external collaborators or have a large base of in-house specialist knowledge beyond initial setup requirements.

8.3 ML Approaches

8.3.1 Introduction to ML in the Genetics of Brain Disorders

Decreasing costs of computation, increasing volumes of data and fundamental advances in ML have converged over the past few years, particularly since 2020, creating a landscape where the potential impact of ML is substantial (Figure 8.2). However, successful practical applications in brain disorders, and biosciences in general, implemented using low risk-of-bias (ROB) procedures in a similar genome-wide manner, are few and far between.

Given the aim of predicting a neurological or psychiatric outcome (a disease or disorder) or identifying novel genetic variants associated with it, the focus here is on supervised methods aiming to estimate a function that maps from predictors to outcome. Unsupervised methods, which include dimensionality reduction approaches like principal

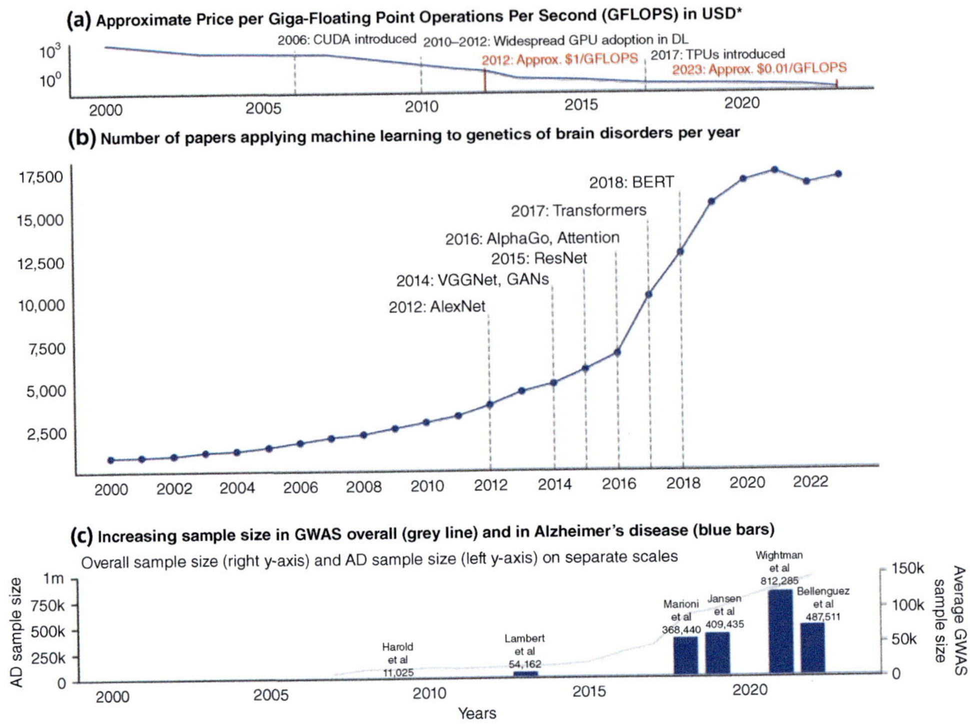

Figure 8.2 Trends in cost of compute, interest in ML and the size of data available. The approximate price (*estimated in 2023 USD) per giga floating point operations per second (GFLOPS) has continuously decreased over the last 20 years (a), as expected, while widespread GPU adoption in DL and free availability of related software have allowed for much larger models. The number of researchers applying ML to brain disorder genetics is also at an all-time high (b), with key advances in DL architectures likely driving some of the increased interest. Finally, the data collected by large consortia has greatly increased for many major psychiatric and neurological disorders, exemplified here by the GWAS sample sizes for AD (c). Price per GFLOPS taken from https://en.wikipedia.org/wiki/FLOPS#Hardware_costs and plotted as the lowest price in each year. Papers applying ML were extracted using Pold87/academic-keyword-occurrence (Volker Strobel, 2018) and Google Scholar for terms related to genetics, ML and brain disorders. AD GWAS sample sizes taken from Escott-Price and Hardy (2022) and average overall GWAS sample sizes from Abdellaoui et al. (2023).

component analysis (PCA), are widespread in genetics and related omics fields and have been prominent since the earliest genome-wide studies to identify population stratification (Price et al., 2006). However, the surge in interest that has occurred has largely been driven by supervised approaches, which are frequently used with an outcome that is binary (disease or no disease), ordered or unordered categorical (e.g., Braak's stages) or continuous (e.g., MMSE score). While there is some debate about which methods strictly fall into ML or statistics, here we include approaches that allow for learning from data with some control for tuning through hyperparameters, and thus include methods such as penalised linear and logistic regression, but not their unpenalised counterparts.

8.3.2 The Scale of the Problem

A key point to note when considering applications of supervised models is that broad interest in genetics is present within ML circles, with many tools and papers made readily available (Kopp et al., 2020). However, these are chiefly focused on problems related to predicting genetic elements or more direct outcomes from genetic structures, such as the effect on DNA conformation, splicing or protein structure. While these are complex problems, they are also fundamentally easier in that predictors and responses are tightly coupled, and an approximate mapping between the two is likely to be learnable with accurate data and a well-designed model. By contrast, the issue of mapping from variants to disease outcome encompasses many biological levels, across wide-ranging physical and temporal scales, from cells to organs and over decades of life, under exposure to different environmental conditions, including *in utero*, all of which finally culminate in disease status. Such a mapping is conceptually a gargantuan task. The methods used to approach this problem have, despite their complexity, been relatively simple.

8.4 Prominent ML Models Applied to Brain Disorders

As with other fields, the application of supervised ML in brain disorders tends to cluster around a few core methods. Briefly, the most common options are penalised regression, random forests (RFs), gradient boosting machines (GBMs), support vector machines (SVMs) and neural networks, with prominent applications of each. Many other approaches have also been used in the literature, including Bayesian networks, naïve Bayes and decision trees (Table 8.1).

Table 8.1 The most frequently used methods in psychiatric genetics and the genetics of AD for phenotype prediction using ML approaches. Counts for the most common methods were collected from two systematic reviews in psychiatry (Bracher-Smith et al., 2020) and AD (Rowe et al., 2021). Boosting and neural network approaches were not used in studies selected based on the review criteria for AD but have been used elsewhere in the literature.

Method	Psychiatric disorders		Alzheimer's disease	
	Studies (n = 13)	Models	Studies (n = 12)	Models
SVMs	8	19	8	44
Neural networks	7	23	—	—
RFs	5	8	3	5
Penalised regression	3	6	4	15
Naïve Bayes	2	2	4	10
Bayesian networks	2	5	2	4
Boosting	4	4	—	—

8.4.1 Penalised Regression

These approaches build on linear or logistic regression models that seek unbiased estimates of the coefficients in the model. They do this by obtaining maximum likelihood estimates (MLEs) by maximising the log-likelihood function $\ell(\beta)$ for each SNP:

$$\ell(\beta; y, X) = \sum_{i=1}^{N} \left[y_i x_i \beta - \log(1 + \exp(x_i \beta)) \right],$$

where N is the total sample size, y_i is the outcome (case/control status), and x_i is the genotype of individual i, $i = 1,\ldots,N$.

However, effect sizes β are often enlarged in small datasets when compared to the true population parameters. A regression model can be penalised, or regularised, by including penalty terms in the loss function to shrink the coefficients of the model towards zero. This reduces overfitting and, on average, results in better predictions.

The penalties applied are in fact the L_1 norm,

$$\|x\|_1 = \sum_{j=1}^{p} |x_j|,$$

where p is the number of predictors, and the L_2 norm,

$$\|x\|_2 = \left(\sum_{j=1}^{p} x_j^2 \right)^{\frac{1}{2}} = \sqrt{x^T x},$$

also known as the Manhattan and Euclidean norms, respectively. These are applied to the log-likelihood to obtain the least absolute shrinkage and selection operator (LASSO) (Tibshirani, 1996) and Ridge (Hoerl & Kennard, 1970) regression models, which shrink coefficients to zero or near-zero, respectively. Penalised regression approaches are some of the most common in genetic epidemiology. They are used routinely in aspects of standard genetic analysis such as fine mapping (Schaid et al., 2018), and several computationally efficient implementations for biobank-scale data have recently been developed (Mbatchou et al., 2021; Qian et al., 2020; Raben et al., 2023). In complex trait prediction in general, they have received widespread interest (Okser et al., 2014), and in the genetics of brain disorders specifically, LASSO has been used in predicting AD (Oriol et al., 2019), anorexia (Guo et al., 2016), SCZ (Trakadis et al., 2019) and BD (Li et al., 2013), while ridge regression has been used in BD and SCZ (Li et al., 2013).

8.4.2 Tree-based Ensembles

Two similar but contrasting approaches aim to ensemble decision trees in order to improve prediction: RFs and GBMs, which use trees as the weak learner. The most popular implementation of decision trees is classification and regression trees (CARTs), developed jointly by Friedman and Breiman following independent efforts (Breiman, 1984), which partition the predictor space through recursive binary splits. When the resulting trees are allowed to continue splitting until no more are possible, we get deep trees that closely fit the training data. These trees both have high variance and struggle to learn additive effects (Hastie et al., 2009), making them unsuitable for highly polygenic traits with a largely additive genetic architecture, such as height. One approach to mitigate this is to average over many trees and reduce the variance. This is implemented in RFs, which were first introduced by Breiman (2001). RFs have a long history of interest in bioinformatics, likely due to their simplicity and relative ease of interpretation. Their ability to learn statistical interactions and their inherently algorithmic nature also make them more adaptable and easier to customise for developing new implementations. Indeed, they have been modified to detect epistasis (Yoshida et al., 2011; Li et al., 2016), include statistical tests (Wei & Lu, 2014) and handle genome-wide data (Schwarz et al., 2010). They have seen some application in predicting brain disorders, including SCZ (Trakadis et al., 2019), BD (Acikel et al., 2016) and autism (Engchuan et al., 2015), but with little of the customisation seen in studies focused on detecting interactions from genotypes in general.

In contrast to RFs, interest in applying GBMs to genetic data has only begun relatively recently. GBMs may be applied with any 'weak' learner, which individually contributes little to prediction and has high bias and low variance, though only the use of decision trees, as this is the most common, is considered here. Instead of learning many deeper trees in parallel and averaging over them to reduce variance as in RFs, here we sequentially learn many shallow trees and iteratively add more to reduce the bias. The most popular implementations are XGBoost, LightGBM and CatBoost. These rely on both the gradient and the Hessian of the loss function, and so can take arbitrary loss functions as long as they are differentiable. This further means that the L_1 and L_2 penalties introduced for regression can be applied here too to regularise the models (in addition to other hyperparameters related to tree construction, step size in gradient-based learning and number of boosting rounds). Taking XGBoost as an example, each split is then decided by the Gain in a level-wise manner, defined as

$$\text{Gain} = \frac{1}{2}\left[\frac{G_{jmL}^2}{H_{jmL}+\lambda} + \frac{G_{jmR}^2}{H_{jmR}+\lambda} - \frac{\left(G_{jmL}+G_{jmR}\right)^2}{H_{jmL}+H_{jmR}+\lambda}\right] - \gamma,$$

where the gradient and Hessian for the m-th iteration (tree) in node j have been combined to give G_{jm} and H_{jm} for the left (L) and right (R) branches from a node split j. The size of the tree is partly capped using γ, which sets the minimum gain required for splitting, while the strength of the L_2 penalty is adjusted using λ.

A boosting approach was recently adapted for large-scale genetic data, though using linear base learners (Klinkhammer et al., 2023). Gradient boosting has also been applied to predict SCZ by combining MRI and genetic data (Yang et al., 2010), from exome data (Trakadis et al., 2019), and by combining PRS with environmental factors (Bracher-Smith et al., 2022), in addition to being used to identify interactions that affect SCZ-related phenotypes (Andreasen et al., 2012). Similar efforts have been made in the prediction of AD using WGS data (Ahmed et al., 2022).

In addition to handling binary outcomes, both RFs and GBMs are able to learn time-to-event or 'survival' models where data are right-censored. For RFs, this is implemented in survival forests (Ishwaran et al., 2008), while for GBMs, approaches like XGBoost can take an arbitrary loss function as long as it is differentiable. This means the hazard survival function can be used as the loss, allowing time-to-event gradient boosting models to be trained. However, despite the prevalence of right-censored data in neurodegenerative disorders like dementia, we are only aware of a single application in each for genetic prediction using gradient boosting with a hazard loss (Khajehpiri et al., 2022) or random survival forests (Sarica et al., 2023), though the latter only uses the *APOE* genotype as a genetic predictor.

8.4.3 Neural Networks and Deep Learning

Following the increase in available training data and computing power (Figure 8.2) and seminal publications in fields such as computer vision and, more recently, large language models, neural networks have become frequently used in the biosciences. However, the architectures used are typically out-of-the-box approaches or simple perceptrons.

Briefly, neural networks rely on networks of 'neurons', a node in the network that takes a linear combination of inputs and then applies a (often non-linear) function to produce the output. Neurons are connected to each other in layers, such that inputs may be predictors or outputs from a previous layer. The weights for each linear combination at each node are learned through successive updates, which take the loss with respect to the outcome and the final output node, and then propagate this backward through the network. The simplest architecture is to have several layers of nodes, where all nodes within a layer connect to all nodes in the neighbouring layers, producing a feed-forward, fully connected multilayer perceptron (MLP). Though simple MLPs have been applied to AD (Shigemizu et al., 2023), SCZ (Chen et al., 2018) and autism (Engchuan et al., 2015), more often variations are used in the genetics of psychiatric disorders. Architectures have included radial basis function networks (Pirooznia et al., 2012; Aguiar-Pulido et al., 2010), embedding layers (Ghafouri-Fard et al., 2019), restricted Boltzmann machines (Wang et al., 2018), convolutional neural networks (CNNs) (Jo et al., 2022) and graph neural networks (Zhou et al., 2023).

More recent publications have involved modifying the architecture of the network using known information from protein–protein interactions or other networks, such as that implemented in PINNet (Kim & Lee, 2023) and GenNet (van Hilten et al., 2021), which have recently been applied to SCZ (Martins et al., 2024). These are typically used to define a sparse architecture, which results in a slightly constrained network with fewer parameters to train and some semblance of biological knowledge built into its connections to enhance interpretability, though it is not necessarily the case that biologically understandable connections will be prioritised.

8.4.4 Support Vector Machines

Unlike regression models that apply probability theory to assign a likelihood of a particular class, SVMs take a geometrical approach to directly assign class labels. They do this by considering all variables as dimensions in space, with specific observations (individuals in our case) being points in this space, and then learning a maximally separating hyperplane in the training set. The location of the hyperplane is only decided by the 'support vectors' – those that touch the margin, a buffer region defined around the hyperplane. In reality, there is never a perfectly separating hyperplane – some 'negative' class labels (controls) will still fall on the 'positive' (case) side of the hyperplane and vice versa, and so a soft hyperplane is learned while allowing for some number of violations. Though this 'max-margin' perspective may initially seem strange, it can be reformulated as a penalised optimisation problem using an L_2-regularised L_1 hinge loss

$$\min_{w,b} \frac{1}{2} \|w\|_2^2 + C\sum_{i=1}^{N} \ell_{\text{hinge}}(b,\ w; y_i, x_i)$$

where w are the weights and b is the bias, analogous to the betas and intercept in regression. Similar to other penalised approaches, the formulation can be adapted to an L_2-regularised L_2 hinge loss by changing the penalty term on the loss function. To optimise these models, the parameter C, the cost, which defines the upper limit for violations to the margin, must be tuned. However, it is not the hyperplane itself, but the use of kernel functions that makes SVMs an attractive option. These allow us to obtain a non-linear decision boundary in the original space, without explicitly having to transform variables. The most popular kernels are the linear, polynomial and radial basis kernels, with the latter two having their own additional hyperparameter to constrain flexibility of the hyperplane (d and γ, respectively).

SVMs have been used for genetic prediction in SCZ (Yang et al., 2010; Vivian-Griffiths et al., 2019) and AD (Hamid et al., 2016; Zhang et al., 2014; An et al., 2017; Zhou et al., 2019; Bi et al., 2019; Hamid et al., 2019), but more widespread adoption has been limited for several reasons. Training SVMs with a non-linear kernel on a large number of predictors is typically slower than RFs or GBMs, and less customisable than neural networks. They have also historically fallen behind in interpretability, as it is not possible to comprehend relationships between predictors in the kernel space, and the method does not have a built-in simple interpretation like coefficients in regression models, importance scores in gradient boosting or the host of explainable artificial intelligence (XAI) methods now available for neural networks, though general methods like Shapley additive explanations (SHAP) values can also be applied to SVMs. Finally, SVMs rely on a good choice of kernel, though in practice, linear or radial basis kernels are often the best performing with appropriate hyperparameter tuning.

8.5 Challenges and Opportunities in ML Applications

8.5.1 Common Pitfalls in Applied ML

Use of ML methods to predict psychiatric and neurodegenerative disorders from genetic data has been reviewed recently (e.g. Bracher-Smith et al., 2020). Systematic reviews that assess ROB – the likelihood that performance estimates are inflated due to choices in model development or validation – have highlighted widespread issues in the field, as the majority of studies have high ROB with inadequate reporting on standard modelling steps. This is not just limited to the genetics of brain disorders – high ROB (Navarro et al., 2021) and biased reporting (Andaur Navarro et al., 2023) are prevalent in applied ML. The issues can be broken down into those surrounding participants, predictors, outcome and analysis, as is done in the Prediction model Risk of Bias Assessment Tool (PROBAST) framework for assessing ROB (Moons et al., 2019). For genetic prediction of brain disorders, problems often arise due to methodological differences between the fields of clinical prediction modelling, ML and genetic association studies, wherein a study that correctly implements cross-validation may fail to perform adequate QC beforehand or handle population stratification; conversely, a study with stringent QC may inadvertently use the test set when selecting predictors to go into the model.

For participants, genetic studies frequently use a case-control design, which is effective for identifying genetic risk factors for rare outcomes but inadequate for clinical prediction modelling as it does not permit the estimation of absolute risks. Instead, designs such as case-cohort, nested case-control or prospective cohort are preferred. Case-cohort and nested case-control designs involve sampling from an existing cohort and can be used for prediction models if the sampling proportion in controls is accounted for in the analysis (Biesheuvel et al., 2008). In order to project the prediction (tested in case-control study) to the whole population, the prediction accuracy should be adjusted based on the disease prevalence in the population and the ratio of cases to controls in the sample (Kallner, 2018). Final prediction models can be adjusted for the base rate (prevalence) differences from nested case-control designs using Elkan's transformation (Elkan, 2001) or further recalibrated with Platt scaling if needed (Platt, 1999).

Predictors and outcome measures are typically marked as high ROB due to smaller issues that are easily addressed in study design or reporting. Handling of predictors often involves bias, as univariable tests of association are applied routinely in GWAS and are often used in selection of predictors for genetic prediction models. However, their use in prediction modelling is discouraged because predictors may differ in importance when evaluated in isolation compared to when considered alongside other variables (Sun et al., 1996). Outcome handling also often falls foul of poor reporting, as many studies fail to clearly note that outcome measurement (e.g., through clinical assessment) and predictor measurement (e.g., through genotyping and QC) were blind to each other.

ROB concerns around analysis are more serious and require re-analysis. Most studies are assigned high ROB here due to failing to adhere to appropriate procedures for ML, with many 'leaking' information between train and test sets by not applying predictor manipulations or selection exclusively to the training set (or fold if within cross-validation), or by using the test set to adjust model hyperparameters. This can significantly bias the estimates of prediction performance (Vabalas et al., 2019).

Analysis efforts also frequently stumble on achieving thorough hyperparameter tuning, optimal internal validation or reporting model calibration alongside discrimination and classification metrics (Table 8.2). All of these have well-documented solutions, and readers are encouraged to follow the recently developed Transparent Reporting of a multivariable prediction model for Individual Prognosis Or Diagnosis (TRIPOD)+AI guidelines (Collins et al., 2024), and consider guidelines specific to genetic data, such as GRIPS (Janssens et al., 2011).

Perhaps the largest issue that lacks a simple one-size-fits-all solution is handling population structure. This is a specific source of bias in genetic studies, wherein the association of genetic ancestry with predictors and the outcome can confound both associations (Marchini et al., 2004) and predictions (Martin et al., 2017) from genetic data. Application

Table 8.2 Common pitfalls in developing and validating ML models of brain disorders.

Step	Issue	Description	Solution
Model development	Leakage	Selection or transformation of predictions prior to CV; use of the test split to inform training	Clearly separate train and test data before analysis, and use standard pipeline implementations (such as in scikit-learn) to avoid inadvertent leakage
Model development	Hyperparameter tuning	Small, arbitrary handful of options reported for HP search, which were likely to be partially chosen through trial and error, unclear reporting	Systematically search using cross-validated randomised search or more recent approaches in the train/validation splits only; report all distributions or grids searched
Internal validation	Split-sample or apparent validation	Prediction accuracies overly optimistic for split-sample approaches with small sample sizes, and from hyperparameter tuning through repeated rounds of CV	Nested cross-validation, which optimises HPs in an inner fold and evaluates the model on the outer fold (Vabalas et al., 2019), or split-sample approaches with large samples and tests of predictor and covariate distributions
Internal/external validation	Model calibration	Comparing observed and predicted probabilities of the outcome occurring; often absent in genetic prediction (Janssens et al., 2011)	Recalibrate using Platt scaling (Platt, 1999); assess calibration graphically using loess smoothers (Austin and Steyerberg, 2014)
Internal/external validation	Population stratification	Genetic ancestry is associated with predictors and outcome, confounding associations	Exclude participants in outlier groups or adjust prior to or within ML models

of supervised ML with the aim of detecting ancestry has proven particularly effective (Flagel et al., 2019; Schrider & Kern, 2018), raising the concern that flexible models may inadvertently learn complex patterns related to ancestry. Addressing this by excluding outlying populations before analysis risks losing many participants, and even within a single population, lower-level population stratification still impacts analyses (Cook et al., 2020). Instead, solutions implemented during modelling appear to work best but depend on the model itself. Modifications to RFs have been proposed (Stephan et al., 2015; Zhao et al., 2012), and neural networks can include the covariates in the training model but only predict from other nodes (van Hilten et al., 2021). However, there is currently no consensus approach for gradient boosting, no systematic comparison of techniques, and no clear evidence that adjusting using principal components (a linear combination of variants) prevents further learning of non-linear ancestry-based effects.

The potential for ML methods to deliver enhanced and robust predictions is high. To realise this potential, it is essential to establish and adhere to rigorous standards for reporting on implementation related to handling of samples, predictors, hyperparameters and prevention of information leakage. Achieving successful outcomes is contingent upon strict adherence to these standards throughout the development, validation and reporting of models. Without such adherence, the promise of improved prediction through ML cannot be fulfilled. However, while some critical challenges have not been fully addressed, there are also significant opportunities waiting to be explored.

8.5.2 Under-utilised and Unexplored Approaches

As discussed so far, a broad range of methods have been applied to brain disorders, but these mostly focus on out-of-the-box approaches with minimal customisation. By contrast, the largest successes in DL have been from highly customised models that embed assumptions about the data into the model, such as the use of convolutions and pooling in CNNs, which capture the translational invariance inherent in images, or the use of transformers in large language models (LLMs), which excel at incorporating distant relationships in sequence data that are essential for understanding language. Customising neural networks by integrating biological domain knowledge – beyond the pathway information that has been recently incorporated into models – could yield networks with improved predictive accuracy and greater biological interpretability. There is an immense wealth of bioinformatics databases that could be leveraged to inform the design of neural network architectures. Custom layers that encode our understanding of how different genetic features map to disease status, similar to how convolutions capture spatial hierarchies in computer vision, have the potential to significantly advance the field. However, it is important to recognise that features from other domain-specific models cannot be directly transplanted with the same level of success. Taking convolutional layers again as an example, though they are effective in other contexts, they may be more prone to identifying correlations within LD blocks when applied to genotype–phenotype relationships, potentially resulting in population-specific findings rather than universal genetic associations.

Outside of DL, RF and gradient boosting approaches offer promising avenues for further customisation. A large number of studies have extensively adapted RFs in various ways to handle genetic data (Botta et al., 2014; Nguyen et al., 2015; te Beest et al., 2017; Deng & Runger, 2013). Most of these custom implementations were originally designed for detecting interactions and managing high-dimensional data. However, these methods have not yet been widely applied to recent large-scale datasets from GWAS consortia working on brain disorders, as other techniques have become more prevalent in recent years. Similarly, gradient boosting methods, which have demonstrated their effectiveness in other areas of ML, could be further adapted to better address the specific challenges posed by genetic data. Implementations that natively handle LD, incorporate techniques from RFs to embed biological knowledge or enable batchwise data loading from standard genotype file formats for distributed training in low-memory environments would be particularly valuable and warrant further exploration.

The challenge of population stratification, a well-known source of confounding in genetic association studies, is not adequately addressed in most ML applications in the genetics of brain disorders. Traditional methods like PCA have been useful, but they are limited to the linear effects of ancestry and cannot always be included in the model itself without also biasing the final prediction, such as in SVMs. There is a pressing need for a clear comparison of approaches that can adjust for population stratification more effectively within ML models, potentially through modifications to existing methods like RFs and gradient boosting or through novel strategies within neural networks. Overall, the field has significant room for growth in the customisation and refinement of ML approaches.

8.5.3 Emerging Strengths of a Federated Approach

Though the number of studies and participants in meta-analyses from consortia continues to increase, many genetic datasets remain siloed in computing centres in universities, hospitals or companies, sometimes despite a willingness to cooperate, because data access agreements do not allow for the transfer of raw data. The recent development of FL allows for collaborative learning on private biological data, such as electronic health records, neuroimaging data, and genomic or other 'omics' data, to be analysed on separate servers, potentially in separate countries, without compromising on data privacy. FL was introduced by Google (Konečný et al., 2016a; Konečný et al., 2016b) as an alternative to CL to allow model training to be coordinated across local edge devices or clients and then aggregated as a single global model on a central server. There are a number of variants of FL, but the core principle is that a global model is created by combining local models trained in each separate server, maintaining security and privacy. Local weight updates from learning, and not any raw data, are all that is shared across sites. The exact details of the aggregation method, initialisation of the network, and where the local and updated global parameters are held are some of the features that define the different implementations of FL.

In addition to addressing privacy and data-sharing concerns, FL may also mitigate issues related to divergent predictor distributions across cohorts. In centralised ML, data are typically assumed to be independent and identically distributed (IID), but this assumption is often violated in large meta-analyses of genetic data, particularly when samples from different global populations are combined. Traditionally, studies have not explicitly addressed these violations, adhering to the standard practises in the field. In contrast, FL treats each site separately, with weight updates being iteratively combined. This approach allows for the aggregation of local models in a way that can accommodate non-IID data, similar to the techniques used in classical meta-analyses (Table 8.3).

Applications of FL to brain disorders have largely focused on non-genetic data. To date, only a handful have investigated FL for disease prediction from genomic data. Kolobkov et al. (2024) applied FL approaches to data from the UK Biobank for the prediction of diseases and ancestry from genetics. They found that the accuracy between CL and FL paradigms was similar, particularly when the client–server communication was increased (i.e., fewer local epochs before updating global parameters). Importantly, they show that federated PCA performs as well as standard centralised PCA for dimensionality reduction, opening the possibility of entire federated pipelines from quality control to disease

Table 8.3 Common aggregation methods in FL and their similarities with classical meta-analysis techniques.

Method	Description	Relation to classical meta-analysis
FedAvg	A weighted combination of local model parameters, with weights proportional to local data size	Similar to a fixed-effect meta-analysis, but iteratively combines model updates to maintain a global model, rather than taking the local MLEs and averaging once
FedProx	Similar to FedAvg, but uses a modified objective function with a proximal term to better handle heterogeneous data	Similar to a random-effect meta-analysis, which adjusts weights to account for within-study heterogeneity
FedOpt	Adaptively aggregates local model updates to scale and adjust them due to data size and model performance	Does not have a direct comparison in classic meta-analysis, but in principle is similar to iteratively trying alternative weights for studies based on additional factors

prediction. Work by Raimondi et al. (2023), this time using FL for analysing the genetics of Crohn's disease, also demonstrated similar performance between FL and CL. Applications in brain disorders specifically have focused on other omics datasets. Detailed experiments by Danek et al. (2024), here applied to biomarkers in PD, again highlight the similar performance achieved between CL and FL implementations, but note that the higher number of FL nodes can degrade performance. Collectively, these studies have established a solid foundation for expanding the use of FL in the genetics of brain disorders, paving the way for future research to leverage its full potential.

8.6 Conclusion

This chapter has explored the growing intersection of ML and the genetics of brain disorders, highlighting both the opportunities and challenges inherent in this multidisciplinary field. While traditional methods like PRS have provided a foundation for genetic prediction, they fall short in accounting for the complex, non-linear relationships between genetic variants and disease outcomes. ML, with its flexibility and capacity for handling large-scale data, offers a promising alternative to enhance predictive accuracy and uncover novel genetic insights. However, the successful application of ML in this domain requires careful consideration of issues such as noisy labels, population stratification and the high-dimensional nature of genetic data.

Looking ahead, the continued integration of ML into the field will likely depend on the development of more customised models and the adoption of innovative approaches like FL, which can address the privacy and data-sharing limitations that currently hinder large-scale collaborative research. By adhering to rigorous methodological standards and embracing the unique complexities of genetic data, ML has the potential to significantly advance our understanding of brain disorders and pave the way for more personalised and effective interventions.

References

Abdellaoui, A., Yengo, L., Verweij, K.J.H. et al. (2023). 15 years of GWAS discovery: realizing the promise. *The American Journal of Human Genetics* 110 (2): 179–194.

Acikel, C., Aydin Son, Y., Celik, C. et al. (2016). Evaluation of potential novel variations and their interactions related to bipolar disorders: analysis of genome-wide association study data. *Neuropsychiatric Disease and Treatment* 12: 2997–3004.

Agerbo, E., Sullivan, P.F., Vilhjalmsson, B.J. et al. (2015). Polygenic risk score, parental socioeconomic status, family history of psychiatric disorders, and the risk for schizophrenia: a Danish population-based study and meta-analysis. *JAMA Psychiatry* 72 (7): 635–641.

Aguiar-Pulido, V., Seoane, J.A., Rabuñal, J.R. et al. (2010). Machine learning techniques for single nucleotide polymorphism—disease classification models in schizophrenia. *Molecules* 15: 4875–4889.

Ahmed, H., Soliman, H., and Elmogy, M. (2022). Early detection of Alzheimer's disease using single nucleotide polymorphisms analysis based on gradient boosting tree. *Computers in Biology and Medicine* 146: 105622.

An, L., Adeli, E., Liu, M. et al. (2017). A hierarchical feature and sample selection framework and its application for Alzheimer's disease diagnosis. *Scientific Reports* 7 (1): 45269.

Andaur Navarro, C.L., Damen, J.A.A., Takada, T. et al. (2023). Systematic review finds "spin" practices and poor reporting standards in studies on machine learning-based prediction models. *Journal of Clinical Epidemiology* 158: 99–110.

Andreasen, N.C., Wilcox, M.A., Ho, B.-C. et al. (2012). Statistical epistasis and progressive brain change in schizophrenia: an approach for examining the relationships between multiple genes. *Molecular Psychiatry* 17 (11): 1093–1102.

Austin, P.C. and Steyerberg, E.W. (2014). Graphical assessment of internal and external calibration of logistic regression models by using loess smoothers. *Statistics in Medicine* 33 (3): 517–535.

Baselmans, B.M.L., Yengo, L., van Rheenen, W. et al. (2021). Risk in Relatives, Heritability, SNP-based heritability, and genetic correlations in psychiatric disorders: a review. *Biological Psychiatry* 89 (1): 11–19.

Bature, F., Guinn, B.A., Pang, D. et al. (2017). Signs and symptoms preceding the diagnosis of Alzheimer's disease: a systematic scoping review of literature from 1937 to 2016. *BMJ Open* 7 (8): e015746.

Bellenguez, C., Grenier-Boley, B., and Lambert, J.C. (2020). Genetics of Alzheimer's disease: where we are, and where we are going. *Current Opinion in Neurobiology* 61: 40–48.

Bellou, E., Stevenson-Hoare, J., and Escott-Price, V. (2020). Polygenic risk and pleiotropy in neurodegenerative diseases. *Neurobiology of Disease* 142: 104953.

Bi, X.-a., Cai, R., Wang, Y. et al. (2019). Frontiers | Effective diagnosis of Alzheimer's disease via multimodal fusion analysis framework. *Frontiers in Genetics* 10: 976.

Biesheuvel, C.J., Vergouwe, Y., Oudega, R. et al. (2008). Advantages of the nested case-control design in diagnostic research. *BMC Medical Research Methodology* 8: 48.

Bodmer, W., Bonilla, C., Bodmer, W. et al. (2008). Common and rare variants in multifactorial susceptibility to common diseases. *Nature Genetics* 40 (6): 695–701.

Botta, V., Louppe, G., Geurts, P. et al. (2014). Exploiting SNP correlations within random forest for genome-wide association studies. *Plos One* 9 (4): e93379.

Bracher-Smith, M., Crawford, K., Escott-Price, V. et al. (2020). Machine learning for genetic prediction of psychiatric disorders: a systematic review. *Molecular Psychiatry* 26 (1): 1–10.

Bracher-Smith, M., Rees, E., Menzies, G. (2022). Machine learning for prediction of schizophrenia using genetic and demographic factors in the UK biobank. *Schizophrenia Research* 246: 156–164.

Breiman, L. (1984). *Classification and Regression Trees*. Boca Raton: CRC Press.

Breiman, L. (2001). Random forests. *Machine Learning* 45 (1): 5–32.

Chen, J., Wu, J.-s., Mize, T. et al. (2018). Prediction of schizophrenia diagnosis by integration of genetically correlated conditions and traits. *Journal of Neuroimmune Pharmacology* 13 (4): 532–540.

Collins, G.S., Moons, K.G.M., Dhiman, P. et al. (2024). TRIPOD+AI statement: updated guidance for reporting clinical prediction models that use regression or machine learning methods. *BMJ* 385: q902.

Brainstorm Consortium, Anttila, V., Bulik-Sullivan, B. et al. (2018). Analysis of shared heritability in common disorders of the brain. *Science* 360 (6395): eaap8757.

Cook, J.P., Mahajan, A., and Morris, A.P. (2020). Fine-scale population structure in the UK Biobank: implications for genome-wide association studies. *Human Molecular Genetics*. 29 (16): 2803–2811.

Corder, E.H., Saunders, A.M., Strittmatter, W.J. et al. (1993). Gene dose of apolipoprotein E type 4 allele and the risk of Alzheimer's disease in late onset families. *Science* 261 (5123): 921–923.

Danek, B.P., Makarious, M.B., Dadu, A. et al. (2024). Federated learning for multi-omics: a performance evaluation in Parkinson's disease. *Patterns* 5 (3): 100945.

Deelen, J., Evans, D.S., Arking, D.E. et al. (2019). A meta-analysis of genome-wide association studies identifies multiple longevity genes. *Nature Communications* 10 (1): 3669.

Demirkan, A., Penninx, B.W., Hek, K. et al. (2011). Genetic risk profiles for depression and anxiety in adult and elderly cohorts. *Molecular Psychiatry* 16 (7): 773–783.

Deng, H. and Runger, G. (2013). Gene selection with guided regularized random forest. *Pattern Recognition*. 46 (12): 3483–3489.

Duong, S., Patel, T., and Chang, F. (2017). Dementia: what pharmacists need to know. *Canadian Pharmacists Journal* 150 (2): 118–129.

Dyck, C.H.v., Swanson, C.J., Aisen, P. et al. (2023). Lecanemab in early Alzheimer's disease. *New England Journal of Medicine* 388 (1): 9–21.

Elkan, C. (2001). The foundations of cost-sensitive learning. Proceedings of the 17th international joint conference on Artificial intelligence - Volume 2.

Engchuan, W., Dhindsa, K., Lionel, A.C. et al. (2015). Performance of case-control rare copy number variation annotation in classification of autism. *BMC Medical Genomics* 8 (Suppl 1): S7.

Escott-Price, V. and Hardy, J. (2022). Genome-wide association studies for Alzheimer's disease: bigger is not always better. *Brain Communications* 4 (3): fcac125.

Escott-Price, V., IPDGC, Nalls, M. et al. (2015). Polygenic risk of Parkinson disease is correlated with disease age at onset. *Annals of Neurology* 77 (4): 582–591.

Escott-Price, V., Myers, A.J., Huentelman, M. et al. (2017a). Polygenic risk score analysis of pathologically confirmed Alzheimer disease. *Annuals of Neurology* 82 (2): 311–314.

Escott-Price, V., Shoai, M., Pither, R. et al. (2017b). Polygenic score prediction captures nearly all common genetic risk for Alzheimer's disease. *Neurobiology of* Ageing 49: 214 e7–e11.

Escott-Price, V., Sims, R., Bannister, C. et al. (2015). Common polygenic variation enhances risk prediction for Alzheimer's disease. *Brain* 138 (12): 3673–3684.

Falconer, D.S. (1996). *Introduction to Quantitative Genetics*. Pearson.

Fisher, R.A. (1919). XV.— The correlation between relatives on the supposition of Mendelian inheritance. | Earth and environmental science transactions of the royal society of Edinburgh | Cambridge core. *Earth and Environmental Science Transactions of The Royal Society of Edinburgh* 52 (2).

Flagel, L., Brandvain, Y., and Schrider, D.R. (2019). The unreasonable effectiveness of convolutional neural networks in population genetic inference. *Molecular Biology and Evolution* 36 (2): 220–238.

FLOPS [Website]. Wikipedia https://en.wikipedia.org/wiki/FLOPS#Hardware_costs. (accessed 09 August 2024).

Foley, K., Borchian, P., Garceau, D. et al. (2020). Determining the role of ApoE4 in age-related cerebrovascular decline. *Innovation in Aging* 4 (1): 118–119.

Fortea, J., Pegueroles, J., Alcolea, D. et al. (2024). APOE4 homozygosity represents a distinct genetic form of Alzheimer's disease. *Nature Medicine* 30 (5): 2093.

Ge, T., Chen, C.Y., Ni, Y. et al. (2019). Polygenic prediction via Bayesian regression and continuous shrinkage priors. *Nature Communications* 10 (1): 1776.

Ghafouri-Fard, S., Taheri, M., Omrani, M.D. et al. (2019). Application of single-nucleotide polymorphisms in the diagnosis of autism spectrum disorders: a preliminary study with artificial neural networks. *Journal of Molecular Neuroscience* 68 (4): 515–521.

Grotzinger, A.D., Mallard, T.T., Akingbuwa, W.A. et al. (2022). Genetic architecture of 11 major psychiatric disorders at biobehavioral, functional genomic and molecular genetic levels of analysis. *Nature Genetics* 54 (5): 548–559.

Guo, Y., Wei, Z., Keating, B.J. et al. (2016). Machine learning derived risk prediction of anorexia nervosa. *BMC Medical Genomics* 9 (1): 4.

Hamid, M.M.A.E., Mabrouk, M.S., and Omar, Y.M.K. (2019). Developing an early predictive system for identifying genetic biomarkers associated to Alzheimer's disease using machine learning techniques. *Biomedical Engineering: Applications, Basis and Communications* 31 (05): 1950040.

Hamid, M.M.A.E., Omar, Y.M.K., and Mabrouk, M.S. (2016). Identifying genetic biomarkers associated to Alzheimer's disease using Support Vector Machine | IEEE Conference Publication | IEEE Xplore. *2016 8th Cairo International Biomedical Engineering Conference (CIBEC).*

Hara, T., Owada, Y., Takata, A. et al. (2022). Genetics of bipolar disorder: insights into its complex architecture and biology from common and rare variants. *Journal of Human Genetics* 68 (3): 183–191.

Hardy, J. and Escott-Price, V. (2019). Genes, pathways and risk prediction in Alzheimer's disease. *Human Molecular Genetics* 28 (R2): R235–R240.

Harrison, J.R., Mistry, S., Muskett, N. et al. (2020). From polygenic scores to precision medicine in Alzheimer's disease: a systematic review. *Journal of Alzheimer's Disease* 74 (4): 1271–1283.

Hastie, T., Tibshirani, R., and Friedman, J. (2009). *The Elements of Statistical Learning: Data Mining, Inference, and Prediction*. Springer Science & Business Media.

Healy, D.G., Falchi, M., O'Sullivan, S.S. et al. (2008). Phenotype, genotype, and worldwide genetic penetrance of LRRK2-associated Parkinson's disease: a case-control study. *The Lancet Neurology* 7 (7): 583–590.

Hoerl, A.E. and Kennard, R.W. (1970). Ridge regression: biased estimation for nonorthogonal problems. *Technometrics* 12 (1): 55–67.

International Schizophrenia Consortium; Purcell, S.M., Wray, N.R. (2009). Common polygenic variation contributes to risk of schizophrenia and bipolar disorder. *Nature* 460 (7256): 748–752.

Ishwaran, H., Kogalur, U.B., Blackstone, E.H. et al. (2008). Random survival forests. *The Annals of Applied Statistics* 2 (3).

Janssens, A.C., Ioannidis, J.P., Bedrosian, S. et al. (2011). Strengthening the reporting of Genetic RIsk Prediction Studies (GRIPS): explanation and elaboration. *Journal of Clinical Epidemiology* 64 (8): e1–e22.

Jo, T., Nho, K., Bice, P. et al. (2022). Initiative FTAsDN. Deep learning-based identification of genetic variants: application to Alzheimer's disease classification. *Briefings in Bioinformatics* 23 (2): bbac022.

Kallner, A. (2018). Bayes' theorem, the ROC diagram and reference values: definition and use in clinical diagnosis. *Biochemia Medica (Zagreb)* 28 (1): 010101.

Karantzoulis, S. and Galvin, J.E. (2011). Distinguishing Alzheimer's disease from other major forms of dementia. *Expert Review of Neurother* 11 (11): 1579–1591.

Karlsson, I.K., Escott-Price, V., Gatz, M. et al. (2022). Measuring heritable contributions to Alzheimer's disease: polygenic risk score analysis with twins. *Brain Communications* 4 (1): fcab308.

Khajehpiri, B., Moghaddam, H.A., Forouzanfar, M. et al. (2022). Survival analysis in cognitively normal subjects and in patients with mild cognitive impairment using a proportional hazards model with extreme gradient boosting regression. *Journal of Alzheimer's Disease: JAD* 85 (2): 837–850.

Kim, Y. and Lee, H. (2023). PINNet: a deep neural network with pathway prior knowledge for Alzheimer's disease. *Frontiers in Aging Neuroscience* 15: 1126156.

Klinkhammer, H, Staerk, C., Maj, C. et al. (2023). A statistical boosting framework for polygenic risk scores based on large-scale genotype data. *Frontiers in Genetics* 13: 1076440.

Koch, S., Schmidtke, J., Krawczak, M. et al. (2023). Clinical utility of polygenic risk scores: a critical 2023 appraisal. *Journal of Community Genetics* 14 (5): 471–487.

Kolobkov, D., Mishra Sharma, S., Medvedev, A. et al. (2024). Efficacy of federated learning on genomic data: a study on the UK Biobank and the 1000 Genomes Project. *Frontiers in Big Data* 7: 1266031.

Konečný, J., Brendan McMahan, H., Ramage, D. et al. (2016a). Federated optimization: distributed machine learning for on-device intelligence. arXiv e-prints.

Konečný, J., Brendan McMahan, H., Yu, F.X. et al. (2016b). Federated learning: strategies for improving communication efficiency. arXiv e-prints.

Kopp, W., Monti, R., Tamburrini, A. et al. (2020). Deep learning for genomics using janggu. *Nature Communications* 11 (1): 3488.

Lall, K., Magi, R., Morris, A. et al. (2017). Personalized risk prediction for type 2 diabetes: the potential of genetic risk scores. *Genetics in Medicine* 19 (3): 322–329.

Lewis, C.M. and Vassos, E. (2017). Prospects for using risk scores in polygenic medicine. *Genome Medicine* 9 (1): 96.

Li, C., Yang, C., Gelernter, J. et al. (2013). Improving genetic risk prediction by leveraging pleiotropy. *Human Genetics* 133 (5): 639–650.

Li, J., Malley, J.D., Andrew, A.S. et al. (2016). Detecting gene-gene interactions using a permutation-based random forest method. *BioData Mining* 9 (1): 14.

Li, W. and Reich, J.A. (2000). Complete enumeration and classification of two-locus disease models. *Human Heredity* 50 (6): 334–349.

Lloyd-Jones, L.R., Zeng, J., Sidorenko, J. et al. (2019). Improved polygenic prediction by Bayesian multiple regression on summary statistics. *Nature Communications* 10 (1): 5086.

Marchini, J., Cardon, L.R., Phillips, M.S. et al. (2004). The effects of human population structure on large genetic association studies. *Nature Genetics* 36 (5): 512–517.

Martin, A.R., Gignoux, C.R., Walters, R.K. et al. (2017). Human demographic history impacts genetic risk prediction across diverse populations. *The American Journal of Human Genetics* 100 (4): 635–649.

Martins, D., Abbasi, M., Egas, C. et al. (2024). Enhancing schizophrenia phenotype prediction from genotype data through knowledge-driven deep neural network models. *Genomics* 116 (5): 110910.

Mbatchou, J., Barnard, L., Backman, J. et al. (2021). Computationally efficient whole-genome regression for quantitative and binary traits. *Nature Genetics* 53 (7): 1097–1103.

Moons, K.G.M., Wolff, R.F., Riley, R.D. et al. (2019). PROBAST: A tool to assess the risk of bias and applicability of prediction model studies. *Annals of Internal Medicine* 170 (1): 51–58.

Mullins, N., Forstner, A.J., O'Connell, K.S. et al. (2021). Genome-wide association study of more than 40,000 bipolar disorder cases provides new insights into the underlying biology. *Nature Genetics* 53 (6): 817–829.

Natarajan, P., Young, R., Stitziel, N.O. et al. (2017). Polygenic risk score identifies subgroup with higher burden of atherosclerosis and greater relative benefit from statin therapy in the primary prevention setting. *Circulation* 135 (22): 2091–2101.

Navarro, C.L.A., Damen, J.A.A., Takada, T. et al. (2021). Risk of bias in studies on prediction models developed using supervised machine learning techniques: systematic review. *BMJ* 375: n2281.

Nguyen, T.-T., Huang, J.Z., Wu, Q. et al. (2015). Genome-wide association data classification and SNPs selection using two-stage quality-based Random Forests. *BMC Genomics* 16 (2): S5.

Nichols, E., Steinmetz, J.D., Vollset, S.E. et al. (2022). Estimation of the global prevalence of dementia in 2019 and forecasted prevalence in 2050: an analysis for the global burden of disease study 2019. *The Lancet Public Health* 7 (2): e105–e125.

Okser, S., Pahikkala, T., Airola, A. et al. (2014). Regularized machine learning in the genetic prediction of complex traits. *PLOS Genetics* 10 (11): e1004754.

Oriol, J.D.V., Vallejo, E.E., Estrada, K. et al. (2019). Initiative TAsDN. Benchmarking machine learning models for late-onset Alzheimer's disease prediction from genomic data. *BMC Bioinformatics* 20 (1): 709.

Owen Michael, J. (2014). New approaches to psychiatric diagnostic classification. *Neuron* 84 (3): 564–571.

Paraskevaidi, M., Martin-Hirsch, P.L., and Martin, F.L. (2018). Progress and challenges in the diagnosis of dementia: a critical review. *ACS Chemical Neuroscience* 9 (3): 446–461.

Pate, A., Emsley, R., Ashcroft, D.M. et al. (2019). The uncertainty with using risk prediction models for individual decision making: an exemplar cohort study examining the prediction of cardiovascular disease in English primary care. *BMC Medicine* 17 (1): 134.

Pirooznia, M., Seifuddin, F., Judy, J. et al. (2012). Data mining approaches for genome-wide association of mood disorders. *Psychiatric Genetics* 22 (2): 55–61.

Platt, J. (1999). Probabilistic outputs for support vector machines and comparisons to regularized likelihood methods. *Advances in Large Margin Classifiers* 10 (3): 61–74.

Volker Strobel. (2018). Pold87/academic-keyword-occurrence: First release. (v1.0.0). Zenodo. https://doi.org/10.5281/zenodo.1218409

Price, A.L., Patterson, N.J., Plenge, R.M. et al. (2006). Principal components analysis corrects for stratification in genome-wide association studies. *Nature Genetics* 38 (8): 904–909.

Qian, J., Tanigawa, Y., Du, W. et al. (2020). A fast and scalable framework for large-scale and ultrahigh-dimensional sparse regression with application to the UK Biobank. *PLoS Genetics* 16 (10): e1009141.

Raben, T.G., Lello, L., Widen, E. et al. (2023). Biobank-scale methods and projections for sparse polygenic prediction from machine learning. *Scientific Reports* 13 (1): 11662.

Rahimi, J., Kovacs, G.G., Rahimi, J. et al. (2014). Prevalence of mixed pathologies in the aging brain. *Alzheimer's Research & Therapy* 6 (9): 82.

Raimondi, D., Chizari, H., Verplaetse, N. et al. (2023). Genome interpretation in a federated learning context allows the multi-centre exome-based risk prediction of Crohn's disease patients. *Scientific Reports* 13 (1): 19449.

Rees, E. and Kirov, G. (2021). Copy number variation and neuropsychiatric illness. *Current Opinion in Genetics & Development* 68: 57–63.

Rowe, T.W., Katzourou, I.K., Stevenson-Hoare, J.O. et al. (2021). Escott-Price V. Machine learning for the life-time risk prediction of Alzheimer's disease: a systematic review. *Brain Communications* 3 (4): fcab246.

Sarica, A., Aracri, F., Bianco, M.G. et al. (2023). Explainability of random survival forests in predicting conversion risk from mild cognitive impairment to Alzheimer's disease. *Brain Informatics* 10 (1): 31.

Schaid, D.J., Chen, W., Larson, N.B. et al. (2018). From genome-wide associations to candidate causal variants by statistical fine-mapping. *Nature Reviews Genetics* 19 (8): 491–504.

Scheltens, P., Strooper, B.D., Kivipelto, M. et al. (2021). Alzheimer's disease. *The Lancet* 397 (10284): 1577–1590.

Schrider, D.R. and Kern, A.D. (2018). Supervised machine learning for population genetics: a new paradigm. *Trends Genetics* 34 (4): 301–312.

Schwarz, D.F., König, I.R., and Ziegler, A. (2010). On safari to Random Jungle: a fast implementation of random forests for high-dimensional data. *Bioinformatics (Oxford, England)* 26 (14): 1752–1758.

Shigemizu, D., Akiyama, S., Suganuma, M. et al. (2023). Classification and deep-learning–based prediction of Alzheimer disease subtypes by using genomic data. *Translational Psychiatry* 13 (1): 232.

Sims, J.R., Zimmer, J.A., Evans, C.D. et al. (2023). Donanemab in Early Symptomatic Alzheimer Disease: The TRAILBLAZER-ALZ 2 Randomized Clinical Trial. *JAMA* 330 (6): 512–527.

Singh, T., Walters, J.T.R., Johnstone, M. et al. (2017). The contribution of rare variants to risk of schizophrenia in individuals with and without intellectual disability. *Nature Genetics* 49 (8): 1167–1173.

Song, H., Kim, M., Park, D. et al. (2020). Learning from noisy labels with deep neural networks: a survey. arXiv e-prints.

Speed, D., Cai, N., Consortium, U. et al. (2017). Reevaluation of SNP heritability in complex human traits. *Nature Genetics* 49 (7): 986–992.

Speed, D., Holmes, J., and Balding, D.J. (2020). Evaluating and improving heritability models using summary statistics. *Nature Genetics* 52 (4): 458–462.

Stephan, J., Stegle, O., and Beyer A. (2015). A random forest approach to capture genetic effects in the presence of population structure. *Nature Communications* 6: 7432.

Sun, G.W., Shook, T.L., and Kay, G.L. (1996). Inappropriate use of bivariable analysis to screen risk factors for use in multivariable analysis. *Journal of Clinical Epidemiology* 49 (8): 907–916.

te Beest, D.E., Mes, S.W., Wilting, S.M. et al. (2017). Improved high-dimensional prediction with Random Forests by the use of co-data. *BMC Bioinformatics* 18 (1).

Tibshirani, R. (1996). Regression shrinkage and selection via the lasso. *Journal of the Royal Statistical Society Series B: Statistical Methodology*. 58 (1): 267–288.

Trakadis, Y.J., Sardaar, S., Chen, A. et al. (2019). Machine learning in schizophrenia genomics, a case-control study using 5,090 exomes. *American Journal of Medical Genetics Part B: Neuropsychiatric Genetics* 180 (2): 103–112.

Vabalas, A., Gowen, E., Poliakoff, E. et al. (2019). Machine learning algorithm validation with a limited sample size. *PLoS One* 14 (11): e0224365.

van Hilten, A., Kushner, S.A., Kayser, M. et al. (2021). GenNet framework: interpretable deep learning for predicting phenotypes from genetic data. *Communications Biology* 4 (1): 1094.

Vivian-Griffiths, T., Baker, E., Schmidt, K.M. et al. (2019). Predictive modelling of schizophrenia from genomic data: comparison of polygenic risk score with kernel support vector machines approach. *American Journal of Medical Genetics Part B: Neuropsychiatric Genetics* 180 (1): 80–85.

Wang, D., Liu, S., Warrell, J. et al. (2018). Comprehensive functional genomic resource and integrative model for the human brain. *Science* 362 (6420): eaat8464.

Wei, C. and Lu, Q. (2014). GWGGI: software for genome-wide gene-gene interaction analysis. *BMC Genetics* 15 (1): 101.

Wray, N.R., Goddard, M.E., and Visscher, P.M. (2007). Prediction of individual genetic risk to disease from genome-wide association studies. *Genome Research* 17 (10): 1520–1528.

Yang, H., Liu, J., Sui, J. et al. (2010). A hybrid machine learning method for fusing fMRI and genetic data: combining both improves classification of schizophrenia. *Frontiers in Human Neuroscience* 4: 192.

Yoshida, M., Koike, A., Yoshida, M. et al. (2011). SNPInterForest: A new method for detecting epistatic interactions. *BMC Bioinformatics* 12 (1): 469.

Zhang, Q., Prive, F., Vilhjalmsson, B. et al. (2021). Improved genetic prediction of complex traits from individual-level data or summary statistics. *Nature Communications* 12 (1): 4192.

Zhang, Y. and Liu, J.S. (2007). Bayesian inference of epistatic interactions in case-control studies. *Nature Genetics* 39 (9): 1167–1173.

Zhang, Z., Huang, H., Shen, D. et al. (2014). Integrative analysis of multi-dimensional imaging genomics data for Alzheimer's disease prediction. *Frontiers in Aging Neuroscience* 6: 260.

Zhao, Y., Chen, F., Zhai, R. et al. (2012). Correction for population stratification in random forest analysis. *Journal of International Epidemiology* 41 (6): 1798–1806.

Zhou, T., Liu, M., Thung, K.-H. et al. (2019). Latent representation learning for Alzheimer's disease diagnosis with incomplete multi-modality neuroimaging and genetic data. *IEEE Transactions on Medical Imaging* 38 (10): 2411–2422.

Zhou, X., Chen, Y., Ip, F.C.F. et al. (2023). Deep learning-based polygenic risk analysis for Alzheimer's disease prediction. *Communications Medicine* 3 (1): 49.

9

Multiscale AI for Precision Neuro-oncology

*Ruodan Yan[1], Xiaofei Wang[2], Hao Chen[2], and Chao Li[1,2,3]**

[1]*Department of Applied Mathematics and Theoretical Physics, University of Cambridge, Cambridge, UK*
[2]*Department of Clinical Neurosciences, University of Cambridge, Cambridge, UK*
[3]*School of Science and Engineering, School of Medicine, University of Dundee, Dundee, UK*

*Corresponding author: cl647@cam.ac.uk

9.1 Introduction

Gliomas are a heterogeneous group of primary brain tumours originating from glial cells, the supportive cells of the central nervous system. They represent the most common and aggressive form of brain cancer in adults (Ostrom et al. 2022). The current standard of care for gliomas typically involves maximal surgical resection followed by adjuvant treatment, e.g., radiotherapy and chemotherapy (Stupp et al. 2005). Despite advancements in these treatments, outcomes for high-grade gliomas remain dismal. Glioblastoma (GBM), the most aggressive form of glioma, has a median survival of approximately 15 months following surgical diagnosis and a 5-year survival rate less than 8% (Salvalaggio et al. 2024).

Multiple factors contribute to the poor prognosis associated with gliomas. Among the most significant is their highly infiltrative growth pattern, which allows tumour cells to migrate diffusely into surrounding brain tissue, often beyond what is visible on standard imaging. This makes complete surgical resection nearly impossible and leads to inevitable recurrence. Compounding this challenge is the pronounced intra- and intertumoral heterogeneity observed in gliomas. Within a single tumour, diverse cellular subpopulations can exhibit varying degrees of malignancy, treatment resistance and metabolic profiles. Between patients, gliomas differ substantially in their genetic mutations, epigenetic landscapes, microenvironmental interactions and response to therapy. Such heterogeneity complicates treatment planning and limits the effectiveness of standard therapies, which are typically guided by pathological and radiological assessments. Conventional treatment strategies often treat gliomas as a homogeneous entity, failing to account for their diverse molecular and phenotypic characteristics. This has led to a growing consensus that static, one-size-fits-all protocols are insufficient for managing these complex tumours.

Glioma biology spans multiple spatial scales, from macroscale brain network disruptions captured by magnetic resonance imaging (MRI) to cellular morphology observed in histology, and down to spatial gene expression patterns resolved by spatial transcriptomics (ST). Integrating these multiscale data sources offers a promising strategy to improve understanding and clinical management of glioma. Imaging techniques such as MRI and histological imaging provide multifaceted information that spans structural, functional and cellular features on microscopic to macroscopic scales. Integrating these multiscale data sources offers a powerful strategy to improve our understanding and clinical management of glioma.

In parallel, advances in image-based artificial intelligence (AI) provide the computational foundation to analyse and integrate these high-dimensional datasets. By leveraging these multiscale inputs, AI enables a more comprehensive understanding of glioma biology, linking brain network alterations, cellular morphology and molecular signatures into a unified predictive framework. Collectively, these AI techniques offer robust tools for discovering latent patterns across scales, improving classification of glioma subtypes, and predicting treatment response and patient survival, thereby advancing precision neuro-oncology from concept to clinical reality.

This chapter is organised into three main sections to explore the transformative potential of AI in neuro-oncology using image-based tools: Section 9.2 delves into how advanced MRI techniques and AI can characterise intricate brain connectivity affected by gliomas. Section 9.3 examines the role of AI in enhancing pathological assessments and uncovering new biomarkers. Section 9.4 addresses the critical aspect of translating AI approaches into real-world clinical datasets, ensuring that AI-driven solutions are robust, reliable and applicable.

Through these sections, we aim to highlight some emerging perspectives on leveraging AI in various diagnostic imaging approaches to transform glioma management and pave the way for a new era of personalised neuro-oncology.

9.2 AI-based Brain Mapping for Neuro-oncology

While gliomas arise primarily as focal lesions, accumulating evidence shows that they infiltrate surrounding tissue and induce widespread structural and functional alterations beyond the visible lesion. In particular, recent research has revealed that these tumours can cause a profound reorganisation of neural connectivity; reciprocally, neuronal activity has been shown to promote tumour proliferation and progression (Krishna et al. 2023). This complex and bidirectional interaction between gliomas and neural networks highlights the importance of studying neural connectivity to gain a deeper understanding of glioma invasion, which could provide a non-invasive tool for more precise patient stratification and personalised management.

The concept of the brain network, or connectome, refers to the complex network of cortical and subcortical regions and the pathways that structurally and functionally connect them (Bullmore and Sporns 2009). Unlike traditional imaging approaches that assess local features in isolation, connectome analysis offers a system-level view of how gliomas disrupt global brain organisation (Mandal et al. 2023). This approach can uncover widespread network perturbations that may not be evident in conventional imaging. The overarching goal of studying the brain connectome in glioma is to develop network-based tools to characterise tumour invasion and its broader impact on brain structure and function, ultimately enabling more precise diagnosis, prognostication and treatment planning.

The brain regions are structurally connected to others via white matter tracts, which are typically assessed using **diffusion MRI (dMRI)**. Structural connectivity is commonly quantified as the strength or density of white matter fibres that connect two regions (Wei et al. 2021). In parallel, brain regions can be functionally connected through synchronised neuronal activity, which can be indirectly measured using the blood oxygen level-dependent (BOLD) signal in **functional MRI (fMRI)**. Functional connectivity is defined as the statistical similarity between the time series of neuronal activity across different regions (Mandal et al. 2023). Another form of connectivity, vascular connectivity, inferred from dynamic perfusion patterns captured by **perfusion MRI (pMRI)**, reflects inter-regional haemodynamic coupling and is used to measure cerebral properties such as blood volume, flow and permeability (Yan et al. 2024).

By integrating these imaging modalities, researchers can construct comprehensive maps of the structural, functional and vascular networks of the brain. However, the high dimensionality and complexity of the resulting data pose significant challenges in identifying clinically significant patterns and understanding the organisation of these networks (Wei et al. 2023a). To address this, advanced network modelling and graph-theoretical analysis techniques have been proposed, drawing on methodologies from systems neuroscience and complex network theory, to characterise connectivity patterns and quantify alterations associated with disease (Van den Heuvel and Sporns 2013; Derks et al. 2021; Zhang and Hong 2024).

In this section, we will explore five main areas. Specifically, Section 9.2.1 outlines the methods for constructing brain connectomes from imaging modalities, followed by Section 9.2.2, which discusses tumour-induced network disruptions. Next, Section 9.2.3 introduces graph-theoretical analysis and connectivity quantification in tumour-bearing brains. Then, Section 9.2.4 highlights how connectome-based methods inform brain tumour characterisation. Finally, Section 9.2.5 concludes with future directions for network-based approaches in neuro-oncology. These discussions aim to elucidate the potential of connectome analysis in enhancing our understanding of glioma biology and its implications for clinical practice.

9.2.1 Constructing the Brain Connectome

Brain networks can be derived from anatomical or physiological data, giving rise to structural and functional networks, respectively. It is essential to maintain this distinction when interpreting brain network datasets, as the underlying biological substrates and analytical methods differ between the two.

The construction of a brain network involves transforming the imaging data into a graph-based representation through a series of systematic steps, beginning with the identification of the network nodes (Figure 9.1). Defining the nodes is the first critical step towards producing a network model. Nodes are typically defined by parcellating the cortical and subcortical grey matter into discrete regions based on anatomical landmarks or atlas-based boundaries, such as those defined by widely used atlases like the Automated Anatomical Labeling (AAL) Rolls et al. (2020) or Desikan et al. (2006), implemented in tools like FreeSurfer (Fischl, 2012). As an alternative to atlas-based methods, clustering

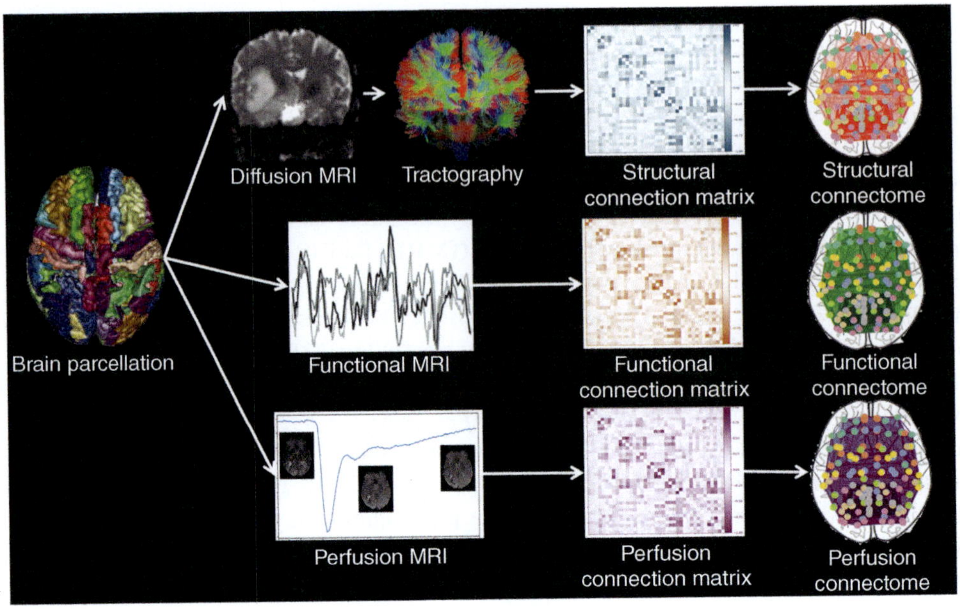

Figure 9.1 Illustration of the brain network construction. *Source*: Adapted from Torsten Rohlfing et al., 2009,

algorithms can be used to generate voxel-wise parcellations with uniform size and spatial coherence. Data-driven parcellation approaches, including spectral clustering, independent component analysis (ICA) or hierarchical clustering, are often implemented using neuroimaging software packages like FSL (Smith et al. 2004), SPM (Ashburner, 2012) or Connectome Workbench (https://www.humanconnectome.org/software/connectome-workbench). Parcellations such as the Human Connectome Project's Multi-Modal Parcellation (HCP-MMP) (Glasser et al. 2016) provide refined, functionally informed subdivisions of the brain.

Once nodes are defined, edges are quantified by measuring the structural or functional relationships between each pair of regions. These pairwise connections are then assembled into a connectivity matrix, which encapsulates the topology of the brain network and serves as the basis for downstream graph-theoretical analysis.

9.2.1.1 Functional Connectome

The functional connectome models brain connectivity based on the temporal correlations of neural activity, commonly derived from BOLD fMRI. Consequently, functional connectivity can be computed through metrics such as cross-correlation, mutual information or spectral coherence.

In gliomas, functional connectivity is disrupted due to both tumour-induced alterations in neural activity and changes in local haemodynamics, including increased oxygenated blood flow from angiogenesis. Incorporating the assessment of the resting-state fMRI (rs-fMRI) network into the traditional glioma mapping approaches could be a versatile method for anatomical and functional localisation and evaluation of molecular parameters.

The exploration of large-scale functional connectivity networks uses two complementary approaches. The first approach is called seed-based methods. They compute the similarity between the time series of a predefined region and all other voxels, revealing network structures like the canonical 7- and 17-network parcellations proposed by Thomas Yeo et al. (2011). However, the tight dependence on seed location is also a limitation of the seed-based approach, since slight differences in the registration of a seed between subjects can result in the identification of distinct networks. Alternatively, ICA decomposes fMRI data into components of signal and noise. ICA is used both to denoise data and to identify rs-functional networks, which naturally manifest as the signal components from the analysis. ICA requires fewer *a priori* assumptions than seed-based methods, as it identifies components in a data-driven manner without predefined regions of interest.

Studies have shown that tumours can disrupt functional organisation. Hart et al. (2019) investigated the global impact of GBM on brain function using rs-fMRI. The authors found a pattern of reduced activity near the tumour, followed by overactivity farther away, forming a gradient. Beyond this, Nenning et al. (2020) found that tumour-induced disruptions were more strongly related to functional proximity than physical distance. These findings highlight the global impact of focal lesions on brain-wide connectivity.

Other work has proposed mechanisms for how tumour infiltration interferes with cognitive processing, demonstrating that gliomas can alter the information content of electrical brain activity. Aabedi et al. (2021) used electrocorticography to compare speech-related brain activity in glioma-infiltrated and normal-appearing cortex. While glioma-infiltrated regions could still show coordinated activity, they had reduced capacity to encode information and relied on broader cortical networks. This suggests a compensatory mechanism in which nearby or distant regions support compromised functions.

9.2.1.2 Structural Connectome

Diffusion MRI enables non-invasive assessment of brain tissue microstructure (Wei et al. 2023a; Wang et al. 2023c). By capturing the diffusion behaviour of water molecules, dMRI reveals important characteristics such as white matter integrity, extracellular volume and cellular density. These features are particularly important in neuro-oncology, as brain tumours often disrupt normal white matter architecture.

One of the major applications of dMRI is the construction of the structural connectome, which models white matter fibre pathways as a brain network. This enables a deeper understanding of how tumours alter structural connectivity and contribute to cognitive dysfunction (Wei et al. 2023b).

Constructing a structural connectome involves several steps (see Figure 9.1). First, network nodes are defined by a brain atlas built based on prior anatomical knowledge. Then, structural connectivity between region pairs is estimated using fibre tractography, which traces white matter pathways by tracking anisotropic water diffusion along axons (Rolls et al. 2020; Desikan et al. 2006; Desikan et al. 2006; Jeurissen et al. 2019). These pairwise measurements are then aggregated into an adjacency matrix, representing a graph. In dMRI studies, structural connectivity is commonly modelled as undirected edges, as current dMRI methods cannot resolve the directionality of axonal projections.

While this pipeline enables the reconstruction of white matter networks, its accuracy can be compromised in the presence of brain tumours. Mass effects (e.g., oedema, displacement) can distort local diffusion signals and reduce tract reconstruction accuracy (Wei et al. 2021). While streamline tractography remains common, they struggle to resolve crossing fibres or distinguish tumour-infiltrated tracts from displaced pathways (Zhang et al. 2012).

To overcome these limitations, recent studies have proposed template-based frameworks that leverage normative data from healthy populations. For example, Wei et al. (2021) developed an iterative method projecting tumour-affected dMRI data onto a healthy tractography atlas, minimising mass-effect biases and improving infiltration detection. This approach quantifies connectivity loss using graph metrics such as edge density (i.e., number of streamlines between regions) and global efficiency, revealing subnetworks vulnerable to glioma progression.

Advances in multi-shell dMRI have further improved specificity. Models such as neurite orientation dispersion and density imaging (Zhang et al. 2012) distinguish between tumour cellularity (intracellular volume fraction) and oedema (isotropic free water). Fixel-based analysis (Raffelt et al. 2017) provides additional insights by resolving axonal density and fibre morphology in infiltrated tracts.

Studies have shown that gliomas preferentially invade along axonal tracts, leading to network-level disruptions that correlate with cognitive deficits and recurrence risk (Derks et al. 2021). The connectome can thus reconstruct large-scale white matter trajectories to reveal how tumour growth displaces or damages critical tracts (Jeurissen et al. 2019).

To quantify tumour-induced disruptions, graph-theoretical metrics (Section 9.2.3) are applied to structural networks. In healthy brains, studies consistently identify small-world architecture in structural networks, an organisation characterised by high clustering and short path lengths, which balances communication efficiency and wiring cost. Hub regions, densely interconnected through rich-club organisations (i.e., highly connected nodes that preferentially connect to each other) (van den Heuvel and Sporns, 2013; Bullmore and Sporns, 2012), are particularly vulnerable to tumour infiltration, as their disruption disproportionately impacts global network efficiency (Bullmore and Sporns, 2012).

Gliomas degrade this optimised architecture, leading to increased path lengths and reduced modularity (Crossley et al. 2014). These changes correlate with cognitive deficits, such as impaired executive function or memory, and may predict postoperative outcomes (Douw et al. 2009; Derks et al. 2021). For instance, tumours invading rich-club hubs are associated with more severe neurological symptoms, highlighting the clinical value of network-based biomarkers.

9.2.1.3 Perfusion Connectome

Tumour vascularity can be assessed preoperatively using pMRI. The dominant dynamic gadolinium-based contrast agent injection pMRI techniques include (1) T1-weighted dynamic contrast-enhanced MRI and (2) T2- or T2*-weighted dynamic susceptibility contrast (DSC) MRI. In brain tumour studies, DSC-MRI is more commonly used (Shiroishi et al. 2023). On the other hand, arterial spin labelling (ASL) is a non-invasive pMRI technique, without the need of contrast agent injection. Instead, ASL uses magnetically labelled arterial blood water to provide haemodynamic information (Hashido et al. 2020). All three methods use indicators (either contrast agents or magnetically labelled water) that modify the MRI signal in proportion to their concentration. Rapid dynamic MRI is then used to track the spatiotemporal variations in signal induced by the indicator. In neuro-oncology, pMRI plays a critical role in assessing tumour aggressiveness due to angiogenesis, a hallmark of cancer.

Previous studies have predominantly relied on predefined pharmacokinetic models (Li et al. 2019), leveraging mathematical representations to calculate perfusion parameters from the kinetic curve, such as relative cerebral blood volume. However, the resulting parametric maps are inherently simplified and limited in modelling complex physiological kinetics. Although various attempts have been made to develop machine learning and deep learning approaches (Pinto et al. 2018; Heo et al. 2023) for extracting more comprehensive features from parametric maps, accurate and interpretable modelling of the full spatiotemporal perfusion signal remains an unmet challenge.

Unlike conventional analysis, network-based approaches leverage the temporal and spatial correlations of perfusion signals across regions to uncover system-level alterations in vascular dynamics driven by tumour-induced disruption. Similar to functional connectome, the perfusion connectivity between two brain regions is typically defined as the statistical similarity between their respective time–signal curve (Figure 9.1). Previous studies have quantified network perfusion in ASL using both seed-based and ICA methods, showing important correlation of perfusion information such as cerebral blood flow (CBF) with network topology (Jann et al. 2013; Liang et al. 2014).

Brain networks provide a promising framework for comprehensively characterising global and regional brain organisation. Based on fMRI data, graph neural networks (GNNs) have been successfully applied to model functional brain networks and associated disorders (Li et al. 2021). However, very few studies have explored network modelling of pMRI data. Yan et al. (2024) were among the first to apply a similar framework to pMRI, using predefined brain regions as nodes and temporal correlations between their perfusion signals as edges to model the spatiotemporal relationships within brain vascularity. This approach represents a novel application of connectomic analysis to the vascular domain, offering new insights into tumour-induced disruptions in cerebral perfusion dynamics.

9.2.2 Connectome Disruptions from Tumour Invasion

Understanding brain connectome disruptions caused by gliomas provides valuable insights into their impact on brain function and clinical outcomes. Wei et al. (2023b) systematically quantified tumour invasion by comparing patient structural connectome disruptions with those of healthy controls. This approach involved constructing tractography templates from healthy subjects, generating fractional anisotropy (FA) skeletons and comparing the connectivity strengths of patients against age-matched norms. The study found that GBM-induced connectome disruptions extended into the normal-appearing brain and were significantly associated with lower preoperative performance, impaired cognition and poorer survival. These findings underscore the prognostic value of structural connectivity metrics in GBM.

Building on this, recent large-scale tractography analyses have further clarified how different tumour types uniquely affect white matter pathways. For instance, Altieri et al. (2025) revealed that gliomas induce histotype-specific alterations in the inferior-fronto-occipital fascicle, with GBMs and low-grade gliomas significantly reducing fascicle volume, length, and FA—consistent with both infiltrative and destructive processes. Complementarily,

Armocida et al. (2025) found that diffusion-based markers—specifically FA, mean diffusivity and tract irregularity (TI)—in the brain adjacent to tumours were found to correlate closely with glioma grade, Ki-67 proliferation index and resection extent. Among these, TI stood out as an independent indicator of tumour aggressiveness, emphasising the value of advanced diffusion tractography in preoperative network assessment and surgical planning.

Complementary findings from fMRI studies further illustrate how tumours disrupt resting-state network dynamics, particularly within networks like the default mode and salience networks. For example, Hadjiabadi et al. (2018) showed that gliomas located in hub regions (e.g., frontal and temporal lobes) disrupt functional connectivity within the default mode network (DMN). Using rs-fMRI, they observed reduced connectivity between the DMN and the frontoparietal control network, which correlated with deficits in attention and executive function. Importantly, these disruptions were not simply attributable to tumour size, suggesting that functional network reconfiguration—rather than mass effect alone—contributes to cognitive impairment.

This interplay between structural and functional disruption is further demonstrated in multimodal studies. Derks et al. (2021) investigated multimodal connectivity changes by combining dMRI and fMRI. Their findings revealed that tumours affecting salience network hubs (e.g., anterior insula and dorsal anterior cingulate cortex) induced aberrant connectivity in task-positive networks, impairing the brain's ability to shift between internally and externally directed cognition. This 'network collapse' mechanism explains why severe cognitive deficits may persist even when structural connectivity appears intact, highlighting the importance of incorporating functional metrics into clinical assessments.

Further integrating biological context, Mandal et al. (2020) linked glioma spatial distribution to underlying biological and connectomic features. Their model explained 58% of the variation in glioma location frequency, demonstrating strong associations with brain regions enriched in glioma-initiating cells such as neural stem cells and oligodendrocyte precursor cells. Additionally, they found that gliomas preferentially occupy and disrupt functional hubs, resulting in reduced global small-world clustering.

Taken together, these findings reveal that glioma-induced connectome disruption is not merely local but reflects a complex interaction of biological predisposition, network topology and tumour behaviour. This underscores the need to integrate structural, functional and molecular perspectives to fully understand and manage glioma impact on the brain.

9.2.3 Graph Theory in Neuro-oncology

Graph theory has emerged as an essential analytical framework for exploring the intricate connectivity patterns within the brain (Zhang and Hong 2024). In neuro-oncology, it plays a pivotal role in guiding tumour resections and in deciphering the network-level dynamics that support glioma development and progression. At its core, graph theory represents complex systems as mathematical structures composed of nodes (or vertices) and edges (or links) that connect pairs of nodes. These graphs serve as abstractions of real-world networks. In brain network analysis, nodes typically correspond to distinct cortical or subcortical regions, while edges represent structural, functional or vascular interactions between these regions.

In the context of brain networks, three fundamental properties are commonly assessed: integration, segregation and centrality (Mandal et al. 2023). Table 9.1 (Mandal et al. 2023) summarises commonly used graph-theoretical metrics for quantifying these properties.

Integration refers to the efficiency of information transfer across the network, often measured by **global efficiency** or **characteristic path length**. A highly integrated network enables rapid and direct communication between distant brain regions. **Segregation**,

Table 9.1 Descriptions of graph theory measures.

Term	Description	Global or Local?
Measures of integration		
Global efficiency	Average inverse shortest path length. Path length is operationalised as the inverse of the weights in a weighted network. A high global efficiency denotes a well-integrated network.	Global
Characteristic path length	Average shortest path length. A low characteristic path length denotes a well-integrated network.	Global
Measures of segregation		
Clustering coefficient	Fraction of a node's neighbours that are also neighbours. It can be averaged across all nodes in the network to provide a global measure of segregation.	Local
Modularity	Degree to which a network can be subdivided into groups of nodes (i.e., modules) with maximal intraconnectivity and minimal interconnectivity. It cannot be calculated directly, but rather after the optimal modular structure has been determined by an optimisation algorithm.	Global
Measures of centrality		
Degree	Number of edges involving a given node. Applied to binary graphs. A node with a high degree can be interpreted as a hub.	Local
Strength	Sum of weighted edges involving a given node. Applied to weighted graphs. A node with high strength can be interpreted as a hub.	Local
Betweenness centrality	Fraction of all shortest paths to pass through a given node. A node with a high betweenness centrality can be interpreted as a hub.	Local
Local network statistics		
Participation coefficient	Distribution of a node's edges with other nodes in different modules versus the same module. A high participation coefficient denotes a 'connector hub,' which helps link distinct modules.	Local
Within-module Z-score	Z-score of nodal strength within a given node's module. A high within-module Z-score denotes a 'provincial hub,' important for integration within a given module.	Local

Source: Mandal et al. (2023)/Oxford University Press/CC BY 4.0.

in contrast, reflects the tendency of the network to form specialised, densely interconnected communities. Measures of segregation differ based on scale; for instance, the clustering coefficient captures local connectivity by counting closed triangles (i.e., fully connected subgraphs of three nodes), while modularity assesses the strength of community structure by comparing within-module to between-module connectivity (Watts and Strogatz, 1998; Girvan and Newman, 2002; Guimera and Amaral, 2005). **Centrality** broadly refers to the importance or influence of a node within a network. This influence can be defined in multiple ways. The most basic centrality measure is **degree**, which counts the number of direct connections a node has. For weighted graphs, **strength** is used to quantify the total weight of a node's connections. Other centrality metrics include **betweenness centrality**, which captures the proportion of shortest paths that pass through a node and reflects its role as a connector in global communication (Freeman, 2002). **Local efficiency**, another related measure, reflects how efficiently a node communicates with its neighbours. Intermodular influence is captured by metrics such as the **participation coefficient**, which reflects the extent to which a node links across different modules, and the **within-module Z-score**, which measures its connectedness within its own module (Guimera and Amaral, 2005; van den Heuvel and Sporns, 2013). Collectively, these measures help identify **hubs**, highly influential nodes critical for information flow, which play vital roles in both healthy brain function and disease.

In neuro-oncology, graph-theoretical methods are widely used to assess alterations in structural and functional brain networks (Bullmore and Sporns, 2009). By quantifying network features such as degree, clustering, path length and efficiency, researchers gain insights into how gliomas disrupt global brain connectivity (Wei et al. 2023b; Salvalaggio et al. 2024). Such disruptions are frequently associated with cognitive and neurological impairments in brain tumour patients (Dou et al. 2019; Aerts et al. 2016).

Moreover, graph theory provides a framework for identifying biomarkers of tumour progression and treatment response. For instance, changes in network topology may correlate with clinical severity or predict postoperative outcomes. In addition, this approach supports investigations into **brain plasticity**, shedding light on how networks reorganise in response to tumour invasion or surgical resection.

9.2.4 AI Approaches for Modelling Connectome in Neuro-oncology

9.2.4.1 Connectome-based Characterisation of Brain Tumours

Building on graph theory, which provides a robust framework for exploring brain connectivity patterns (Zhang and Hong, 2024), recent efforts have focused on understanding how tumours disrupt these complex topologies, including hubs, modules, small-worldness and core–periphery structure, and how such disruptions affect brain dynamics (Farahani et al. 2019). This provides a foundation for approaches to model, predict and interpret tumour-related network alterations and their clinical implications.

Cognition and functional mapping: In dMRI studies, Friedrich et al. (2024) employed connectome-based predictive modelling to link structural connectivity loss (e.g., reduced fibre numbers in frontoparietal tracts) with post-treatment cognitive decline. Their findings indicate that in glioma patients, cognitive performance after treatment relies heavily on the integrity of key network hubs. In fMRI studies, Niu et al. (2021) demonstrated that a neural network approach using rs-fMRI effectively predicted motor activation maps in patients with perirolandic brain tumours, outperforming ICA through higher spatial correlations and Dice coefficients relative to task-based fMRI, particularly in gliomas and on non-tumour sides. This highlights its clinical potential for presurgical mapping, especially in contexts of neurovascular uncoupling and functional reorganisation.

Genotype prediction: Significant progress in understanding intertumoral glioma heterogeneity has come from identifying key genetic distinctions between low-grade gliomas and GBMs (Mandal et al. 2023). Among these, the most clinically relevant prognostic marker is the isocitrate dehydrogenase (IDH) mutation, which is associated with a less aggressive disease course and serves as a diagnostic hallmark for diffuse low-grade gliomas.

Increasingly, imaging-derived brain networks are being leveraged to complement molecular diagnostics by capturing spatially distributed tumour effects on brain structure and function. Recent studies have demonstrated the utility of the brain connectome, constructed from modalities such as dMRI and pMRI, in predicting key molecular markers like IDH mutation (Wei et al. 2023a; Yan et al. 2024). These approaches highlight that tumours exert widespread influence beyond their anatomical boundaries, disrupting white matter integrity and reconfiguring vascular connectivity across distant brain regions.

For example, Yan et al. (2024) incorporated spatiotemporal graph structure learning and dual-attention mechanisms to optimise feature extraction from pMRI, enhancing the characterisation of tumour-related brain regions. They highlight the utility of pMRI in predicting the *IDH* mutation status in gliomas, which is a crucial marker of tumour malignancy. By modelling 4D pMRI using GNNs, this study demonstrates improved predictive capabilities for glioma genotyping, offering a non-invasive alternative to traditional methods.

Tumour progression and treatment response: Beyond prediction, graph theory serves as a powerful tool for identifying biomarkers of tumour progression and treatment response. Previous work suggests that network metric changes, such as node degree and clustering coefficient, can indicate tumour progression and patient outcomes (Semmel et al. 2022). Additionally, this approach enables the study of brain plasticity, where the brain reorganises its networks to compensate for the presence of a tumour. Krishna et al. (Krishna et al. 2023) have demonstrated that brain network reorganisation occurs in response to tumours, underscoring the brain's adaptive mechanisms.

Finally, biophysical models can provide simplified simulations for understanding biophysical processes inside the human body and the physiological mechanisms underlying medical imaging. For example, Zhang et al. (2023) used Physics-Informed Neural Networks (PINNs) for GBM growth modelling, estimating patient-specific parameters in a reaction–diffusion PDE model by integrating structural MRI with tumour proliferation and diffusion dynamics.

9.2.4.2 GNNs for Connectome Modelling

The aforementioned clinical applications rely heavily on the ability to accurately model complex brain networks and their disruptions by gliomas. To this end, advanced machine learning models, particularly GNNs, have been introduced for more precise modelling of tumour-induced brain network disruptions. Unlike traditional machine learning methods with limited ability to handle the complex, non-Euclidean nature of brain connectivity data, GNNs inherently learn from graph-structured data and provide an elegant solution to this challenge.

GNNs operate on a message-passing mechanism, where each node represents a brain region, and they communicate with their neighbouring nodes via edges that encode inter-regional connectivity. This process allows the GNNs to capture complex spatial–temporal relationships across modalities (Bessadok et al. 2023). By allowing for end-to-end learning and automatic feature extraction from neuroimaging data, they have shown promise in analysing bidirectional tumour–brain interactions, uncovering imaging biomarkers and supporting presurgical decision-making (Wein et al. 2021; Kim et al. 2021; Yang et al. 2024; Hu et al. 2024; Wang et al. 2024a).

Several GNN architectures have been proposed, each employing distinct aggregation strategies. Notable examples include graph convolutional networks, graph attention networks and GraphSAGE. These models differ in their design and assumptions, making them suitable for a wide range of applications, such as brain connectivity analysis. Table 9.2 (Mohammadi and Karwowski, 2025) summarises various GNN types used in connectome research, along with their characteristics and application areas.

GNNs have demonstrated strong potential for modelling dynamic brain networks in neuro-oncology, particularly for capturing temporal changes in connectivity associated with tumour invasion and angiogenesis. Unlike traditional static graph models that assume fixed relationships between regions, GNNs can incorporate temporal edges to represent time-varying interactions. This allows for a more precise analysis of how brain connectivity is altered by tumours or treatments. Dynamic GNNs, which process sequences of graphs, are particularly effective for analysing fMRI and pMRI data, where inter-regional connections shift in response to varying external stimuli or contrast agent injection (Zhao et al. 2022a; Yan et al. 2024).

Beyond modelling structural and functional connectivity, GNNs offer a promising approach to tackling inter-subject variability in neuro-oncology. Brain tumours, especially

Table 9.2 Overview of different types of GNNs utilised in brain connectivity studies.

GNN Variant	Key Features	Common Applications in Brain Studies
Graph convolutional networks	Laplacian-based filters, spectral domain	Static connectivity classification
Graph attention networks	Attention mechanisms for edge weighting	Regional feature importance studies
Dynamic graph CNNs (DGCNNs)	Temporal graph analysis	Dynamic connectivity studies
Hierarchical GNNs (Hier GNNs)	Multi-level graph structures, hierarchical learning	Multiscale brain connectivity analysis
Spatiotemporal GNNs	Integration of spatial and temporal data	Spatiotemporal brain network analysis

Source: Mohammadi and Karwowski (2025)/MDPI/CC BY 4.0.

gliomas, exhibit significant heterogeneity in tumour location, grade and biological behaviour, making it challenging to identify consistent patterns across patients. GNNs can learn shared representations across subjects while preserving individual-specific features, enabling the discovery of consistent network patterns within heterogeneous tumour populations. This capacity is especially valuable for identifying reliable imaging biomarkers, understanding treatment-related brain plasticity and supporting personalised clinical decision-making in neuro-oncology. By combining deep learning with graph-based modelling, GNNs help reveal population-level insights that might otherwise be obscured by subject-specific differences.

Though GNN application to neuroimaging data remains challenging: (1) **Accuracy of graph construction**: In brain tumour studies, brain graphs are commonly derived from dMRI, fMRI and pMRI, yet they are susceptible to noise, low spatial resolution and inherent model assumptions, particularly in tumour-infiltrated or peritumoral regions. These factors can compromise the accuracy and reliability of graph representations and cause bias in downstream tasks. (2) **Overfitting and generalisation**: Neuro-oncology datasets often suffer from limited and imbalanced sample sizes. These limitations increase the risk of overfitting, particularly when using complex models like GNNs, which have a high capacity and require substantial data to generalise effectively. (3) **Interpretability in neuro-oncology**: While GNNs excel at modelling complex network interactions, their 'black box' nature still limits biological understanding of tumour invasion and clinical trust. In the context of glioma, understanding how tumours disrupt brain networks is essential for diagnosis, prognosis and treatment planning. Improving interpretability is therefore critical to ensure that GNN-based tools can support both clinical decision-making and neuroscientific discovery.

9.2.4.3 Multimodal Learning

Multimodal learning shows excellent performance at integrating multiple data modalities while minimising the domain gap between them (Meng et al. 2024). Technically, multimodal learning frameworks integrate data from different imaging modalities by aligning their representations into a shared feature space. This can be achieved using dual-branch or cross-modal GNNs, where each branch encodes modality-specific features before fusing them through attention or contrastive learning mechanisms (Ektefaie et al. 2023).

By leveraging complementary information from structural MRI, dMRI and fMRI, these models can better capture the complex, spatially distributed effects of tumours on the brain network. For instance, combining structural and functional connectivity through GNNs can help reveal how anatomical changes in the brain may lead to alterations in functional connectivity, providing deeper insights into the pathophysiology of neurological disorders (Xu et al. 2023; Falcó-Roget et al. 2024).

Wei et al. (2023a) developed a multimodal learning framework that combines anatomical MRI with dMRI to reconstruct brain networks and predict glioma genotype. This approach employs hierarchical attention mechanisms to extract relevant features from the brain network, addressing the gradient of tumour invasion and improving the model's robustness against pathological confounders. The proposed methods have the potential for automated, rapid diagnosis and prognosis in glioma patients based on pre-treatment MRI, essential for patient risk stratification and treatment planning towards precision medicine.

Similarly, Falcó-Roget et al. (2024) investigated functional and structural connectivity within brain tumour regions, identifying altered resting-state signal dynamics and developing a tumour-inclusive fibre-tracking pipeline. Their machine learning framework predicts postoperative structural reorganisation and disentangles tumour-type-specific connectivity patterns. These findings underscore the critical need to incorporate tumour-affected areas in neuroimaging to elucidate complex functional and structural (dis)connection profiles. The results emphasise that damaged tissues exhibit non-trivial connectivity reorganisation, necessitating their inclusion in neuroimaging analyses to advance surgical planning and improve understanding of tumour-induced brain network alterations.

Further, Sun and Peng (2024) developed a dual-GNN using multimodal MRI and clinical features for GBM survival prediction. By leveraging both radiomic and textual features (e.g., age), their method achieves improved accuracy in classifying short-, medium-, and long-term survival.

Together, these studies highlight that multimodal learning not only improves predictive performance but also provides a more complete view of tumour–brain interactions. This is especially valuable for advancing personalised diagnosis, treatment planning and monitoring of glioma patients in clinical settings.

9.2.5 Future Directions

Multimodal MRI connectome studies, encompassing dMRI, pMRI and AI-driven analytical methods, provide critical insights into the structural and functional alterations caused by gliomas. Given the complexity and heterogeneity of gliomas, traditional clinical models often oversimplify the intricate interactions between tumour growth, microenvironmental changes and brain network disruptions. Integrating multimodal connectome with AI offers a transformative opportunity to bridge this gap by modelling the complex interplay of tumour biology and connectomic alterations. As imaging and computational technologies continue to advance, these integrative approaches are expected to become increasingly central to the comprehensive management of glioma patients.

Therefore, clinicians may be able to move beyond static diagnostic snapshots towards dynamic, system-level models. Such models may enable clinicians to track disease progression, monitor treatment effects and personalise interventions based on the evolving network state of the brain. Future research should focus on refining AI algorithms, discovering novel imaging-genomic biomarkers and validating these methods across larger, multi-centre cohorts to support robust clinical translation.

Additionally, while genotyping a single gene, such as *IDH1*, is now relatively straightforward and cost-effective, multimodal learning that integrates different imaging types with genomic data or more comprehensive multi-omics data is of significant promise to better characterise gliomas. This integrative paradigm could reduce reliance on invasive tissue sampling and offer a non-invasive surrogate for molecular diagnosis and prognosis.

A major question of interest is how structural and functional brain network topologies relate to one another. Functional connections are supported by the underlying structural architecture that mediates communication between distant brain regions. As such, structural and functional networks are often correlated, with both exhibiting canonical features such as small-worldness and hub-centric organisation. However, functional connectivity is inherently more dynamic, capturing transient and higher-order interactions mediated by multi-synaptic pathways. Given that many pathologies, including glioma, can affect both network types differently, understanding their interdependence remains a key challenge.

Combining multimodal connectome approaches further enriches our understanding of glioma-related brain alterations. These modalities offer complementary perspectives on tumour behaviour, integrating vascular, anatomical and functional dimensions. For instance, regions exhibiting elevated cerebral blood volume and CBF often co-localise with disrupted white matter tracts, indicating that gliomas affect not only localised haemodynamics but also broader networks of neural communication. This integrative framework supports a more holistic characterisation of tumour impact on brain structure and function and may inform the development of network-guided treatment strategies and personalised therapeutic planning.

9.3 AI-based Tissue Mapping for Neuro-oncology

Histopathology remains the gold standard for cancer diagnosis, offering rich tissue- and cellular-level information for both clinical assessment and biological discovery (Madabhushi and Lee, 2016; Bera et al. 2019; Acs et al. 2020). However, interpreting histological slides, such as haematoxylin and eosin (H&E)-stained sections, requires substantial time and expertise, limiting the effective implementation of precision oncology.

Recent advances in digital pathology have shown great potential in automating and scaling tissue-based analysis (Bankhead, 2020). Deep learning-based digital pathology methods now support not only diagnostic workflows but also computational biomarker discovery, uncovering spatial patterns related to prognosis, treatment response and tumour biology (Echle et al. 2021; Shmatko et al. 2022).

Meanwhile, the emergence of ST technologies has enabled mapping gene expression onto the tissue context, creating a powerful bridge between molecular biology and histopathology. These spatially resolved omics data enable high-resolution modelling of the tumour microenvironment, cellular interactions and tissue-level heterogeneity, especially relevant for gliomas.

In this section, we first review general AI-driven approaches for cancer diagnosis using histological images. Then, we focus on applications in glioma diagnosis, highlighting disease-specific challenges and solutions. In addition, we further explore how AI refines ST in both spatial resolution and gene coverage, enabling downstream molecular biology for cancer analysis. Finally, we discuss how learning-based approaches increasingly enable insights into cancer evolution and intratumoral heterogeneity.

9.3.1 AI-based Digital Pathology

Deep learning-based digital pathology methods can typically be divided into cell-level and tissue-level modelling. These methods are used for various downstream tasks, such as cell segmentation (Hatipoglu and Bilgin, 2017) and tissue phenotyping (Nalisnik et al. 2017). Here, we introduce representative methods for both cell- and tissue-level modelling using histopathology images, particularly the routinely acquired H&E images.

9.3.1.1 Cell-level Modelling

Cell segmentation is an important task in cell-level modelling, which is defined as assigning pixel-level labels to cells and nuclei (Hatipoglu and Bilgin, 2017). Mahmood et al. (2020) presented a multistage mitotic-cell-detection method based on the faster region-based convolutional neural network (Faster R-CNN) for efficient detection of mitotic cells based on histology images. After generating the candidate image patches, a ResNet-50-based (He et al. 2016) classifier was then utilised for final classification. Meanwhile, Rauf et al. (2023) designed an attention-guided multiscale deep object detection framework, called DC-Lym-AF, for lymphocyte analysis based on immunohistochemistry images. Specifically, DC-Lym-AF comprises four stages, i.e., pre-processing phase, screening phase, localisation phase and post-processing. Of note, in the screening phase, a custom convolutional neural network (CNN) is developed to screen lymphocytic regions by performing a patch-level classification. In addition, Wang et al. (2023d) further proposed a generalisable and robust mitosis detection algorithm (called FMDet) to improve the model generalisability across image domains. Specifically, in FMDet, a robust feature extractor is developed to capture the appearance variations of mitotic cells, constructed by integrating a channel-wise multiscale attention mechanism into a fully convolutional network structure.

Despite success, cell patch detection only provides coarse-grained locations of cells and is thereby incapable of generating fine-grained pixel-wise locations of individual cells, hindering the extraction of valuable morphological features of cells. To solve this problem, many works (Hatipoglu and Bilgin, 2017; Albayrak and Bilgin, 2019; Ahmad et al. 2023) have recently been proposed to achieve pixel-level segmentation of cells based on histology images. For instance, Hatipoglu et al. (2017) proposed to incorporate cellular and extracellular spatial relationships into CNNs, via windowing with small patches of various sizes. Similarly, Albayrak et al. (2019) devised a two-stage segmentation method to obtain cellular structures in high-dimensional histology images of renal cell carcinoma. Specifically, the image is first segmented into superpixels, which are then clustered to find similar superpixels that compose the cell nuclei.

In addition, Ahmad et al. (2023) proposed a novel spatial-channel attention-based modified UNet (Ronneberger et al. 2015) architecture with ResNet (He et al. 2016) blocks in encoder layers. Moreover, most recently, some works (Graham et al. 2019; Hörst et al. 2024; Jaisakthi et al. 2023; Graham et al. 2023) have further proposed to achieve joint learning of cell segmentation and downstream tasks, such as disease classification. For example, Graham et al. (2019) introduced Hover-Net, a model that simultaneously segments and classifies nuclei by leveraging the distances of nuclear pixels from their centre of mass. In addition, Graham et al. (2019) present a multi-task learning approach for segmentation and classification of nuclei, glands, lumina and different tissue regions.

9.3.1.2 Tissue-level Modelling

Unlike cell-level modelling, tissue-level modelling excels at capturing global phenotypic features, offering a more comprehensive view of tissue architecture and pathology. Recent tissue-level modelling approaches have significantly advanced computer-aided classification of both cancer (Xiang et al. 2019; Wang et al. 2023c) and non-cancer diseases (Pan et al. 2021; Douget et al. 2023). For instance, Jiang et al. (2021) proposed a light-weight attention mechanism-based framework, namely, DRANet, to differentiate 11 types of skin diseases based on a real histopathological image set. Meanwhile, Huang et al. (2023) developed a deep learning-based workflow for in vitro and intraoperative pathological diagnosis of liver cancer using Raman spectroscopy. Specifically, they distinguished carcinoma tissues from adjacent non-tumour tissues in a rapid, non-disruptive and label-free manner by using Raman spectroscopy combined with deep learning.

In addition, Khazaee and Rezaee (2023) proposed an ensemble-based multi-tissue classification approach using a novel hybrid deep learning framework for colorectal cancer diagnosis. Similarly, Yang et al. (2022b) introduced FA-MSCN, a feature-aligned multiscale convolutional network designed for liver tumour diagnosis. FA-MSCN enhances diagnostic accuracy by integrating features from multiple magnifications, allowing the model to incorporate more contextual information from neighbouring regions. While deep learning-based algorithms have shown considerable success in various cancers, they still face substantial challenges in the context of brain tumours, especially gliomas. This difficulty arises from the pronounced heterogeneity of gliomas, the intricate diagnostic criteria and the limited availability of sample data (Perry and Wesseling, 2016; Louis et al. 2021).

9.3.2 AI-based Glioma Classification

Gliomas are highly aggressive brain tumours, requiring accurate grading for effective treatment. Recently, the diagnostic criteria have shifted from purely relying on histology features to integrating histology features and molecular markers. This integration of histological and molecular data reflects the growing complexity of glioma diagnosis and the need for precision-driven tools. Advances in AI and digital pathology have enabled automated glioma grading, making the analysis of histology images more efficient and reliable. These methods not only improve diagnostic accuracy but also help connect traditional histopathology with molecular pathology. In this section, we review current glioma classification methods and examine how combining histology and molecular data is advancing precision oncology.

9.3.2.1 Classifying Glioma via Histopathology

In recent years, with the rapid development of deep learning, there have been many works for the diagnosis of glioma based on histopathological images (Zhang et al. 2022, Truong et al. 2020; Jose et al. 2022; Yang et al. 2023). Specifically, for glioma grading, Zhang et al. (2022) proposed a mutual contrastive learning method, in which histological features of Formalin-Fixed, Paraffin-Embedded (FFPE) and frozen sections are learned together for better performance. Similarly, Truong et al. (2020) devised a ResNet-based method for the grading of gliomas, with the network output visualised for analysing the tumour heterogeneity. Of note, they found that data augmentation did not improve model training, while a smaller batch size helped to prevent overfitting and led to improved model performance.

Additionally, Pei et al. (2021) proposed a novel deep learning-based classification method that fused genomic features with cellularity features for glioma grading. They devised a novel over-segmentation strategy for region-of-interest (ROI) selection in large histopathology images. More recently, Elazab et al. (2024) proposed to utilise the colour and texture features from histology images, which were then fused and screened to reduce the computation time, beneficial for the real-world deployment of the deep learning system.

On the other hand, many works have also been proposed in the glioma subtyping (Jose et al. 2022; Yang et al. 2023). For instance, Jose et al. (2022) designed an ensemble method by cascading the deep learning networks of VGG-Net (Simonyan and Zisserman, 2014) and ResNet (He et al. 2016). In this proposed network, GBM was first identified, and then astrocytoma and oligodendroglioma were further distinguished in the second stage. Similarly, Yang et al. (2023) proposed a trainable, prototype-enhanced multiple-instance learning network, called TPMIL, to distinguish between GBM, astrocytoma and oligodendroglioma. Different from conventional methods that rely on a certain number of selected patches for feature refinement, TPMIL softly clustered all the patches of the histology images by allocating them to their corresponding prototypes. Moreover, TPMIL also enabled a more accurate interpretability based on the distance of the instances from the trained prototypes, which served as an alternative to the conventional attention score-based interpretability.

More recently, Yuan et al. (2024) proposed using a fine-tuned CNN model to extract deep learning features, followed by biclustering to identify glioma subtypes. Of note, the prognostic value of the predicted image subtypes was assessed via Cox regression on survival outcomes and validated with 189 samples from the Clinical Proteomic Tumor Analysis Consortium (CPTAC) dataset, revealing that the GBM image subtypes were significantly associated with overall survival and progression-free survival.

However, despite success, most existing deep learning methods are based on the obsolete glioma taxonomy criteria (Louis et al. 2007, 2016), which only focus on histology features, such as necrosis and microvascular proliferation. As a result, there is a growing need to develop deep learning-based methods that align with the latest glioma diagnostic criteria (Louis et al. 2021), which incorporate both histology and molecular markers in the classification process. The detailed WHO 2021 classification pipeline and deep learning methods targeting the current pipeline are introduced as follows.

9.3.2.2 Classifying Glioma via Molecular Pathology

In the latest WHO Classification of CNS tumours (Louis et al. 2021), molecular markers, e.g., IDH mutations, for the first time, are assigned as the primary indicator of gliomas taxonomy, while some histological features remain as a reference, e.g., necrosis. The hierarchical schematic diagram for glioma classification is shown in Figure 9.2. As shown, there are six classes of gliomas in total in this version. Specifically, IDH wildtype is directly diagnosed as Grade 4 GBM, while typical GBM histological features (i.e., necrosis, microvascular proliferation and active mitosis) and/or molecular markers (i.e., TERT promoter [TERTp] mutation and/or EGFR gene amplification and/or gain of entire chromosome 7 and loss of entire chromosome 10 [+7/10]) can be of diagnostic value. On the other hand, when IDH mutates, the patients will then be diagnosed as oligodendroglioma or astrocytoma if tested as chromosome 1p/19q co-deletion (CODEL) or not, respectively. For further grading of astrocytoma, both histological features and molecular markers are leveraged, e.g., diagnosing Grade 4 astrocytoma if typical GBM histological features and/or homozygous deletion (HOMDEL) of cyclin-dependent kinase inhibitor 2A/B (CDKN2A/B) are observed. Moreover, of note, there is no evident modification in distinguishing Grade 2 from Grade 3 gliomas in the new version of the WHO classification criteria, either in astrocytoma or oligodendroglioma. For instance, if there exist signals of hypercellularity or anaplasia in histologic images of astrocytoma, the tumour can then be diagnosed as Grade 3 astrocytoma.

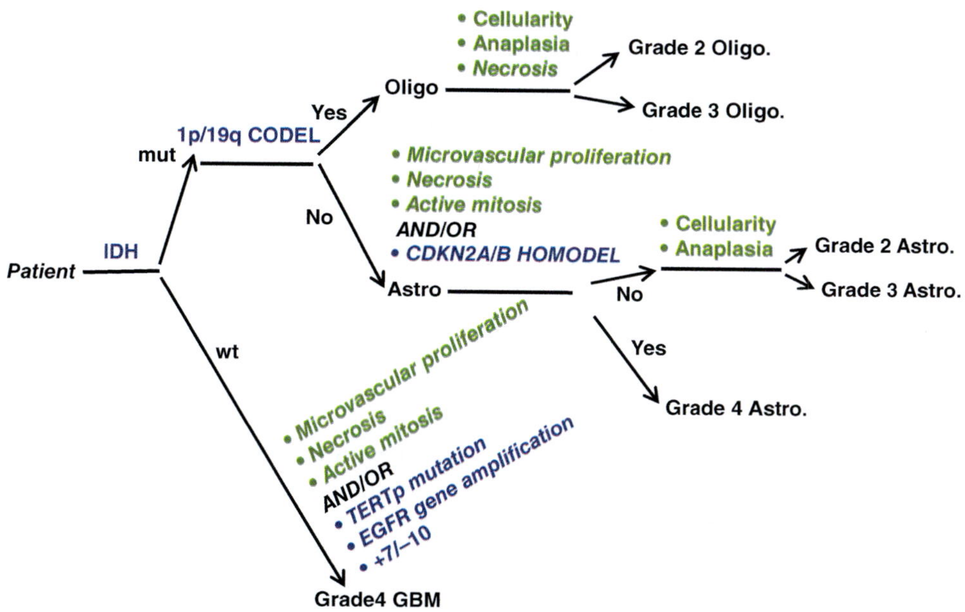

Figure 9.2 Latest WHO CNS tumour classification pipeline. Of note, wt denotes wildtype; mut denotes mutated; Astro. denotes astrocytoma; Oligo. denotes oligodendroglioma.

Most recently, a few deep learning-based methods (Hollon et al. 2023; Nasrallah et al. 2023; Wang et al. 2023b; Jiang et al. 2024) have achieved automatic glioma diagnosis using the up-to-date criteria. For example, Nasrallah et al. (2023) proposed a hierarchical vision transformer-based method based on cryosection slides, called CHARM, to separately predict molecular markers and histology features, which are then combined for final diagnosis. Despite the pioneering attempt, CHARM suffers from an incomplete WHO 2021 diagnostic scheme, i.e., Grade 4 astrocytoma, a newly defined class in the latest criteria, was not included in the training process, due to limited patients in the training cohort.

Similarly, Hollon et al. (2023) developed DeepGlioma, an AI diagnostic screening system to predict the molecular markers (i.e., IDH, 1p/19q and ATRX) for the glioma diagnosis based on stimulated Raman histology images. Specifically, DeepGlioma utilises the automated tumour segmentation as a preprocessing step for the detection of ROI region, which is then cropped to patches and fed into a transformer-based network for molecular marker prediction. Despite good performance, DeepGlioma only focuses on the prediction of molecular markers and ignores the diagnostically important histology features, e.g., necrosis, mitosis, cellularity and anaplasia, which are indispensable in the latest diagnostic criteria. Besides, the small scale of the included glioma patients (i.e., n = 153) further restricts the significance of the study.

Furthermore, Wang et al. (2023b) devised a ResNet-50-based deep learning structure with a novel patch selection strategy for classifying glioma based on FFPE sections. Despite success, they suffer from the incorrect taxonomy process of Grade 4 GBM, which is an important high-grade glioma class with poor survival. In addition, Wang et al. (2023b) can only generate the final glioma class, incapable of simultaneously predicting the molecular markers and histology features that are essential in the current integrated diagnostic pipeline, thereby reducing its practical application value and clinical interpretability of the deep learning prediction system. Another work (Jiang et al. 2024) has recently been proposed for the tasks of glioma subtyping and predicting the molecular markers of IDH mutation and MGMT (O-6-methylguanine-DNA methyltransferase) methylation; yet, this work followed the previous WHO 2016 diagnostic criteria and thus is not applicable for the current integrated scheme.

Despite the potential of existing works, significant limitations remain in existing methods: (1) Current methods are incapable of achieving efficient interaction between the histology features and molecular markers for a clinically interpretable diagnosis. The interaction mentioned involves incorporating clinical multimodal relationship priors into the deep learning model through specific design strategies. For instance, this can be achieved by cross-modally constraining the feature consistency between the IDH marker and high-grade histological features, aligning with the clinical observation that most patients diagnosed histologically with GBM are still classified as Grade 4 GBM in the current integrated diagnostic framework. (2) Existing methods are validated in small real-world cohorts diagnosed with WHO 2021 criteria, as the guideline was published relatively recently, leaving insufficient time for the existing methods to include extensive patients. (3) The diagnosis scheme used in existing works is incomplete or inconsistent with the latest criteria, as aforementioned. (4) Most models were trained with a single type of whole slide images (WSIs), i.e., either FFPE or frozen tissue sections, whereas in real-world practice, both FFPE- and frozen-based WSIs are commonly used for diagnosis in (1) clinical scenarios.

Next, we review the SOTA methods for integrated learning with both molecular markers and histology features for precision oncology.

9.3.2.3 Integrating Histology and Molecular Pathology

Recently, joint modelling of multimodal data of histology and molecular markers has achieved progress in precision oncology, advancing AI towards practical clinical application in various cancers (Zhang et al. 2024b; Wei et al. 2023a). Multimodal algorithms are generally categorised into three types: early, intermediate and late fusion. Early fusion integrates raw data or extracted features before inputting them into the proposed algorithms. Intermediate fusion (Zhou and Chen, 2023; Jaume et al. 2024) modulates multimodal feature embeddings at specific fusion modules or different layers within the model. Late fusion (Shao et al. 2019) involves modality-specific models that aggregate predictions from each modality for the final classification.

Zhou et al. (2023) proposed an intermediate fusion framework (CMAT), which explores intrinsic associations between histopathology images and gene expression profiles through a cross-modal attention module and a modality alignment and fusion module. To explore the intrinsic cross-modal correlations, they further designed a cross-modal attention module as the information bridge between different modalities to perform cross-modal interactions and transfer complementary information. Of note, the authors found that taking the generated cross-modal representation to enhance and recalibrate intramodal representation can significantly improve its discrimination for comprehensive survival analysis. Similarly, Jaume et al. (2024) developed a late fusion framework named SURVPATH, a memory-efficient multimodal transformer that integrates gene pathways and histopathology images for pan-cancer survival prediction. Specifically, they fused both modalities using a memory-efficient multimodal transformer that can model interactions between pathway and histology patch tokens.

For glioma, several multimodal data fusion methods have been proposed (Chen et al. 2020; Qiu et al. 2024; Zhang et al. 2024b; Zhang et al. 2024a; Yi et al. 2024). Specifically, Chen et al. (2020) introduced the Pathomic Fusion method that integrates gene expression and WSI for jointly predicting survival and glioma classification. Qiu et al. (2024) devised a disentangled multimodal framework that can handle both complete and incomplete samples for glioma diagnosis and prognosis. Most recently, Zhang et al. (2024a) designed a general-purpose multimodal learning framework for both cancer diagnosis and prognosis tasks. Specifically, they proposed a novel knowledge-driven subspace fusion (KS-Fusion) scheme to effectively enhance the multimodal interactions. Specifically, in KS-Fusion, they presented a cross-modal deformable attention (CM-Deform) module with a gene-guided consistency strategy to capture specific morphological features of genes.

Similarly, Yi et al. (2024) proposed a Unified Modeling Enhanced Multimodal Learning (UMEML) framework that employs a hierarchical attention structure to effectively leverage shared and complementary features of both modalities of histology and genomics. Specifically, to mitigate unimodal bias from modality imbalance, they utilise a query-based cross-attention mechanism for prototype clustering in the pathology encoder. The prototype assignment and modularity strategy are designed to align shared features and minimise modality gaps. An additional registration mechanism with learnable tokens is introduced to enhance cross-modal feature integration and robustness in multimodal unified modelling. Although effective, current multimodal methods depend on molecular data, which is expensive, time-consuming and relies on tissue availability. These limitations reduce their practicality and hinder widespread clinical adoption.

To tackle this challenge, Xing et al. (2022) proposed a novel discrepancy and gradient-guided modality distillation framework. In this approach, WSIs and gene expressions are combined during training, while the branch for gene expressions is distilled during testing. However, despite this innovative approach, modality distillation-based multimodal learning methods fall short in generating molecular markers. This limitation renders them impractical for clinical use according to the latest diagnosis criteria. Multi-task learning shows promise to improve on existing methods, which can predict both histology features and molecular markers only using WSIs as input. For instance, Coudray et al. (2018) proposed an Inception-v3-based deep learning method (Szegedy et al. 2016) to separately predict both classification and mutation from non-small-cell lung cancer histopathology images. However, most existing methods perform multiple tasks separately, without explicit interaction between histology and molecular marker predictions. Most recently, Wang et al. (2023c) proposed a novel multi-task learning method, called DeepMO-Glioma, with efficient modality interaction via the dynamic confidence constraint loss and the specially designed gradient modulation strategy. This approach facilitates explicit task interplay, enhancing the predictive power and clinical applicability of the model by integrating histology and molecular markers more effectively.

9.3.2.4 Limitations

Existing glioma classification methods, particularly those integrating histology and molecular pathology, face several key limitations. First, current fusion strategies often struggle with the granularity mismatch between modalities. Specifically, WSIs capture millions of cellular-level details, whereas molecular data (e.g., IDH mutation status) typically reflect tumour-level information. This discrepancy makes it challenging to accurately align and integrate the two data types within a unified model.

Second, most existing methods are trained and evaluated on a single dataset, such as TCGA, without external validation. This significantly limits their generalisability and clinical applicability across diverse patient populations and pathology workflows.

Finally, a major limitation is the lack of interpretability in current AI models. These models often function as 'black boxes', providing little insight into the specific histological features that drive predictions of molecular markers. This opacity poses a challenge in clinical settings, where understanding the rationale behind a diagnosis is essential for trust, verification and decision-making.

9.3.3 AI-based Spatial Transcriptomics

While bulk molecular markers have been successfully utilised, we further explore the integration of spatial gene profiling for more effective neuro-oncology modelling. ST is inherently multimodal, combining spatial gene expression maps with histology images and holding promise for precision neuro-oncology. However, despite their rich information, current ST platforms face challenges in limited spatial resolution and gene profiling coverage. To address these challenges, various computer vision-based computational methods have emerged to analyse and enhance ST data, enabling better feature extraction for neuro-oncology modelling. Here, we first introduce the fundamentals of ST techniques in Section 9.3.3.1. Next, we review computational approaches for super-resolving ST maps in Section 9.3.3.2. Finally, in Section 9.3.3.3, we explore techniques for imputing under-profiled genes in ST, enabling downstream tasks such as cell-type deconvolution. An illustration of the ST super-resolution and gene imputation process is shown in Figure 9.4.

9.3.3.1 Spatial Transcriptomics: A Brief Introduction

Human tissues are composed of various cell subclones in which biological functions are highly correlated with their structural distributions and cellular interactions. An in-depth understanding of disease mechanisms, such as heterogeneity, requires molecular profiling with advanced techniques, e.g., next-generation sequencing (NGS)-based approaches (Atta and Fan, 2021). As a representative technique of NGS, recently developed single-cell RNA sequencing (scRNA-seq) (Kharchenko, 2021) can provide extensive information about transcriptomes at the single-cell level. Nevertheless, current segregation-based protocols inevitably dissociate cells from tissues, so that the important spatial information of sequencing is lost (Du et al. 2023). To preserve spatial information, spatially resolved transcriptomics techniques have therefore been proposed.

Techniques for obtaining the spatial transcriptome can generally be classified into **imaging-based methods**, e.g., ouroboros single-molecule fluorescent in situ hybridisation (osmFISH; Codeluppi et al. 2018), and **sequencing-based methods**, e.g., Visium Spatial

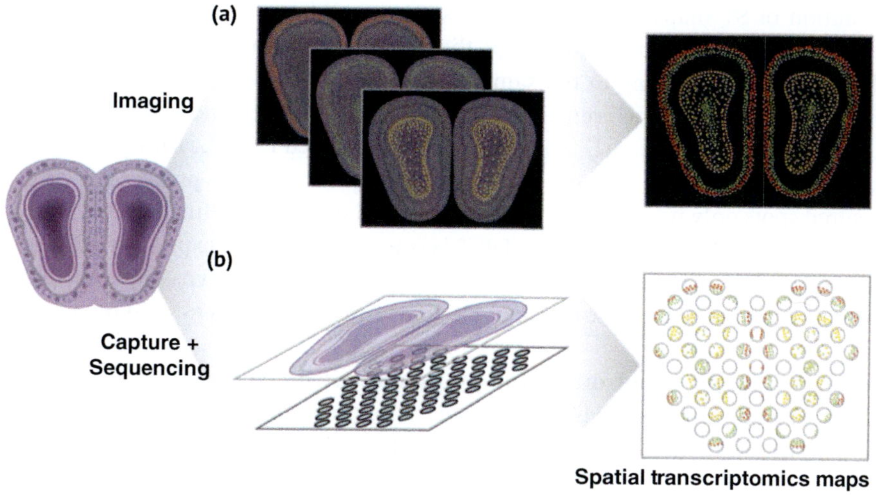

Figure 9.3 An example of imaging-based and sequencing-based ST maps. (a) Imaging-based transcriptomic profiling result. (b) Sequencing-based ST. *Source*: Atta and Fan (2021)/SpringerNature/CC BY 4.0.

Gene Expression used by 10× Genomics (Rao et al. 2020). Generally, there is a trade-off between experimental throughput versus spatial resolution, i.e., imaging-based methods usually offer higher spatial resolution but fewer genes profiled, while most sequencing-based methods generally provide higher (sometimes nearly genome-wide) experimental throughput but lower spatial resolution (Shen et al. 2022). Of note, among the ST techniques, 10× Visium ST is now one of the most popular technical solutions due to its practicality and commercialisation (Moses and Pachter, 2022). An example of imaging- and sequencing-based ST maps is shown in Figure 9.3.

9.3.3.2 AI Approaches for Enhancing ST Spatial Resolution

Popular ST techniques, such as Visium (Ståhl et al. 2016) and SLIDE-seqV2 (Stickels et al. 2021), measure gene expression in tissue spots but suffer from low spatial resolution (e.g., 100 μm px^{-1} for Visium), which limits their ability to capture gene expression at the cellular level (10 μm px^{-1}). Emerging high-resolution ST methods, such as Xenium (Salas et al. 2023) and Visium HD (Polanski et al. 2024), aim to address this limitation. However, these advanced techniques are expensive, time-consuming and constrained by the technical challenge of low capture sensitivity. Computational methods offer the potential to improve the spatial resolution of ST maps. In general, existing computational methods can be divided into two groups in terms of the use of paired histology images: expression prior-based methods and histology image-based methods. A detailed review of the two groups of ST resolution enhancement approaches is introduced as follows.

Super-resolving ST maps based on spatial expression prior: The expression prior-based methods (Zhao et al. 2023; Zhao et al. 2021) try to exploit the spatial patterns of gene expressions for ST super-resolution. One representative method is BayesSpace (Zhao et al. 2021), a Bayesian statistical method using spot neighbourhood information for enhancing the resolution of ST maps. Specifically, BayesSpace enables spatial clustering by modelling a low-dimensional representation of the gene expression matrix and encouraging neighbouring spots to belong to the same cluster via a spatial prior. Hence, BayesSpace does not rely on any extra reference data, e.g., histology images and scRNA-seq data, in real-world applications. Similarly, Zhao et al. (2023) proposed Deep Learning-based Spatial Transcriptomics (DIST), a method that imputes gene expression profiles on new and unmeasured spots only using spatial gene expression data to get a refined spatial map with a higher resolution. The whole idea of DIST is imputing the gene expression of unmeasured spots by self-supervised learning. Specifically, the authors downsample the original gene expression maps and create artificial LR gene expression maps to train the model.

Predicting ST maps based on histology images: Histology features observed from tissue sections are enriched with phenotypic structure and morphology information. Previous studies show that image-level histology features are associated with tissue gene expression (Schmauch et al. 2020; Badea and Stănescu, 2020). This intrinsic link establishes the feasibility of predicting ST maps using histology features extracted from tissue images as guidance. Therefore, several studies (He et al. 2020; Xie et al. 2024; Jia et al. 2024) have made efforts in this direction. For example, as a pioneering attempt, He et al. (2020) devised ST-Net, utilising ImageNet-pretrained DenseNet-121 (Huang et al. 2017) to successfully predict the spatial expression of 250 genes of breast cancer. The ST-Net can predict substantial variation in expression within regions that are all labelled as tumour by clinicians, demonstrating that the model captures intratumour heterogeneity. Similarly, Jia et al. (2024) devised THItoGene, a hybrid neural network that utilises dynamic convolutional and capsule networks to adaptively sense potential molecular signals in histological images for exploring the relationship between high-resolution pathology image phenotypes and the regulation of gene expression. A comprehensive benchmark evaluation using datasets from human breast cancer and cutaneous squamous cell carcinoma has demonstrated the great performance of THItoGene in spatial gene expression prediction. Specifically, THItoGene reaches Pearson correlation coefficients of 0.747, 0.711, 0.672 and 0.452 for four commonly used breast tumour-related genes of FN1, SCD, IGKC and FASN, respectively.

More recently, Xie et al. (2024) proposed a bi-modal contrastive-based framework (BLEEP) for predicting expression from histology images. Specifically, BLEEP uses contrastive learning to construct a low-dimensional joint embedding space from a reference dataset using paired image and expression profiles at micrometre resolution. With this approach, the gene expression of any query image patch can be imputed using the expression profiles from the reference dataset. However, as these approaches only focus on predicting spot-based ST maps, they are incapable of enhancing the resolution of ST maps.

To enhance the resolution of ST maps, recent studies (Bergenstråhle et al. 2022; Hu et al. 2023; Zhang et al. 2024; Wang et al. 2024b) have further proposed to super-resolve ST maps using histology images. For example, Hu et al. (2023) proposed a ST super-resolution approach, TESLA, which combines the colour space of histology images and the position features of spot-based ST maps for ST enhancement. Similarly, Hou et al. (2022) proposed to first generate the cell segmentation map from the histology image and then fuse the cell segmentation map with spot-based ST map for generating a single-cell resolution ST map.

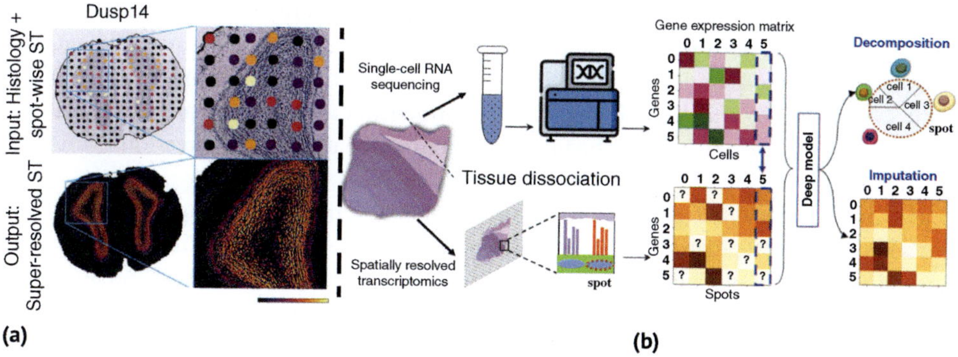

Figure 9.4 (a) Inferred super-resolved expression of the input paired low-resolution ST and histologic image for XFuse method. *Source*: Adapted from Bergenstråhle et al. (2022)/Springer Nature. (b) Illustration of the gene imputation and cell-type deconvolution process. *Source*: Nasab et al. (2022)/Oxford University Press/CC BY 4.0.

However, one drawback of these methods is the dependency on the reference ST maps when deployed in the real world, which hinders their application in scenarios without ST availability. Hence, some recent works (Bergenstråhle et al. 2022; Zhang et al. 2024) have been proposed to remove the dependency of the reference spot-based ST maps in model inference. For example, Bergenstråhle et al. (2022) proposed a transcriptome-wide ST super-resolution model, called XFuse, by integrating low-resolution gene expression from in situ sequencing and high-resolution histologic images. In model training, XFuse assumes that the gene expression data and histologic images conform to the negative binomial and Gaussian distributions, respectively. Then, XFuse casts spatial gene expression and histologic image data as observable effects of a latent tissue state, which is designed to capture both global and local anatomical features for ST super-resolution. In model validation, XFuse can use either (1) the paired low-resolution ST and histologic image (see an example in Figure 9.4a) or (2) standalone histologic images with reference ST from other parts of the same tissue. Similarly, Zhang et al. (2024) devised a Vision Transformers-based method (Chen et al. 2022), iStar, to infer high-resolution ST maps. However, as these methods only use the spot ST as weak supervision, they are less capable of modelling the cross-modal interactions between high-resolution ST and histology images. Most recently, Wang et al. (2024b) proposed a diffusion-based model (Diff-ST) for ST enhancement, explicitly integrating histology and high-resolution ST maps to enhance the spot-based ST resolution.

9.3.3.3 Imputing Under-profiled Genes in ST

To overcome the under-profiling limitation in ST, some computational methods (Lopez et al. 2019; Abdelaal et al. 2020; Shengquan et al. 2021) have recently been proposed to impute the missing genes (see an example in Figure 9.4b). Specifically, to impute missing genes, Lopez et al. (2019) proposed the variational inference (gimVI), a deep generative model for integrating ST data and unpaired standard scRNA-seq data obtained from the same biological tissue. In addition, Abdelaal et al. (2020) devised the spatial gene enhancement (SpaGE) method, which integrates spatial and scRNA-seq datasets to predict whole-transcriptome expressions in their spatial configuration. Specifically, SpaGE relies on domain adaptation learning paradigms to correct for differences in the sensitivity of transcript detection between both single-cell and ST technologies. Moreover, Shengquan et al. (2021) proposed a reference-based method, namely stPlus, that utilises information in scRNA-seq data to enhance ST. Based on an autoencoder with a specially designed loss function, stPlus performs joint embedding and predicts spatial gene expression through a weighted k-nearest-neighbour network.

Based on the imputed gene profiling, some tailored methods (Song and Su, 2021; Biancalani et al. 2021; Liao et al. 2022) have further proposed to uncover the cell compositions using ST maps, which are further utilised to study the spatial molecular features of tissue at the single-cell level. For instance, Song et al. (2021) proposed a graph-based convolutional network, called DSTG, to deconvolute ST with the integration of scRNA-seq data. In addition, Biancalani et al. (2021) presented Tangram, an algorithm that aligns scRNA-seq data with various forms of ST data and further achieves single-cell deconvolution of ST. While most deconvolution methods require paired ST and scRNA-seq data as input, Liao et al. (2022) tried to predict single-cell ST data from bulk transcriptomes with existing high-quality scRNA-seq data and ST as references. The key idea of Liao et al. (2022) is the development of a beta variational autoencoder (β-VAE; Higgins et al. 2017), which is used to generate scRNA-seq directly from the bulk transcriptomes with the standard reference scRNA-seq.

9.3.3.4 Limitations

Despite success, existing AI methods for enhancing ST resolution and profiling coverage are still facing challenges.

For ST super-resolution, both expression prior- and histology image-based AI methods have limitations. Specifically, expression prior-based methods rely solely on ST, without utilising the histological information at the tissue and cellular levels provided by ST maps. This reliance on genomic data alone challenges the biological confidence of the imputed fine-grained ST resolution. Meanwhile, histology image-based ST resolution enhancement methods draw expression information exclusively from paired low-resolution ST maps, which inherently suffer from low sequencing sensitivity (up to hundreds of times lower than scRNA-seq) and a limited gene panel (also significantly smaller than that of scRNA-seq).

For ST imputation, current AI-based imputation methods still face notable limitations. First, many rely heavily on paired scRNA-seq data, which is often impractical due to tissue constraints and high costs. Additionally, most existing approaches overlook the valuable information embedded in corresponding histological images.

9.3.4 AI Modelling Cancer Evolution and Tumour Heterogeneity

9.3.4.1 Cancer Evolution and Tumour Heterogeneity

In cancer diagnosis and treatment, tumour heterogeneity is a major factor that influences clinical outcomes. In general, tumour heterogeneity describes the observation that different tumour cells can show distinct morphological characteristics, such as cellularity, gene expression and motility (Fittall and Van Loo, 2019). This observation occurs both among tumours, i.e., intertumor heterogeneity, and within tumours, i.e., intratumour heterogeneity. Tumour heterogeneity is widely recognised as the result of tumour evolution, which is

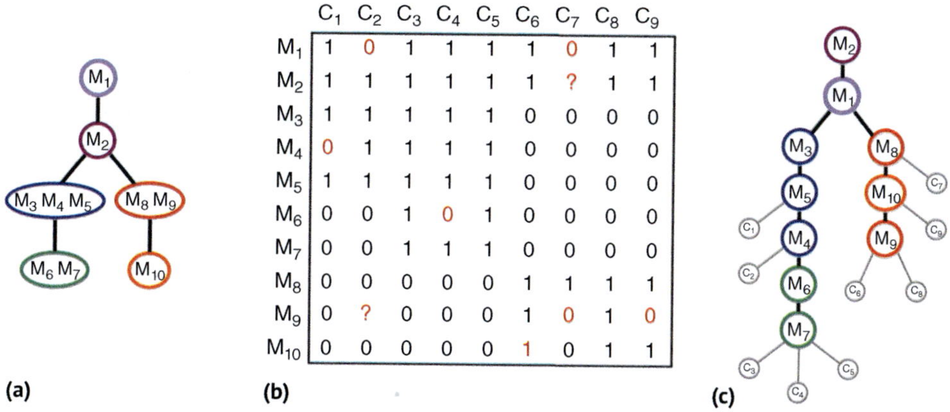

Figure 9.5 An example of evolutionary history and subclonal composition prediction based on the single-cell mutation profile. (a) Ground truth evolution tree with mutations of M_1 to M_{10}: different subclones are represented in different colours, with the tree structure indicating the temporal order in which the subclones evolved. (b) Cell mutation profiles derived from the single-cell sequencing data for nine cells of c_1 to c_9, where '1' indicates the observed presence of a mutation and '0' denotes absence. A '?' indicates a missing data point. c) Inferred single-cell mutation tree annotated with single-cell placements. Not all cells can be placed such that their observed mutation profile matches the mutations acquired along the lineage from the root to their attachment point. The branching point of the ground truth tree is inferred correctly, due to the strong signal that the red/orange and blue/green mutations do not occur in the same cell. *Source*: Malikic et al. (2019)/Springer Nature/CC BY 4.0.

driven by mutations that cause growth advantages to different subpopulations of cancer cells (Zellmer and Zhang, 2014). Tumour evolution is identified by the arising of genetically distinct subclones via driver mutations and cellular interaction within the microenvironment.

Generally, the clonal evolution models can be divided into four groups (Manini and López, 2023), i.e., linear evolution, punctuated evolution, neutral evolution and branch evolution. These models are based on different assumptions; for example, in branch evolution, it is assumed that the clonal and subclonal diversification develops following Darwinian patterns. To predict the evolutionary dynamics of individual tumours, there are mainly two steps: (1) evolutionary history prediction and (2) identification of its subclonal composition (Malikic et al. 2019). Among recent techniques in tumour evolution prediction, bulk sequencing and single-cell DNA sequencing results are mostly used. An example of evolutionary history and subclonal composition prediction based on the single-cell mutation profile (Malikic et al. 2019) is shown in Figure 9.5. Finally, regarding the meaning of tumour evolution analysis, understanding the heterogeneity existing within tumours and their evolution features in response to therapy may allow early cancer diagnosis and more optimised treatment strategies (Fittall and Van Loo, 2019).

9.3.4.2 Learning-based Methods

In recent years, many learning-based computational methods (Fischer et al. 2014; Durante et al. 2020; Ouellette and Awadalla, 2022) have been proposed for the prediction and analysis of phylogenetic trees. Among the existing methods, the main research directions can be divided into two groups, i.e., designing efficient tumour evolution inference algorithms and exploring evolution information from various omics. Specifically, in the field of inference algorithm design, existing papers (Jahn et al. 2016; Durante et al. 2020; Ouellette and Awadalla, 2022) mainly focus on specific tasks of evolutionary history prediction and subclonal composition identification. Besides, the most common data used for evolutionary inference are single-nucleotide variants (SNVs) and copy number variation (CNV), either from bulk or single-cell sequencing. For instance, Jahn et al. (2016) proposed a stochastic search algorithm, called SCITE, to identify the evolutionary history of a tumour from noisy and incomplete mutation profiles of single cells. SCITE implements a Markov-chain Monte Carlo-based Bayesian inference scheme to identify a phylogenetic tree that ideally fits the observed single-cell genotypes. Durante et al. (2020) utilised inferCNV (https://github.com/broadinstitute/inferCNV/wiki:) method to generate the CNV content from scRNA-seq data, which is then leveraged in a hidden Markov model to determine subclonal CNV events. This work paves a new way to infer CNV from the downstream scRNA-seq data, which is inspiring. This work paves a new path for inferring CNVs from downstream scRNA-seq data, offering an inspiring and innovative approach. Moreover, Ouellette et al. (2022) proposed a proof-of-principle synthetic supervised learning method, called TumE, which integrates simulated models of tumour evolution with Bayesian neural networks, to infer ongoing selection based on variant allele frequencies from bulk DNA sequencing.

On the other hand, many efforts have also been made to explore evolutionary dynamics from multiple omics or sequencing techniques. For example, Fischer et al. (2014) devised a probabilistic algorithm, cloneHD, to realise subclone reconstruction from both CNV and SNV data, so that it is robust to both CNV and SNV-driven tumour evolution inference. Specifically, in a study by Fischer et al. (2014), inferences are performed with a set of coupled hidden Markov models jointly across CNV and SNV from single-cell sequencing. Besides, Gaiti et al. (2019) performed single-cell multi-omics to capture transcriptome and DNA methylation information and performed lineage tracing through DNA methylation data.

9.3.4.3 Limitations

Despite important progress, current tumour evolution studies still face notable limitations. Most approaches rely on either bulk or single-cell sequencing data, each with inherent drawbacks. Bulk sequencing provides only a coarse view of mutation history and lacks resolution to identify subclonal populations. Single-cell methods, while offering higher resolution, are prone to sequencing noise introduced during DNA amplification (Kalisky and Quake, 2011).

In addition, while histology images, e.g., WSIs, offer rich spatial context and morphological detail, their integration into AI-based cancer evolution models remains underdeveloped. One major limitation is the lack of direct ground truth linking microscopic histological features to evolutionary events such as subclonal expansions or temporal mutation trajectories.

Moreover, although computational techniques for modelling cancer evolution have advanced significantly, their application to gliomas remains limited. This gap is likely due to the complex nature of neuro-oncology. Gliomas are highly heterogeneous and interact in complex ways with the brain microenvironment. These factors make analysis especially challenging.

9.3.5 Future Directions

In this section, we outline future directions for AI-based digital pathology, ST enhancement and cancer evolution with tumour heterogeneity analysis.

For AI-based digital pathology, given the impracticality of exhaustively annotating gigapixel pathology slides, weakly supervised learning will remain essential and be further improved. Future studies can explore more sophisticated multi-instance learning algorithms that handle noisy slide-level labels, especially for cases with partially available molecular markers. In addition, to improve the interpretability, future models can be designed to provide the histological patterns of the predicted molecular markers. Moreover, the clinical value, e.g., the prognostic value, should be validated in future works, thus proving the translational value of the proposed models. Finally, the designed AI models should be validated in multi-centre datasets, thus verifying the real-world generalisability of the AI models.

For future directions in AI-based ST enhancement, several key opportunities emerge. Current ST super-resolution methods remain limited by sequencing sensitivity and restricted gene coverage, while many imputation approaches fail to leverage the rich contextual information embedded in paired histological images. A promising direction is the integration of multimodal data, including histology images, scRNA-seq and ST maps, to jointly perform prediction, super-resolution and imputation, enabling more accurate and comprehensive spatial gene profiling.

Furthermore, most existing methods primarily focus on reconstructing gene expression within cellular regions, often overlooking non-cellular niches such as the extracellular matrix. These regions, however, play a critical role in cancer progression and tumour–microenvironment interactions. Future AI models should explicitly model both cellular and non-cellular compartments to fully capture the spatial complexity of tumour biology.

Finally, for AI-based cancer evolution methods, future research should explore integrative AI models that combine histology images with omics data. In addition, advances in spatial omics and multiplex imaging could also enable new multimodal datasets that capture both phenotypic and genotypic heterogeneity across tumour regions. Together, these innovations may pave the way for AI systems capable of reconstructing cancer evolution from tissue slides, ultimately improving our understanding of glioma progression and therapeutic resistance.

9.4 Generalisable AI for Clinical Translation

AI has rapidly transformed patient care, promising to automate time-consuming tasks. AI models are now used for tumour segmentation (delineating tumour boundaries on MRI), tumour classification/grading (inferring tumour type or genetics from images), intraoperative imaging (e.g., interpreting confocal endomicroscopy or hyperspectral images) and surgical decision support (predicting resectability, survival or aiding navigation). In research settings, these algorithms have achieved high performance in critical tasks such as tumour segmentation and classification, fuelling optimism for a paradigm shift in diagnostics, prognostics and treatment planning in neuro-oncology with AI support.

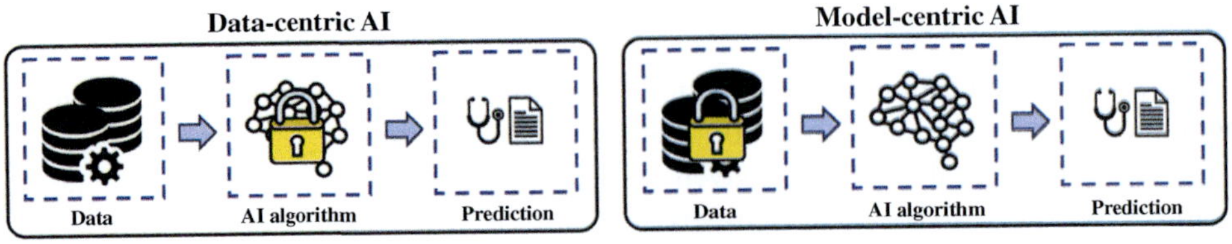

Figure 9.6 The development of AI generalisation paradigms can be broadly categorised into data-centric and model-centric approaches. The 'lock' represents the fixed or unchanged element within the paradigm.

Despite the promising potential and research success, the routine use of these AI tools in clinical practice remains challenging. This is largely due to a key limitation: most models are trained and validated on curated datasets, which do not reflect the variability of real-world clinical environments. In practice, models that perform well in retrospective studies often fail to generalise when applied in real clinical settings, due to factors such as differences in imaging equipment, patient demographics and tumour heterogeneity.

Although generalisability has been widely studied across machine learning research, it is particularly critical and unresolved in high-stakes applications such as neuro-oncology. At its core, an AI system processes input data through computational models to produce clinically meaningful outputs. The quality and trustworthiness of these outputs depend on two principal factors: (1) the integrity and representativeness of the input data and (2) the robustness and adaptability of the model architecture. In real-world clinical settings, both factors are frequently compromised.

The following sections (Sections 9.4.1 and 9.4.2) describe the radiological imaging used in brain tumours and highlight the associated challenges, respectively. It then introduces current strategies to improve generalisation, viewed from two complementary perspectives: data-centric approaches (Section 9.4.3), which aim to enhance the quality, diversity and harmonisation of input data; and model-centric approaches (Section 9.4.4), which focus on developing architectures and learning paradigms capable of robust generalisation. These perspectives are conceptually illustrated in Figure 9.6.

9.4.1 Radiological Imaging for Neuro-oncology

MRI is a cornerstone of brain tumour diagnosis and plays a critical role in surgical planning, image-guided therapy and longitudinal monitoring. The most commonly used MRI modalities: T1-weighted (T1W), T2-weighted (T2W), contrast-enhanced T1 (T1C) and fluid-attenuated inversion recovery (FLAIR), offering complementary tissue contrasts that highlight different pathological features. The examples of these modalities are presented in Figure 9.7:

1. **T1W imaging:** T1W sequences reflect the longitudinal relaxation time (T1) of hydrogen protons, emphasising anatomical detail and fluid compartments. They are especially effective for visualising brain structure, fat-containing lesions and haemorrhages.
2. **T2W imaging:** T2W sequences emphasise transverse relaxation (T2), providing high sensitivity to water content. They are useful for identifying oedema, inflammation and cystic tumour components, offering strong contrast between normal and pathological tissues.

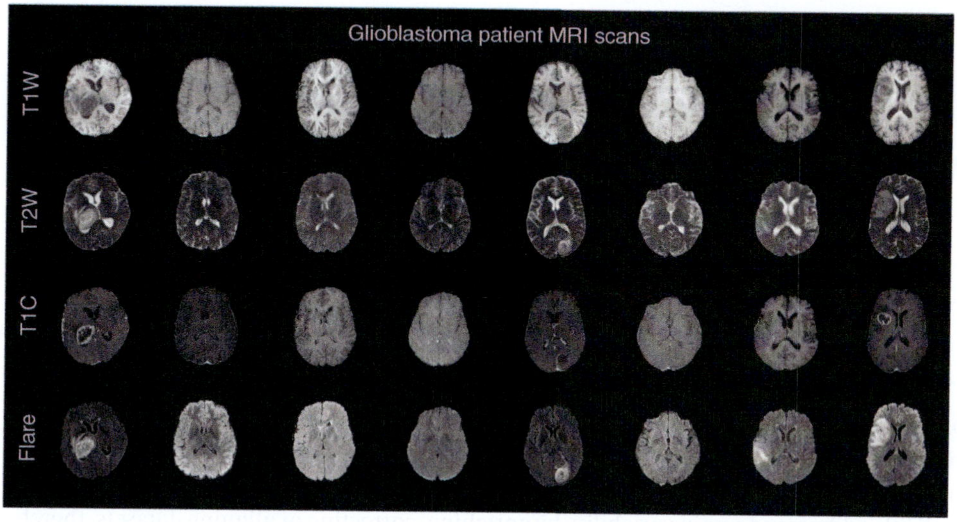

Figure 9.7 Representative MRI scans across four modalities (T1W, T2W, T1C and FLAIR) from GBM patients. These sequences are commonly used in diagnosis, treatment planning and follow-up. *Source*: Baid et al. (2021)/arXiv/CC BY 4.0.

3. **T1C imaging:** T1C uses gadolinium-based contrast agents to highlight areas with a disrupted blood–brain barrier. This modality is essential for delineating tumour margins, detecting active tumour regions and differentiating recurrence from treatment-induced changes.
4. **FLAIR imaging:** FLAIR is a sequence that suppresses cerebrospinal fluid (CSF) signals. This makes it particularly useful for visualising periventricular and cortical lesions, demyelination, gliosis and small infiltrative tumours that may be masked in standard T2W images.

9.4.2 MRI Data Heterogeneity

Curated datasets used for training AI models are typically well-standardised and often sourced from a limited number of clinical sites. In contrast, clinical environments are highly variable, introducing a wide range of heterogeneity in MRI data, including limitations arising from data availability (Challenge I) and missing or incomplete data (Challenge II). The main sources of variability include the following:

1. **Scanner and protocol variabilities (Challenge I):** Differences between MRI scanners (e.g., Siemens vs. GE) and coil configurations can lead to variations in image contrast, resolution and slice thickness. These factors may critically impair the performance of AI-based systems.
2. **Scanning artefacts (Challenge I):** Scanning artefacts can arise from various sources, such as patient movement, metallic implants or anatomical factors like proximity to the sinuses. These artefacts may obscure lesion borders or degrade overall image quality. Additionally, different acquisition protocols (e.g., 1.5T vs. 3.0T) exhibit varying susceptibility to artefacts and influence the visibility of lesion margins, resulting in different signal-to-noise ratios.

3. **Incomplete datasets (Challenge II):** MRI acquisition is inherently time-consuming. Factors such as limited scanner availability, patient tolerance, concerns about contrast agents, and the duration and cost of scanning often lead to incomplete acquisition. A comparison of typical MRI sequence parameters is provided in Table 9.3.
4. **Demographic diversity (Challenge II):** Curated datasets often lack diversity in terms of age, sex, ethnicity and comorbidities, leading to biased model behaviour. For example, paediatric brain tumours are biologically distinct, often located in different regions, and may appear differently on MRI. Applying adult-trained models to paediatric cases can lead to misdiagnosis or reduced accuracy.

9.4.3 Data-centric Generalisation

Data-centric generalisation focuses on enhancing the robustness and adaptability of AI models by improving the quality, completeness and representativeness of training data, rather than relying solely on architectural or algorithmic modifications. This approach explicitly addresses domain variability by enriching, correcting or aligning datasets, thereby promoting more consistent model performance across diverse clinical settings.

In the context of the aforementioned challenges, two primary research directions have emerged to mitigate issues related to both acquired and missing imaging data: medical image harmonisation and medical image synthesis. Harmonisation seeks to reduce inter-scanner and inter-protocol variability in available images, directly targeting Challenge I (see Section 9.4.3.1). In parallel, synthesis techniques aim to mitigate Challenge II by generating missing or underrepresented modalities and sequences, thereby enhancing dataset completeness and diversity (see Section 9.4.3.2).

9.4.3.1 Medical Imaging Harmonisation

Image super-resolution: High-resolution MRI scans with sufficient tissue contrast are often unavailable in clinical practice due to long acquisition times and patient motion. To address this limitation, numerous deep learning methods (Zakka et al. 2020; Esser et al. 2021; Wang et al. 2023a; Mao et al. 2023) have been proposed to enhance resolution computationally.

For example, Esser et al. (2021) propose a hybrid architecture that combines CNNs for learning local image components with transformers to capture long-range spatial dependencies in high-resolution MRI reconstruction. InverseSR (Wang et al. 2023a) leverages a latent diffusion model, pretrained on the UK Biobank dataset (Sudlow et al. 2015), to upsample low-resolution clinical MRI scans. The model explores the latent space of three-dimensional (3D) T1-weighted brain MRI to infer optimal high-resolution mappings, enabling improved delineation of subtle tissue structures in time-constrained clinical workflows. Mao et al. (2023) propose DisC-Diff, which incorporates multi-contrast MRI data to enhance super-resolution fidelity using a diffusion model-based approach. Their method introduces a disentanglement mechanism with channel attention to capture relevant shared features across modalities while minimising redundancy. In parallel, it isolates contrast-specific features that help preserve structural detail in the final output, supporting accurate diagnosis in cases such as small lesion detection or tumour boundary assessment.

Table 9.3 The MRI scanning time data.

Modality	Protocal	Plane	Mode	FOV	TE	TR	Frequency	Thickness	Scan Time
T1W	1.5T	Sagittal/axial	3D	256 mm	Min	2100	≥172	≤1.5 mm	5–10 min
T1C	1.5T	Sagittal/axial	3D	256 mm	Min	2100	≥172	≤1.5 mm	5–10 min
T2W	1.5T	Axial	2D	240 mm	100–120	>3500	≥256	≤4 mm	3–5 min
FLAIR	1.5T	Axial	2D	240 mm	100–140	>6000	≥256	≤4 mm	4–5 min
DWI	1.5T	Axial	2D	240 mm	Min	>5000	128	≤4 mm	3–5 min
T1W	3.0T	Sagittal/axial	3D	256 mm	Min	2100	256	1 mm	5–8 min
T1C	3.0T	Sagittal/axial	3D	256 mm	Min	2100	256	1 mm	5–8 min
T2W	3.0T	Axial	2D	240 mm	80–120	>2500	≥256	3 mm	3–5 min
FLAIR	3.0T	Axial	2D	240 mm	100–140	>6000	≥256	3 mm	4–5 min
DWI	3.0T	Axial	2D	240 mm	Min	>5000	128	3 mm	3–5 min

Source: Adapted from Ellingson et al. (2015). For T1C, use contrast—0.1 mmol/kg injection with 20 cc of a gadolinium-chelated contrast agent at an injection rate of 3–5 cc/s (single, full dose). Abbreviations: 3D, 3-dimensional; FOV, field of view; TE, echo time; TR, repetition time; DWI, diffusion-weighted imaging; Min, the minimum TE that can be achieved based on the scanner's capabilities and the imaging sequence being used. The data is collected from Siemens and Hitachi scanners.

Although diffusion-based models demonstrate strong generative capacity, they are often sensitive to stochastic sampling, which can reduce output consistency. DisC-Diff mitigates this by conditioning generation on low-resolution, multi-contrast MRI inputs processed through dedicated encoders, resulting in more stable and anatomically consistent reconstructions suitable for clinical use.

Undersample Degradation Removal: Accelerated MRI via k-space undersampling enables faster acquisition but introduces artefacts such as aliasing, noise and information loss, which degrade image quality. Despite MRI's excellent soft tissue contrast, long acquisition times limit its clinical practicality. Reconstruction from undersampled measurements has been extensively studied to mitigate these issues (Yang et al. 2022a; Xie and Li, 2022; Korkmaz et al. 2023; Gunel et al. 2022).

For instance, Xie et al. (2022) propose MC-DDPM, a measurement-conditioned diffusion model that generates reconstructions directly in *k*-space. By incorporating the undersampling mask, the model adapts the denoising process to specific scan protocols, making it well-suited for varied clinical workflows where acquisition settings differ across sites. Extending this, Korkmaz et al. (2023) introduce a self-supervised strategy that removes the need for fully sampled labels. Through cross-attention transformers and data consistency constraints, it supports deployment in settings where reference scans are unavailable, such as emergency care or low-resource clinics. To improve generalisability across anatomies and scanner types, Gunel et al. (2022) introduce scale-equivariant unrolled networks, enhancing robustness across diverse patient populations without retraining.

Beyond artefacts, random noise presents an additional challenge by obscuring fine-grained anatomical features essential for preserving lesion boundaries and subtle structural details, which are critical for early diagnosis and disease monitoring. Yang et al. (2022a) address this with a hybrid denoising framework that combines multi-layer perceptrons and CNNs to jointly model global context and local structure. Their method effectively enhances image quality while maintaining clinically relevant details.

9.4.3.2 Medical Imaging Synthesis

Synthesising missing modalities: In brain tumour diagnosis, acquiring multimodal MRI scans (e.g., T1, T1C, T2, FLAIR) is a standard practice, as each modality provides complementary structural and pathological information. However, missing modalities are common in clinical settings due to scanner limitations, time constraints or concerns regarding gadolinium-based contrast agents.

MRI synthesis has emerged as a practical solution to fill these gaps. Pinetz et al. (2023) synthesise full-dose contrast-enhanced images from low-dose scans using a conditional GAN, reducing contrast-agent exposure without sacrificing diagnostic value. This is particularly relevant for patients with renal impairment or paediatric populations, where minimising gadolinium use is essential for long-term safety. Recent methods such as SynDiff (Özbey et al. 2023) and UNIT-DDPM (Sasaki et al. 2021) use diffusion models to reconstruct missing sequences from available ones, enabling consistent tumour visualisation across incomplete scans. CoLa-Diff (Jiang et al. 2023) further improves anatomical accuracy by conditioning on structural brain regions (e.g., grey matter, CSF, tumour) during synthesis.

Such anatomical conditioning helps preserve lesion morphology, supporting more reliable treatment planning and tumour grading.

While most prior work assumes one-to-one synthesis, clinical cases often involve multiple missing and multiple observed modalities. MM-SYN (Chartsias et al. 2017) addresses this by fusing independently encoded feature maps via max pooling. MM-GAN (Sharma and Hamarneh, 2019) and ReMIC (Shen et al. 2020) pre-impute missing modalities with zero matrices and synthesise in a single step. However, these approaches suffer from limited fusion quality and domain gap. To address these limitations, Meng et al. (2024) proposed M2DN, which conditions on available modalities and performs progressive denoising from Gaussian noise. A binary modality mask explicitly encodes which inputs are missing, enabling joint synthesis across any combination of present/absent modalities. This flexibility is valuable in clinical workflows where input availability varies across patients or scan sessions.

The model initialises missing modalities with random Gaussian noise, enabling joint synthesis across multiple modalities. A binary modality mask is incorporated into the denoising process as a conditioning mechanism to explicitly indicate which modalities are missing.

Despite recent progress, synthesised MRI still falls short of clinical standards. In practice, images must be clear, accurate and reliable, yet many synthesised scans show blurring or artefacts, limiting clinician trust. Current methods also struggle with noisy, incomplete or variable data, which are common in real-world settings. Even when more modalities are available, gains in image quality remain modest, highlighting limitations in current techniques. To be clinically useful, future models must handle real-world data variability, improve anatomical fidelity and consistently deliver trustworthy outputs that support confident diagnosis and treatment planning.

Generating more data: Medical image datasets are often limited in size and diversity due to high acquisition costs and patient variability, making it difficult for AI models to generalise across clinical populations and disease types. Synthetic data offers a solution. Recent work by Dorjsembe et al. (2024) shows that adding synthetic images to training datasets improves model performance, especially when real data is scarce.

Generating synthetic data that reflects specific anatomical structures or clinical conditions can help overcome the limitations of small or imbalanced medical datasets and improve the reliability of AI models. For example, Costa et al. (2017) and Zhang et al. (2019) simulate anatomical variation using segmentation maps and sketch-based inputs. More recent approaches, such as those by Blattmann et al. (2022) and Daum et al. (2024), generate realistic 3D medical images by conditioning generative models on text descriptions or patient attributes. These approaches enable large-scale dataset generation while safeguarding sensitive health data, supporting broader access to AI development in clinical research.

Synthesising lesions in healthy scans is another effective way to expand disease representation in training data, especially for rare or early-stage tumours. Niu et al. (2022) introduced a method using pathological descriptors to control lesion location, size and type, allowing targeted lesion generation for specific diagnostic needs. Jin et al. (2021) addressed the challenge of modelling tumour uncertainty boundaries by designing a synthesis framework that mimics the gradual transition between tumour and healthy tissue.

More recent methods have incorporated semantic guidance, such as segmentation maps, to improve anatomical accuracy. For instance, Med-DDPM (Dorjsembe et al. 2024) uses segmentation masks to guide lesion placement during synthesis, enabling pixel-level control in 3D brain MRIs. This level of precision supports training and validating diagnostic algorithms that detect small or irregular lesions. Similarly, Chen et al. (2024b) show that diffusion models can synthesise subtle tumours (often too small to detect visually) and assess their applicability across multiple organs. Such synthetic cases can simulate difficult clinical scenarios, helping improve model sensitivity to early disease and reducing the risk of missed diagnoses.

9.4.4 Model-centric Generalisation

Model-centric generalisation aims to improve the robustness and adaptability of AI systems through architectural and algorithmic innovations, without modifying the input data. This paradigm assumes fixed data availability while acknowledging the presence of domain shifts, inconsistencies and missing modalities, which are common challenges in real-world clinical environments. The central objective is to develop models that maintain performance despite variations in input quality, completeness or distribution. Two primary research directions, corresponding to the previously defined Challenge I and Challenge II, are discussed in the following subsections: robustness to distribution shift (Challenge I) in Section 9.4.4.1 and resilience to incomplete data (Challenge II) in Section 9.4.4.2.

In this context, two key concepts are central: the *source domain*, representing the training distribution, and the *target domain*, representing the testing distribution. Empirical studies have shown that dissimilarities between the source and target domains can lead to increased test error due to training bias. This error tends to grow in proportion to the degree of distributional shift between the training and test data, as demonstrated in prior research.

Within the field of model generalisation, two main methodological streams have been developed: *domain generalisation* (DG), which focuses on improving model performance when trained solely on source domain data, and *domain adaptation* (DA), which aims to adapt the model to the target domain during or after training.

It is worth noting that meta-learning and transfer learning are widely discussed concepts in the context of deep learning (Torralba and Efros, 2011; Ben-David et al. 2006). Both approaches are fundamentally linked to the concept of *generalisation*. Specifically, meta-learning can be interpreted as a form of DG, while transfer learning is generally classified under DA.

9.4.4.1 Model Robustness to Distribution Shift

Domain generalisation: To improve model generalisability within the source domain, numerous studies have proposed training strategies that aim to make models less sensitive to domain-specific variations. Central to these approaches is the assumption that each image consists of two components: domain-invariant features (X) and domain-specific features (Δ) (Sun et al. 2021; Dou et al. 2019). Effective DG relies on the ability to extract and utilise the invariant component while minimising the influence of domain-specific noise.

Ouyang et al. (2022) utilise a randomised network functioning as a style generator to create artificial domains with modified Δ, helping the model manage minor distribution shifts. Alternatively, *Domain Shifts with Uncertainty* (Li et al. 2022) integrates feature statistics to model the uncertainty of X, aiming to reduce prediction instability in the target domain. Building on this, BayeSeg (Gao et al. 2023) assumes that Δ follows a normal distribution, $\Delta \in \mathcal{N}(\mu, \text{diag}(\sigma)^{-1})$, where both μ and σ are trainable. By fixing Δ, the model achieves deterministic disentanglement, improving stability under domain variation. Chen et al. (2024a) highlight that domain shifts arise not only from image intensity differences but also from anatomical variability across populations (e.g., age or geography). This is clinically significant, as models trained on one demographic group may misinterpret anatomical features in others, potentially compromising diagnostic accuracy. To address this, the authors propose a method for disentangling X and Δ by capturing invariant style and geometric components, thereby enabling generalisation across diverse patient populations, as illustrated in Figure 9.8.

The generalization capability of the models is shown in Figure 9.9, which compares the segmentation results of different methods on two unseen domains from the BraTS challenges (Baid et al., 2021), thereby assessing their robustness across diverse patient populations.

Domain adaptation: Domain adaptation aims to adapt the model to the imaging distribution of the target domain, particularly in scenarios where labelled data in the target domain is limited or unavailable.

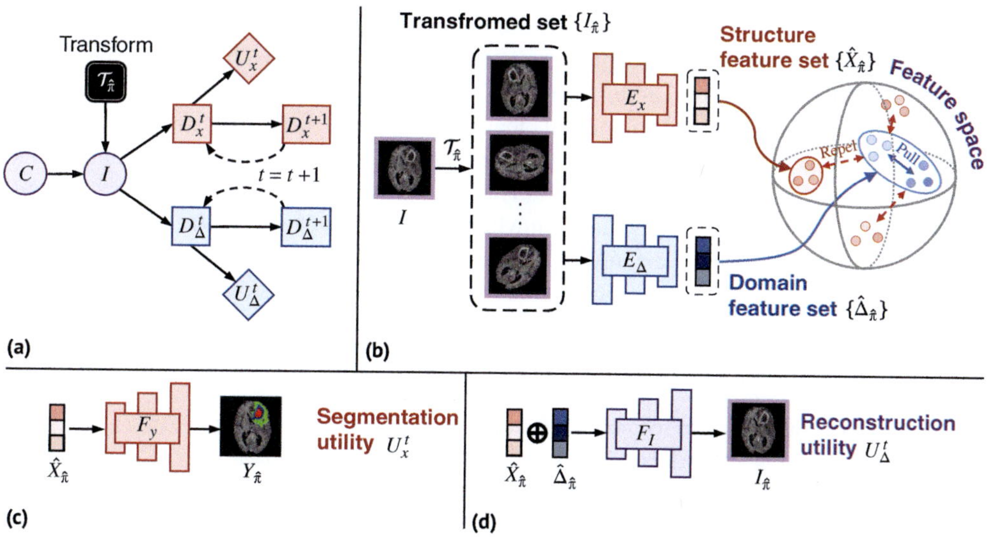

Figure 9.8 Domain game employs a disentanglement framework to separate X and Δ: (a) casual diagram, (b) disentangle feature space, (c) Encoder E_X utility and (d) encoder E_Δ utility. Colours denote associations with distinct features: purple represents the overall input (I), red corresponds to anatomical features (X) and blue indicates domain-specific features (Δ). *Source*: Chen et al. (2024a)/with permission of Springer Nature.

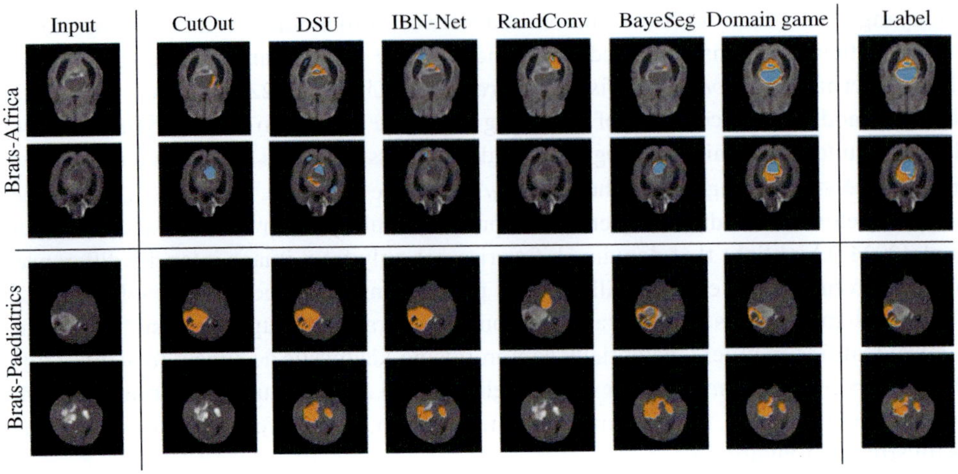

Figure 9.9 Qualitative visualisation of brain tumour segmentation. *Source*: Chen et al. (2024a)/ with permission of Springer Nature.

Manakov et al. (2019) and Yan et al. (2019) propose adapting source models by generating *source-like* images that align with the distribution of the target domain. While these methods aim to produce realistic image representations, Mahmood et al. (2018) and Tian et al. (2021) introduce a reverse translation approach, transforming real images into synthetic ones to preserve diagnostic content while filtering out domain-specific noise. Alternatively, Agarwal et al. (2022) and Wang et al. (2022) apply contrastive objectives to bring semantically similar disease cases closer across domains while pushing dissimilar cases apart. These strategies play a critical role in enhancing model robustness across clinical sites, preserving disease-relevant features while adapting to distributional differences. This enables more consistent and reliable diagnostic performance across diverse healthcare settings.

In addition to image-level alignment, feature alignment strategies aim to map both source and target data into a shared feature space. Kamnitsas et al. (2017) apply this approach using a Siamese network for brain lesion segmentation. Building on this, Bateson et al. (2019) incorporate prior medical knowledge, while Gao et al. (2019) emphasise clinically meaningful, high-level diagnostic features. These methods are especially valuable in clinical practice, where strict privacy regulations often prevent the sharing of patient data across institutions. Aligning features rather than exchanging raw images enables safer and more scalable deployment of AI models across hospitals and healthcare systems.

9.4.4.2 Model Resilience to Incomplete Data

In clinical workflows, models often face incomplete imaging data due to time limits, patient-specific factors or inconsistent scanning protocols across institutions. Ensuring AI systems remain reliable in these settings is crucial for diagnostic consistency. From a model-centric perspective, domain generalisation has been extended to address this type of data incompleteness.

Zhao et al. (2022b) propose a Modality-adaptive Feature Interaction framework for brain tumour segmentation under missing modality conditions. Their method models each imaging modality as a graph node and uses attention-based interaction to extract complementary features. A modality code indicates the presence or absence of each input, allowing the model to adapt dynamically. Wang et al. (2021) introduce the Adversarial Co-training Network, which jointly learns from full and partial modality inputs. The model aligns latent spaces through mutual information transfer to maintain performance when only limited data are available. Wu et al. (2023) present cross-modal shared graph fusion, which learns robust representations that retain core structural features even when some inputs are missing.

These techniques are especially valuable in clinical contexts where complete imaging protocols are often not feasible, such as in emergency care or with paediatric patients who may not tolerate long scan times. Their capacity to maintain diagnostic accuracy despite missing modalities supports safer, more inclusive deployment of AI in real-world healthcare settings.

9.4.5 Future Directions

The preceding subsections have provided an overview of the current state of real-world generalisation in AI applications for neuro-oncology, highlighting both data-centric and model-centric approaches. Despite the remarkable advancements driven by deep learning techniques, their integration into clinical practice remains fraught with significant challenges. Among the most pressing issues are the limited explainability of models, their vulnerability to noise and data distortions, and the difficulty in addressing the inherent heterogeneity of real-world clinical datasets. Furthermore, the evaluation of AI-generated imaging data for authenticity and reliability remains a critical unmet need, which continues to impede the practical adoption of these technologies in clinical settings.

Recent research has sought to mitigate some of these limitations. For example, Han et al. (2023) have identified key shortcomings in existing methods, particularly the inability to assess the quality of synthesised images and quantify the contributions of different input modalities in generating new ones. To address these gaps, they propose an explainable framework that dynamically optimises modality-specific weights during the fusion process. This approach not only enhances the interpretability of the model but also improves its adaptability by prioritising the most relevant input sequences for each task. However, despite these efforts, current solutions still fall short of meeting the rigorous demands of clinical practice. The challenges outlined earlier underscore the need for further commitment to development and comprehensive validation.

9.5 Conclusion

This chapter has explored the transformative potential of multiscale AI modelling in advancing precision neuro-oncology. By integrating structural, functional, cellular and molecular information across diverse imaging modalities, such as MRI, histology and ST, AI offers a powerful framework for capturing the biological and spatial heterogeneity of gliomas. These technologies are not isolated; rather, their convergence through AI enables novel opportunities to stratify patients, personalise treatment strategies and support data-driven clinical decision-making.

Nonetheless, substantial challenges remain in translating these innovations into clinical practice. These include not only ensuring generalisability across centres and patient populations but also building trust through interpretability, transparency and rigorous clinical validation. Furthermore, there is a growing need for robust multimodal learning strategies that can integrate multiscale data while preserving biological coherence, enabling models to capture the underlying mechanisms that drive disease progression and therapeutic response. Overcoming these challenges will require sustained interdisciplinary collaboration and systematic evaluation in prospective, real-world settings. Future work should prioritise collaboration between AI researchers, clinicians and regulatory bodies to ensure that technological advancements align with the practical needs and ethical standards of healthcare.

In summary, AI offers more than analytical power; it reshapes how we understand and treat brain tumours. By aligning computational precision with biological insight, these approaches pave the way towards a new data-driven paradigm in neuro-oncology: one that is personalised, integrative and clinically actionable. As AI technologies continue to evolve, their integration into routine care holds the promise of bringing precision medicine from the research bench to the patient bedside.

References

Aabedi, A.A., Lipkin, B., Kaur, J. et al. (2021). Functional alterations in cortical processing of speech in glioma-infiltrated cortex. *Proceedings of the National Academy of Sciences* 118 (46): e2108959118. https://doi.org.10.1073/pnas. 2108959118. https://www.pnas.org/doi/abs/10.1073/pnas.2108959118._eprint:https://www.pnas.org/doi/pdf/10.1073/pnas.2108959118.

Abdelaal, T., Mourragui, S., Mahfouz, A., and Reinders, M.J. (2020). Spage: spatial gene enhancement using scrna-seq. *Nucleic Acids Research* 48 (18): e107–e107.

Acs, B., Rantalainen, M., and Hartman, J. (2020). Artificial intelligence as the next step towards precision pathology. *Journal of Internal Medicine* 288 (1): 62–81.

Aerts, H., Fias, W., Caeyenberghs, K. et al. (2016). Brain networks under attack: robustness properties and the impact of lesions. *Brain* 139 (12): 3063–3083. https://doi.org/10.1093/brain/aww194.

Agarwal, P., Paudel, D.P., Zaech, J.N. et al. (2022). Unsupervised robust domain adaptation without source data. *Proceedings of the IEEE/CVF Winter Conference on Applications of Computer Vision*, pp. 2009–2018.

Ahmad, I., Xia, Y., Cui, H. et al. (2023). Dan-nucnet: a dual attention based framework for nuclei segmentation in cancer histology images under wild clinical conditions. *Expert Systems with Applications* 213: 118945.

Albayrak, A. and Bilgin, G. (2019). Automatic cell segmentation in histopathological images via two-staged superpixel-based algorithms. *Medical & Biological Engineering & Computing* 57: 653–665.

Altieri, R., Bianconi, A., Caneva, S. et al. (2025). Quantitative evaluation of neuroradiological and morphometric alteration of inferior Fronto-Occipital Fascicle across different brain tumor histotype: an Italian multicentric study. *Acta Neurochirurgica* 167 (1): 71. https://doi.org/10. 1007/s00701-025-06488-6.

Armocida, D., Bianconi, A., Zancana, G.et al. (2025). DTI fiber-tracking parameters adjacent to gliomas: the role of tract irregularity value in operative planning, resection, and outcome. *Journal of Neuro-Oncology* 171 (1): 241–252. https://doi.org/10.1007/s11060-024-04848-3.

Ashburner, J. (2012). SPM: a history. *NeuroImage* 62 (2): 791–800. https://doi.org/10.1016/j.neuroimage.2011.10.025. https://www.sciencedirect.com/science/article/pii/S1053811911011888.

Atta, L. and Fan, J. (2021). Computational challenges and opportunities in spatially resolved transcriptomic data analysis. *Nature Communications* 12 (1): 5283.

Badea, L. and Stănescu, E. (2020). Identifying transcriptomic correlates of histology using deep learning. *PLoS One* 15 (11): e0242858.

Baid, U., Ghodasara, S., Mohan, S. et al. (2021). The rsna-asnr-miccai brats 2021 benchmark on brain tumor segmentation and radiogenomic classification. *arXiv preprint arXiv:2107.02314.*

Bankhead, P. (2020). Developing image analysis methods for digital pathology. *The Journal of Pathology* 257 (4): 391–402.

Bateson, M., Kervadec, H., Dolz, J. et al. (2019). Constrained domain adaptation for segmentation. In: *Medical Image Computing and Computer Assisted Intervention–MICCAI 2019: 22nd International Conference*, Shenzhen, China, (October 13–17, 2019), Proceedings, Part II 22, pp. 326–334. Springer.

Ben-David, S., Blitzer, J., Crammer, K. et al. (2006). Analysis of representations for domain adaptation. *Advances in Neural Information Processing Systems* 19, 137–144.

Bera, K., Schalper, K.A., Rimm, D.L. et al. (2019). Artificial intelligence in digital pathology—new tools for diagnosis and precision oncology. *Nature Reviews Clinical Oncology* 16 (11): 703–715.

Bergenstråhle, L., He, B., Bergenstråhle, J. et al. (2022). Super-resolved spatial transcriptomics by deep data fusion. *Nature Biotechnology* 40 (4): 476–479.

Bessadok, A., Mahjoub, M.A., and Rekik, I. (2023). Graph neural networks in network neuroscience. *IEEE Transactions on Pattern Analysis and Machine Intelligence* 45: 5833–5848. https://doi.org/10.1109/TPAMI.2022.3209686.

Biancalani, T., Scalia, G., Buffoni, L. et al. (2021). Deep learning and alignment of spatially resolved single-cell transcriptomes with tangram. *Nature Methods* 18 (11): 1352–1362.

Blattmann, A., Rombach, R., Oktay, K. et al. (2022). Semi-parametric neural image synthesis. *arXiv preprint arXiv:2204.11824.*

Bullmore, E. and Sporns, O. (2009). Complex brain networks: graph theoretical analysis of structural and functional systems. *Nature Reviews Neuroscience* 10 (3): 186–198. https://doi.org/10.1038/nrn2575.

Bullmore, E. and Sporns, O. (2012). The economy of brain network organization. *Nature Reviews Neuroscience* 13 (5): 336–349. https://doi.org/10.1038/nrn3214.

Chartsias, A., Joyce, T., Giuffrida, M.V. et al. (2017). Multimodal mr synthesis via modality-invariant latent representation. *IEEE Transactions on Medical Imaging* 37 (3): 803–814.

Chen, H., Zhang, H., Chan, U.W. et al. (2024a). Domain game: disentangle anatomical feature for single domain generalized segmentation. *arXiv preprint arXiv:2406.02125.*

Chen, Q., Chen, X., Song, H. et al. (2024b). Towards generalizable tumor synthesis. *Proceedings of the IEEE/CVF Conference on Computer Vision and Pattern Recognition (CVPR)*, pp. 11147–11158.

Chen, R.J., Chen, C., Li, Y. et al. (2022). Scaling vision transformers to gigapixel images via hierarchical self-supervised learning. *Proceedings of the IEEE/CVF Conference on Computer Vision and Pattern Recognition*, pp. 16144–16155.

Chen, R.J., Lu, M.Y., and Wang, J. (2020). Pathomic fusion: an integrated framework for fusing histopathology and genomic features for cancer diagnosis and prognosis. *IEEE Transactions on Medical Imaging* 41 (4): 757–770.

Codeluppi, S., Borm, L.E., and Zeisel, A. (2018). Spatial organization of the somatosensory cortex revealed by osmfish. *Nature Methods* 15 (11): 932–935.

Connectome - Using Connectome Workbench. URL https://www.humanconnectome.org/software/connectome-workbench.

Costa, P., Galdran, A., Meyer, M.I. et al. (2017). Towards adversarial retinal image synthesis (2017). *arXiv preprint arXiv:1701.08974*.

Coudray, N., Ocampo, P.S., Sakellaropoulos, T. et al. (2018). Classification and mutation prediction from non–small cell lung cancer histopathology images using deep learning. *Nature Medicine* 24 (10): 1559–1567.

Crossley, N.A., Mechelli, A., Scott, J. et al. (2014). The hubs of the human connectome are generally implicated in the anatomy of brain disorders. *Brain* 137 (8): 2382–2395. https://doi.org/10.1093/brain/awu132.

Daum, D., Osuala, R., Riess, A. et al. (2024). On differentially private 3D medical image synthesis with controllable latent diffusion models (2024). *arXiv preprint arXiv:2407.16405*.

Derks, J., Kulik, S.D., Numan, T. et al. (2021). Understanding global brain network alterations in glioma patients. *Brain Connectivity* 11 (10): 865–874. https://doi.org/10.1089/brain.2020.0801.

Desikan, R.S., Ségonne, F., Fischl, B. et al. (2006). An automated labeling system for subdividing the human cerebral cortex on MRI scans into gyral based regions of interest. *Neuroimage* 31 (3): 968–980. https://doi.org/10.1016/j.neuroimage.2006.01.021.

Dorjsembe, Z., Pao, H.K., and Odonchimed, S. (2024). Conditional diffusion models for semantic 3D brain MRI synthesis. *IEEE Journal of Biomedical and Health Informatics* 28 (7): 4084–4093.

Dou, Q., Coelho de Castro, D., Kamnitsas, K. et al. (2019). Domain generalization via model-agnostic learning of semantic features. *Advances in Neural Information Processing Systems, 32*: 6447–6458.

Douget, J.E.L., Taïeb, J., Jacob, P. et al. (2023). Improved colorectal cancer survival prediction with deep learning-based WSI analysis on petacc8 and prodige13 cohorts. *Cancer Research* 83 (7_Supplement): 5430–5430.

Douw, L., Klein, M., Fagel, S.S. et al. (2009). Cognitive and radiological effects of radiotherapy in patients with low-grade glioma: long-term follow-up. *The Lancet Neurology* 8 (9): 810–818. https://doi.org/10.1016/S1474-4422(09)70204-2.

Du, J., Yang, Y.C., An, Z.J. et al. (2023). Advances in spatial transcriptomics and related data analysis strategies. *Journal of Translational Medicine* 21 (1): 1–21.

Durante, M.A., Rodriguez, D.A., Kurtenbach, S. et al. (2020). Single-cell analysis reveals new evolutionary complexity in uveal melanoma. *Nature Communications* 11 (1): 496.

Echle, A., Rindtorff, N.T., Brinker, T.J. et al. (2021). Deep learning in cancer pathology: a new generation of clinical biomarkers. *British Journal of Cancer* 124 (4): 686–696.

Ektefaie, Y., Dasoulas, G., Noori, A. et al. (2023). Multimodal learning with graphs. *Nature Machine Intelligence* 5 (4): 340–350. https://doi.org/10.1038/s42256-023-00624-6.

Elazab, N., Gab Allah, W., and Elmogy, M. (2024). Computer-aided diagnosis system for grading brain tumor using histopathology images based on color and texture features. *BMC Medical Imaging* 24 (1): 177.

Ellingson, B.M., Bendszus, M., Boxerman, J. et al. (2015). Consensus recommendations for a standardized brain tumor imaging protocol in clinical trials. *Neuro-oncology* 17 (9): 1188–1198.

Esser, P., Rombach, R., and Ommer, B. (2021). Taming transformers for high-resolution image synthesis. *Proceedings of the IEEE/CVF Conference on Computer Vision and Pattern Recognition (CVPR)*, pp. 12873–12883.

Falcó-Roget, J., Cacciola, A., Sambataro, F. et al. (2017). Functional and structural reorganization in brain tumors: a machine learning approach using desynchronized functional oscillations. *Communications Biology* 7 (1): 1–17. https://doi.org/10.1038/s42003-024-06119-3.

Farahani, F.V., Karwowski, W., and Lighthall, N.R. (2019). Application of graph theory for identifying connectivity patterns in human brain networks: a systematic review. *Frontiers in Neuroscience* 13: 439505. https://doi.org/10.3389/FNINS.2019.00585/BIBTEX. www.frontiersin.org.

Fischer, A., Vázquez-García, I., Illingworth, C.J. et al. (2014). High-definition reconstruction of clonal composition in cancer. *Cell Reports* 7 (5): 1740–1752.

Fischl, B. (2012). FreeSurfer. *NeuroImage* 62 (2): 774–781. https://doi.org/10.1016/j.neuroimage.2012.01.021.

Fittall, M.W. and Van Loo, P. (2019). Translating insights into tumor evolution to clinical practice: promises and challenges. *Genome Medicine* 11 (1): 1–14.

Freeman, L.C. (2002). Centrality in social networks: conceptual clarification. *Social Network: Critical Concepts in Sociology*. 1: 238–263.

Friedrich, M., Filss, C.P., Lohmann, P. et al. (2024). Structural connectome-based predictive modeling of cognitive deficits in treated glioma patients. *Neurooncology Advances* 6 (1): vdad151. https://doi.org/10.1093/noajnl/vdad151.

Gaiti, F., Chaligne, R., Gu, H. et al. (2019). Epigenetic evolution and lineage histories of chronic lymphocytic leukaemia. *Nature* 569 (7757): 576–580.

Gao, S., Zhou, H., Gao, Y. et al. (2023). Bayeseg: Bayesian modeling for medical image segmentation with interpretable generalizability. *Medical Image Analysis* 89: 102889. https://doi.org/10.1016/J.MEDIA.2023.102889.

Gao, Y., Zhang, Y., Cao, Z. et al. (2019). Decoding brain states from fMRI signals by using unsupervised domain adaptation. *IEEE Journal of Biomedical and Health Informatics* 24 (6): 1677–1685.

Girvan, M. and Newman, M.E. (2002). Community structure in social and biological networks. *Proceedings of the National Academy of Sciences of the United States of America* 99 (12): 7821–7826.

Glasser, M.F., Coalson, T.S., Robinson, E.C. et al. (2016). A multi-modal parcellation of human cerebral cortex. *Nature* 536 (7615): 171–178. https://doi.org/10.1038/nature18933.

Graham, S., Vu, Q.D., Jahanifar, M. et al. (2023). One model is all you need: multi-task learning enables simultaneous histology image segmentation and classification. *Medical Image Analysis* 83: 102685.

Graham, S., Vu, Q.D., Raza, S.E.A. et al. (2019). Hover-net: simultaneous segmentation and classification of nuclei in multi-tissue histology images. *Medical Image Analysis* 58: 101563.

Guimera, R. and Amaral, L.A.N. (2005). Cartography of complex networks: modules and universal roles. *Journal of Statistical Mechanics: Theory and Experiment* 2005 (02): P02001.

Gunel, B., Sahiner, A., Desai, A.D. et al. (2022). Scale-equivariant unrolled neural networks for data-efficient accelerated MRI reconstruction. In: Medical Image Computing and Computer Assisted Intervention – MICCAI 2022 (ed. L. Wang, Q. Dou, P.T. Fletcher, S. Speidel, and S. Li), 737–747. Cham: Springer Nature Switzerland.

Hadjiabadi, D.H., Pung, L., Zhang, J. et al. (2018). Brain tumors disrupt the resting-state connectome. *NeuroImage: Clinical* 18: 279–289. https://doi.org/10.1016/j.nicl.2018.01.026.

Han, L., Zhang, T., Huang, Y. et al. (2023). An explainable deep framework: towards task-specific fusion for multi-to-one MRI synthesis. *arXiv preprint arXiv:2307.00885.*

Hart, M.G., Romero-Garcia, R., Price, S.J. et al. (2019). Global effects of focal brain tumors on functional complexity and network robustness: a prospective cohort study. *Neurosurgery* 84 (6): 1201–1213. https://journals.lww.com/neurosurgery/fulltext/2019/06000/global_effects_of_focal_brain_tumors_on_functional.5.aspx.

Hashido, T., Saito, S., and Ishida, T. (2020) A radiomics-based comparative study on arterial spin labeling and dynamic susceptibility contrast perfusion-weighted imaging in gliomas. *Scientific Reports* 10: 1–10. https://doi.org/10.1038/s41598-020-62658-9.

Hatipoglu, N. and Bilgin, G. (2017). Cell segmentation in histopathological images with deep learning algorithms by utilizing spatial relationships. *Medical & Biological Engineering & Computing* 55: 1829–1848.

He, B., Bergenstråhle, L., Stenbeck, L. et al. (2020). Integrating spatial gene expression and breast tumour morphology via deep learning. *Nature Biomedical Engineering* 4 (8): 827–834.

He, K., Zhang, X., Ren, S. et al. (2016). Deep residual learning for image recognition. *Proceedings of the IEEE Conference on Computer Vision and Pattern Recognition*, 770–778.

Heo, D., Lee, J., Yoo, R.E. et al. (2023). Deep learning based on dynamic susceptibility contrast MR imaging for prediction of local progression in adult-type diffuse glioma (grade 4). *Scientific Reports* 13 (1): 13864. https://doi.org/10.1038/s41598-023-41171-9.

Higgins, I., Matthey, L., Pal, A. et al. (2017). Beta-vae: learning basic visual concepts with a constrained variational framework. *International Conference on Learning Representations.*

Hollon, T., Jiang, C., Chowdury, A. et al. (2023). Artificial-intelligence-based molecular classification of diffuse gliomas using rapid, label-free optical imaging. *Nature Medicine* 29 (4): 828–832.

Hörst, F., Rempe, M., Heine, L. et al. (2024). Cellvit: vision transformers for precise cell segmentation and classification. *Medical Image Analysis* 94: 103143.

Hou, S., Tian, K., Yang, S. et al. (2022). St2cell: reconstruction of in situ single-cell spatial transcriptomics by integrating high-resolution histological image. *bioRxiv* pp. 2022–10.

Hu, J., Coleman, K., Zhang, D. et al. (2023). Deciphering tumor ecosystems at super resolution from spatial transcriptomics with tesla. *Cell Systems* 14 (5): 404–417.

Hu, X., Xiao, L., and Wang, Y.P. (2024). A graph neural network based fusion of MRI-derived brain network and clinical data for glioblastoma survival prediction. *ICASSP, IEEE International Conference on Acoustics, Speech and Signal Processing – Proceedings*, 1571–1575 https://doi.org/10.1109/ICASSP48485.2024.10446734.

Huang, G., Liu, Z., Van Der Maaten, L. et al. (2017). Densely connected convolutional networks. *Proceedings of the IEEE Conference on Computer Vision and Pattern Recognition*, 4700–4708.

Huang, L., Sun, H., Sun, L. et al. (2023). Rapid, label-free histopathological diagnosis of liver cancer based on Raman spectroscopy and deep learning. *Nature Communications* 14 (1): 48.

InferCNV: Inferring copy number alterations from tumor single cell RNA-Seq data. https://github.com/broadinstitute/inferCNV/wiki.

Jahn, K., Kuipers, J., and Beerenwinkel, N. (2016). Tree inference for single-cell data. *Genome Biology* 17 (1): 1–17.

Jaisakthi, S., Desingu, K., Mirunalini, P. et al. (2023). A deep learning approach for nucleus segmentation and tumor classification from lung histopathological images. *Network Modeling Analysis in Health Informatics and Bioinformatics* 12 (1): 22.

Jann, K., Orosz, A., Dierks, T. et al. (2013). Quantification of network perfusion in asl cerebral blood flow data with seed based and ica approaches. *Brain Topography* 26: 569–580. https://doi.org/10.1007/S10548-013-0280-3/FIGURES/5.

Jaume, G., Vaidya, A., Chen, R.J. et al. (2024). Modeling dense multimodal interactions between biological pathways and histology for survival prediction. *Proceedings of the IEEE/CVF Conference on Computer Vision and Pattern Recognition*, 11579–11590.

Jeurissen, B., Descoteaux, M., Mori, S. et al. (2019). Diffusion MRI fiber tractography of the brain. *NMR in Biomedicine* 32: e3785. https://doi.org/10.1002/NBM.3785.

Jia, Y., Liu, J., Chen, L. et al. (2024). Thitogene: a deep learning method for predicting spatial transcriptomics from histological images. *Briefings in Bioinformatics* 25 (1): bbad464.

Jiang, L., Mao, Y., Wang, X. et al. (2023). Cola-diff: conditional latent diffusion model for multi-modal MRI synthesis. *International Conference on Medical Image Computing and Computer-Assisted Intervention*, 398–408. Springer.

Jiang, R., Yin, X., Yang, P. et al. (2024). A transformer-based weakly supervised computational pathology method for clinical-grade diagnosis and molecular marker discovery of gliomas. *Nature Machine Intelligence* 6: 876–891.

Jiang, S., Li, H., and Jin, Z. (2021). A visually interpretable deep learning framework for histopathological image-based skin cancer diagnosis. *IEEE Journal of Biomedical and Health Informatics* 25 (5): 1483–1494.

Jin, Q., Cui, H., Sun, C., et al. (2021). Free-form tumor synthesis in computed tomography images via richer generative adversarial network. *Knowledge-Based Systems* 218: 106753.

Jose, L., Liu, S., Russo, C. et al. (2022). Artificial intelligence–assisted classification of gliomas using whole-slide images. *Archives of Pathology & Laboratory Medicine* 147 (8): 916–924.

Kalisky, T. and Quake, S.R. (2011). Single-cell genomics. *Nature Methods* 8 (4): 311–314.

Kamnitsas, K., Baumgartner, C., Ledig, C. et al. (2017). Unsupervised domain adaptation in brain lesion segmentation with adversarial networks. *Information Processing in Medical Imaging: 25th International Conference, IPMI 2017*, Boone, NC, USA, (June 25–30, 2017), Proceedings 25, 597–609. Springer.

Kharchenko, P.V. (2021). The triumphs and limitations of computational methods for scrna-seq. *Nature Methods* 18 (7): 723–732.

Khazaee Fadafen, M. and Rezaee, K. (2023). Ensemble-based multi-tissue classification approach of colorectal cancer histology images using a novel hybrid deep learning framework. *Scientific Reports* 13 (1): 8823.

Kim, B.H., Ye, J.C., and Kim, J.J. (2021). Learning dynamic graph representation of brain connectome with spatio-temporal attention. *Advances in Neural Information Processing Systems*, vol. 34, 4314–4327. https://github.com/egyptdj/stagin.

Korkmaz, Y., Cukur, T., and Patel, V.M. (2023). Self-supervised MRI reconstruction with unrolled diffusion models. In: Medical Image Computing and Computer Assisted Intervention – MICCAI 2023 (ed. H. Greenspan, A. Madabhushi, P. Mousavi et al.), 491–501. Cham: Springer Nature Switzerland.

Krishna, S., Choudhury, A., Keough, M.B. et al. (2023). Glioblastoma remodelling of human neural circuits decreases survival. *Nature* 617 (7961), 599–607. https://doi.org/10.1038/s41586-023-06036-1.

Li, C., Wang, S., Liu, P. et al. (2019). Decoding the interdependence of multiparametric magnetic resonance imaging to reveal patient subgroups correlated with survivals. *Neoplasia* 21 (5): 442–449. https://doi.org/10.1016/j.neo.2019.03.005.

Li, X., Dai, Y., Ge, Y. et al. (2022). Uncertainty modeling for out-of-distribution generalization. *arXiv preprint arXiv:2202.03958*.

Li, X., Zhou, Y., Dvornek, N. et al. (2021). Interpretable brain graph neural network for fMRI analysis. *Medical Image Analysis* 74: 102233. https://doi.org/10.1016/j.media.2021.102233.

Liang, X., Connelly, A., and Calamante, F. (2014). Graph analysis of resting-state asl perfusion MRI data: Nonlinear correlations among cbf and network metrics. *NeuroImage* 87: 265–275. https://doi.org/10.1016/J.NEUROIMAGE.2013.11.013.

Liao, J., Qian, J., Fang, Y. et al. (2022). De novo analysis of bulk rna-seq data at spatially resolved single-cell resolution. *Nature Communications* 13 (1): 6498.

Lopez, R., Nazaret, A., Langevin, M. et al. (2019). A joint model of unpaired data from scrna-seq and spatial transcriptomics for imputing missing gene expression measurements. *arXiv preprint arXiv:1905.02269*.

Louis, D.N., Ohgaki, H., Wiestler, O.D. et al. (2007). The 2007 who classification of tumours of the central nervous system. *Acta Neuropathologica* 114: 97–109.

Louis, D.N., Perry, A., Wesseling P. et al. (2021). The 2021 who classification of tumors of the central nervous system: a summary. *Neuro-oncology* 23 (8): 1231–1251.

Louis, D.N., Perry, A., Reifenberger, G. et al. (2016). The 2016 World Health Organization classification of tumors of the central nervous system: a summary. *Acta Neuropathologica* 131: 803–820.

Madabhushi, A. and Lee, G. (2016). Image analysis and machine learning in digital pathology: Challenges and opportunities. *Medical Image Analysis* 33: 170–175.

Mahmood, F., Chen, R., and Durr, N.J. (2018). Unsupervised reverse domain adaptation for synthetic medical images via adversarial training. *IEEE Transactions on Medical Imaging* 37 (12): 2572–2581.

Mahmood, T., Arsalan, M., Owais, M. et al. (2020). Artificial intelligence-based mitosis detection in breast cancer histopathology images using faster R-CNN and deep CNNs. *Journal of Clinical Medicine* 9 (3): 749.

Malikic, S., Jahn, K., Kuipers, J. et al. (2019). Integrative inference of subclonal tumour evolution from single-cell and bulk sequencing data. *Nature Communications* 10 (1): 2750.

Manakov, I., Rohm, M., Kern, C. et al. (2019). Noise as domain shift: denoising medical images by unpaired image translation. In: *Domain Adaptation and Representation Transfer and Medical Image Learning with Less Labels and Imperfect Data: First MICCAI Workshop, DART 2019, and First International Workshop, MIL3ID 2019, Shenzhen, Held in Conjunction with MICCAI 2019*, Shenzhen, China, (October 13 and 17, 2019), Proceedings 1, 3–10. Springer.

Mandal, A.S., Brem, S., and Suckling, J. (2023). Brain network mapping and glioma pathophysiology. Brain Communications 5 (2): fcad040. https://doi.org/10.1093/braincomms/fcad040.

Mandal, A.S., Romero-Garcia, R., Hart, M.G. et al. (2020). Genetic, cellular, and connectomic characterization of the brain regions commonly plagued by glioma. *Brain* 143 (11): 3294–3307. https://doi.org/10.1093/brain/awaa277.

Manini, C. and López, J.I. (2023). Pathology and the evolutionary dynamics of clear cell renal cell carcinoma. *Translational Oncology* 27: 101601.

Mao, Y., Jiang, L., Chen, X. et al. (2023). Disc-diff: disentangled conditional diffusion model for multi-contrast MRI super-resolution. In: *Medical Image Computing and Computer Assisted Intervention – MICCAI 2023* (ed. Greenspan, H., A. Madabhushi, P. Mousavi et al.). MICCAI 2023. Lecture Notes in Computer Science, Vol 14229. Cham: Springer. https://doi.org/10.1007/978-3-031-43999-5_37.

Meng, X., Sun, K., Xu, J. et al. (2024). Multi-modal modality-masked diffusion network for brain MRI synthesis with random modality missing. IEEE Transactions on Medical Imaging.

Mohammadi, H. and Karwowski, W. (2025). Graph neural networks in brain connectivity studies: Methods, challenges, and future directions. *Brain Sciences* 15 (1): 17. https://doi.org/10.3390/brainsci15010017.

Moses, L. and Pachter, L. (2022). Museum of spatial transcriptomics. *Nature Methods* 19 (5): 534–546.

Nalisnik, M., Amgad, M., Lee, S. et al. (2017). Interactive phenotyping of large-scale histology imaging data with histomicsml. *Scientific Reports* 7 (1): 14588.

Nasab, R.Z., Ghamsari, M.R.E., Argha, A. et al. (2022). Deep learning in spatially resolved transcriptomics: a comprehensive technical view. *arXiv preprint arXiv:2210.04453.*

Nasrallah, M.P., Zhao, J., Tsai, C.C. et al. (2023). Machine learning for cryosection pathology predicts the 2021 who classification of glioma. *Med* 4 (8): 526–540.

Nenning, K.H., Furtner, J., Kiesel, B. et al. (2020). Distributed changes of the functional connectome in patients with glioblastoma. *Scientific Reports* 10 (1): 18312. https://doi.org/10. 1038/s41598-020-74726-1.

Niu, C., Wang, Y., Cohen, A.D. et al. (2021). Machine learning may predict individual hand motor activation from resting-state fMRI in patients with brain tumors in perirolandic cortex. *European Radiology* 31 (7): 5253–5262. https://doi.org/10.1007/s00330-021-07825-w.

Niu, Y., Gu, L., Zhao, Y. et al. (2022). Explainable diabetic retinopathy detection and retinal image generation. *IEEE Journal of Biomedical and Health Informatics* 26 (1): 44–55 https://doi.org/10.1109/JBHI.2021.3110593.

Ostrom, Q.T., Price, M., Neff, C. et al. (2022). CBTRUS statistical report: primary brain and other central nervous system tumors diagnosed in the United States in 2015–2019. *Neuro-oncology* 24 (Supplement_5): v1–v95. https://doi.org/10.1093/neuonc/noac202.

Ouellette, T.W. and Awadalla, P. (2022). Inferring ongoing cancer evolution from single tumour biopsies using synthetic supervised learning. *PLOS Computational Biology* 18 (4): e1010007.

Ouyang, C., Chen, C., Li, S. et al. (2022). Causality-inspired single-source domain generalization for medical image segmentation. *IEEE Transactions on Medical Imaging* 42 (4): 1095–1106.

Özbey, M., Dalmaz, O., Dar, S.U. et al. (2023). Unsupervised medical image translation with adversarial diffusion models. *IEEE Transactions on Medical Imaging* 42 (12): 3524–3539.

Pan, S., Fu, Y., Chen, P. et al. (2021). Multi-task learning-based immunofluorescence classification of kidney disease. *International Journal of Environmental Research and Public Health* 18 (20): 10798.

Pei, L., Jones, K.A., Shboul, Z.A. et al. (2021). Deep neural network analysis of pathology images with integrated molecular data for enhanced glioma classification and grading. *Frontiers in Oncology* 11: 668694.

Perry, A. and Wesseling, P. (2016). Histologic classification of gliomas. *Handbook of Clinical Neurology* 134: 71–95.

Pinetz, T., Kobler, E., Haase, R. et al. (2023). Faithful synthesis of low-dose contrast-enhanced brain MRI scans using noise-preserving conditional gans. In: *Medical Image Computing and Computer Assisted Intervention – MICCAI 2023* (ed. H. Greenspan, A. Madabhushi, P. Mousavi et al.), 607–617. Cham: Springer Nature Switzerland.

Pinto, A., Pereira, S., Meier, R. et al. (2018). Enhancing clinical MRI perfusion maps with data-driven maps of complementary nature for lesion outcome prediction, 107–115. Springer International Publishing. https://doi.org/10.1007/978-3-030-00931-1_13.

Polanski, K., Bartolome-Casado, R., Sarropoulos, I. et al. (2024). Bin2cell reconstructs cells from high resolution Visium HD data. *bioRxiv* 2024–06.

Qiu, L., Zhao, L., Zhao, W. et al. (2024). Dual-space disentangled-multimodal network (ddm-net) for glioma diagnosis and prognosis with incomplete pathology and genomic data. *Physics in Medicine & Biology* 69 (8): 085028.

Raffelt, D.A., Tournier, J.D., Smith, R.E. et al. (2017). Investigating white matter fibre density and morphology using fixel-based analysis. *NeuroImage* 144: 58–73. https://doi.org/10.1016/j.neuroimage.2016.09.029.

Rao, N., Clark, S., and Habern, O. (2020). Bridging genomics and tissue pathology: 10x genomics explores new frontiers with the visium spatial gene expression solution. *Genetic Engineering & Biotechnology News* 40 (2): 50–51.

Rauf, Z., Sohail, A., Khan, S.H. et al. (2023). Attention-guided multi-scale deep object detection framework for lymphocyte analysis in ihc histological images. *Microscopy* 72 (1): 27–42.

Rohlfing, T., Zahr, N.M., Sullivan, E.V. et al. (2010). The SRI24 multichannel atlas of normal adult human brain structure. *Human Brain Mapping*, 31 (5): 798–819. https://pmc.ncbi.nlm.nih.gov/articles/PMC2915788.

Rolls, E.T., Huang, C.C., Lin, C.P. et al. (2020). Automated anatomical labelling atlas 3. *NeuroImage* 206: 116189. https://doi.org/10.1016/j.neuroimage.2019.116189.

Ronneberger, O., Fischer, P., and Brox, T. (2015). U-net: convolutional networks for biomedical image segmentation. *Medical Image Computing and Computer-Assisted Intervention–MICCAI 2015: 18th International Conference*, Munich, Germany, (October 5–9, 2015), Proceedings, Part III 18, 234–241. Springer.

Salas, S.M., Czarnewski, P., Kuemmerle, L.B. et al. (2023). Optimizing xenium in situ data utility by quality assessment and best practice analysis workflows. *BioRxiv* 2023–02.

Salvalaggio, A., Pini, L., Bertoldo, A. et al. (2024). Glioblastoma and brain connectivity: the need for a paradigm shift. *The Lancet Neurology* 23 (7): 740–748. https://doi.org/10.1016/S1474-4422(24)00160-1.

Sasaki, H., Willcocks, C.G., and Breckon, T.P. (2021). Unit-ddpm: unpaired image translation with denoising diffusion probabilistic models. *arXiv preprint arXiv:2104.05358*.

Schmauch, B., Romagnoni, A., Pronier, E. et al. (2020). A deep learning model to predict rna-seq expression of tumours from whole slide images. *Nature Communications* 11 (1): 3877.

Semmel, E.S., Quadri, T.R., and King, T.Z. (2022). Graph theoretical analysis of brain network characteristics in brain tumor patients: a systematic review. *Neuropsychology Review* 32: 651–675. https://doi.org/10.1007/S11065-021-09512-5/METRICS.

Shao, W., Han, Z., Cheng, J. et al. (2019). Integrative analysis of pathological images and multi-dimensional genomic data for early-stage cancer prognosis. *IEEE Transactions on Medical Imaging* 39 (1): 99–110.

Sharma, A. and Hamarneh, G. (2019). Missing MRI pulse sequence synthesis using multi-modal generative adversarial network. *IEEE Transactions on Medical Imaging* 39 (4): 1170–1183.

Shen, L., Zhu, W., Wang, X. et al. (2020). Multi-domain image completion for random missing input data. *IEEE Transactions on Medical Imaging* 40 (4): 1113–1122.

Shen, X., Zhao, Y., Wang, Z. et al. (2022). Recent advances in high-throughput single-cell transcriptomics and spatial transcriptomics. *Lab on a Chip* 22 (24): 4774–4791.

Shengquan, C., Boheng, Z., Xiaoyang, C. et al. (2021). stPlus: a reference-based method for the accurate enhancement of spatial transcriptomics. *Bioinformatics* 37 (Supplement_1): i299–i307.

Shiroishi, M.S., Boxerman, J.L., Quarles, C.C. et al. (2023). Physical principles of dynamic contrast-enhanced and dynamic susceptibility contrast MRI. *Functional Neuroradiology: Principles and Clinical Applications*, 2e (ed. S.H. Faro and F.B. Mohamed), 15–34, Springer. https://doi.org/10.1007/978-3-031-10909-6_2.

Shmatko, A., Ghaffari Laleh, N., Gerstung, M. et al. (2022). Artificial intelligence in histopathology: enhancing cancer research and clinical oncology. *Nature Cancer* 3 (9): 1026–1038.

Simonyan, K. and Zisserman, A. (2014). Very deep convolutional networks for large-scale image recognition. *arXiv preprint arXiv:1409.1556.*

Smith, S.M., Jenkinson, M., Woolrich, M.W. et al. (2004). Advances in functional and structural MR image analysis and implementation as FSL. *Neuroimage* 23 (Suppl 1): S208–S219. https://doi.org/10.1016/j.neuroimage.2004.07.051.

Song, Q. and Su, J. (2021). Dstg: deconvoluting spatial transcriptomics data through graph-based artificial intelligence. *Briefings in Bioinformatics* 22 (5): bbaa414.

Ståhl, P.L., Salmén, F., Vickovic, S. et al. (2016). Visualization and analysis of gene expression in tissue sections by spatial transcriptomics. *Science* 353 (6294): 78–82.

Stickels, R.R., Murray, E., Kumar, P. et al. (2021). Highly sensitive spatial transcriptomics at near-cellular resolution with slide-seqv2. *Nature Biotechnology* 39 (3): 313–319.

Stupp, R., Mason, W.P., van den Bent, M.J. et al. (2005). Radiotherapy plus concomitant and adjuvant temozolomide for glioblastoma. *New England Journal of Medicine* 352 (10): 987–996. https://doi.org/10.1056/NEJMoa043330.

Sudlow, C., Gallacher, J., Allen, N. et al. (2015). UK biobank: an open access resource for identifying the causes of a wide range of complex diseases of middle and old age. *PLoS Medicine* 12 (3): e1001779.

Sun, J. and Peng, Y. (2024). The cross-modality survival prediction method of glioblastoma based on dual-graph neural networks. *Expert Systems with Applications* 254: 124394. https://doi.org/10.1016/j.eswa.2024.124394.

Sun, X., Wu, B., Zheng, X. et al. (2021). Recovering latent causal factor for generalization to distributional shifts. *Advances in Neural Information Processing Systems* 34: 16846–16859.

Szegedy, C., Vanhoucke, V., Ioffe, S. et al. (2016). Rethinking the inception architecture for computer vision. *Proceedings of the IEEE Conference on Computer Vision and Pattern Recognition*, 2818–2826.

Thomas Yeo, B.T., Krienen, F.M., Sepulcre, J. et al. (2011). The organization of the human cerebral cortex estimated by intrinsic functional connectivity. *Journal of Neurophysiology* 106 (3): 1125–1165. https://doi.org/10.1152/jn.00338.2011.

Tian, J., Zhang, J., Li, W. et al. (2021). VDM-DA: virtual domain modeling for source data-free domain adaptation. *IEEE Transactions on Circuits and Systems for Video Technology* 32 (6): 3749–3760.

Torralba, A. and Efros, A.A. (2011). Unbiased look at dataset bias. In: *CVPR 2011*, 1521–1528. IEEE.

Truong, A.H., Sharmanska, V., Limbck-Stanic, C. et al. (2020). Optimization of deep learning methods for visualization of tumor heterogeneity and brain tumor grading through digital pathology. *Neurooncology Advances* 2 (1): vdaa110.

Van den Heuvel, M.P. and Sporns, O. (2013). Network hubs in the human brain. *Trends in Cognitive Sciences* 17 (12): 683–696.

Wang, J., Levman, J., Pinaya, W.H.L. et al. (2023a). Inversesr: 3D brain MRI super-resolution using a latent diffusion model. In: *Medical Image Computing and Computer Assisted Intervention – MICCAI 2023* (ed. H. Greenspan, A. Madabhushi, P. Mousaviet al.), 438–447. Cham: Springer Nature Switzerland.

Wang, K., Shi, J., Tong, X. et al. (2024a). Tg468: a text graph convolutional network for predicting clinical response to immune checkpoint inhibitor therapy. *Briefings in Bioinformatics* 25: 1–15. https://doi.org/10.1093/BIB/BBAE017.

Wang, R., Wu, Z., Weng, Z. et al. (2022). Cross-domain contrastive learning for unsupervised domain adaptation. *IEEE Transactions on Multimedia*.

Wang, W., Zhao, Y., Teng, L. et al. (2023b). Neuropathologist-level integrated classification of adult-type diffuse gliomas using deep learning from whole-slide pathological images. *Nature Communications* 14 (1): 6359.

Wang, X., Huang, X., Price, S.J. et al. (2024b). Cross-modal diffusion modelling for super-resolved spatial transcriptomics. *arXiv preprint arXiv:2404.12973.*

Wang, X., Price, S., and Li, C. (2023c). Multi-task learning of histology and molecular markers for classifying diffuse glioma. *International Conference on Medical Image Computing and Computer-Assisted Intervention*, 551–561. Springer.

Wang, X., Zhang, J., Yang, S. et al. (2023d). A generalizable and robust deep learning algorithm for mitosis detection in multicenter breast histopathological images. *Medical Image Analysis* 84: 102703.

Wang, Y., Zhang, Y., Liu, Y. et al. (2021). Acn: adversarial co-training network for brain tumor segmentation with missing modalities. In: *Medical Image Computing and Computer Assisted Intervention – MICCAI 2021* (ed. M. de Bruijne, P.C. Cattin, S. Cotin, et al.), 410–420. Cham: Springer International Publishing.

Watts, D.J. and Strogatz, S.H. (1998). Collective dynamics of 'small-world' networks. *Nature* 393 (6684): 440–442.

Wei, Y., Chen, X., Zhu, L. et al. (2023a). Multi-modal learning for predicting the genotype of glioma. *IEEE Transactions on Medical Imaging* 42 (11): 3167–3178.

Wei, Y., Li, C., and Price, S.J. (2021). Quantifying structural connectivity in brain tumor patients. In: *Medical Image Computing and Computer Assisted Intervention – MICCAI 2021*, Lecture Notes in Computer Science, pp. 519–529. Springer International Publishing, Cham. https://doi.org/10. 1007/978-3-030-87234-2_49.

Wei, Y., Li, C., Cui, Z. et al. (2023b). Structural connectome quantifies tumour invasion and predicts survival in glioblastoma patients. *Brain* 146 (4): 1714–1727. https://doi.org/10.1093/brain/awac360.

Wein, S., Malloni, W.M., Tomé, A.M. et al. (2021). A graph neural network framework for causal inference in brain networks. Scientific Reports 11: 1–18. https://doi.org/10.1038/s41598-021-87411-8.

Wu, X., Zhong, T., Liang, S. et al. (2023). Collaborative modality generation and tissue segmentation for early-developing macaque brain mr images. In: *Medical Image Computing and Computer Assisted Intervention – MICCAI 2023* (ed. H. Greenspan, A. Madabhushi, P. Mousavi et al.), 470–480. Cham: Springer Nature Switzerland.

Xiang, Z., Ting, Z., Weiyan, F. et al. (2019). Breast cancer diagnosis from histopathological image based on deep learning. *2019 Chinese Control and Decision Conference (CCDC)*, 4616–4619. IEEE.

Xie, R., Pang, K., Chung, S. et al. (2024). Spatially resolved gene expression prediction from histology images via bi-modal contrastive learning. *Advances in Neural Information Processing Systems* 36: 70626–70637.

Xie, Y. and Li, Q. (2022). Measurement-conditioned denoising diffusion probabilistic model for under-sampled medical image reconstruction. *Medical Image Computing and Computer Assisted Intervention – MICCAI 2022* (ed. L. Wang, Q. Dou, P.T. Fletcher et al.), 655–664. Cham: Springer Nature Switzerland.

Xing, X., Chen, Z., Zhu, M. et al. (2022). Discrepancy and gradient-guided multi-modal knowledge distillation for pathological glioma grading. In: *MICCAI 2022*, 636–646. Springer.

Xu, R., Zhu, Q., Li, S. et al. (2023). Mstgc: multi-channel spatio-temporal graph convolution network for multi-modal brain networks fusion. *IEEE Transactions on Neural Systems and Rehabilitation Engineering* 31: 2359–2369. https://doi.org/10.1109/TNSRE.2023.3275608.

Yan, R., Schönlieb, C.B., and Li, C. (2024). Spatiotemporal graph neural network modelling perfusion MRI. In: *Medical Image Computing and Computer Assisted Intervention – MICCAI 2024* (ed. F. Qi, G. Aasa, G. Stamatia et al.), 411–421. Springer Nature Switzerland. https://link.springer.com/chapter/10.1007/978-3-031-72069-7_39.

Yan, W., Wang, Y., Gu, S. et al. (2019). The domain shift problem of medical image segmentation and vendor-adaptation by unet-gan. *Medical Image Computing and Computer Assisted Intervention–MICCAI 2019: 22nd International Conference*, Shenzhen, China, October 13–17, 2019, Proceedings, Part II 22, 623–631. Springer.

Yang, H., Zhang, S., Han, X. et al. (2022a). Denoising of 3D MR images using a voxel-wise hybrid residual mlp-cnn model to improve small lesion diagnostic confidence. In: *Medical Image Computing and Computer Assisted Intervention – MICCAI 2022* (ed. L. Wang, Q. Dou, P.T. Fletcher et al.), 292–302. Cham: Springer Nature Switzerland.

Yang, L., Mehta, D., Liu, S. et al. (2023). Tpmil: trainable prototype enhanced multiple instance learning for whole slide image classification. *arXiv preprint arXiv:2305.00696*.

Yang, T.L., Tsai, H.W., Huang, W.C. et al. (2022b). Pathologic liver tumor detection using feature aligned multi-scale convolutional network. *Artificial Intelligence in Medicine* 125: 102244.

Yang, Y., Ye, C., Cai, G. et al. (2024). Hypercomplex graph neural network: towards deep intersection of multi-modal brain networks. *IEEE Journal of Biomedical and Health Informatics* https://doi.org/10.1109/JBHI.2024.3490664.

Yi, H., Wang, X., Li, K. et al. (2024). Unified modeling enhanced multimodal learning for precision neuro-oncology. *arXiv preprint arXiv:2406.07078.*

Yuan, M., Ding, H., Guo, B. et al. (2024). Image-based subtype classification for glioblastoma using deep learning: prognostic significance and biologic relevance. *JCO Clinical Cancer Informatics* 8: e2300154.

Zakka, C., Saheb, G., Najem, E. et al. (2020). Mammoganesis: controlled generation of high-resolution mammograms for radiology education. *arXiv preprint arXiv:2010.05177.*

Zellmer, V.R. and Zhang, S. (2014). Evolving concepts of tumor heterogeneity. *Cell & Bioscience* 4 (1): 69.

Zhang, D., Schroeder, A., Yan, H. et al. (2024). Inferring super-resolution tissue architecture by integrating spatial transcriptomics with histology. *Nature Biotechnology*, 1–6.

Zhang, H., Schneider, T., Wheeler-Kingshott, C.A. et al. (2012). NODDI: practical in vivo neurite orientation dispersion and density imaging of the human brain. *Neuroimage* 61 (4): 1000–1016. https://doi.org/10.1016/j.neuroimage.2012.03.072.

Zhang, L., Wei, Y., Fu, Y. et al. (2022). Mutual contrastive low-rank learning to disentangle whole slide image representations for glioma grading. *arXiv preprint arXiv:2203.04013.*

Zhang, R.Z., Ezhov, I., Balcerak, M. et al. (2023). Personalized predictions of glioblastoma infiltration: mathematical models, physics-informed neural networks and multimodal scans. *Medical Image Analysis. arXiv preprint arXiv:2311.16536.*

Zhang, T., Fu, H., Zhao, Y. et al. (2019). Skrgan: sketching-rendering unconditional generative adversarial networks for medical image synthesis. *arXiv preprint arXiv:1908.04346.*

Zhang, W. and Hong, Q. (2024). Modeling brain functional networks using graph neural networks: a review and clinical application. *IEEE Transactions on Intelligent Transportation Systems* 1 (2): 58–68. https://doi.org/10.62762/TIS.2024.680959.

Zhang, Y., Wang, X., Meng, F. et al. (2024a). Knowledge-driven subspace fusion and gradient coordination for multi-modal learning. *arXiv preprint arXiv:2406.13979.*

Zhang, Y., Xu, Y., Chen, J. et al. (2024b). Prototypical information bottlenecking and disentangling for multimodal cancer survival prediction. *arXiv preprint arXiv:2401.01646.*

Zhao, E., Stone, M.R., Ren, X. et al. (2021). Spatial transcriptomics at subspot resolution with BayesSpace. *Nature Biotechnology* 39 (11): 1375–1384.

Zhao, K., Duka, B., Xie, H. et al. (2022a). A dynamic graph convolutional neural network framework reveals new insights into connectome dysfunctions in ADHD. *NeuroImage* 246: 118774. https://doi.org/10.1016/j.neuroimage.2021.118774. https://www.sciencedirect.com/science/article/pii/S1053811921010466.

Zhao, Y., Wang, K., and Hu, G. (2023). Dist: spatial transcriptomics enhancement using deep learning. Briefings in Bioinformatics 24 (2): bbad013.

Zhao, Z., Yang, H., Sun, J. (2022b). Modality-adaptive feature interaction for brain tumor segmentation with missing modalities. In: *Medical Image Computing and Computer Assisted Intervention – MICCAI 2022* (ed. L. Wang, Q. Dou, P.T. Fletcher et al.), 183–192. Cham: Springer Nature Switzerland.

Zhou, F. and Chen, H. (2023). Cross-modal translation and alignment for survival analysis. *Proceedings of the IEEE/CVF International Conference on Computer Vision*, 21485–21494.

10

Using AI to Detect Alzheimer's Disease from Speech: The Ethical Questions

Ulla Petti[1], Rune Nyrup[2], Jeffrey M Skopek[3], and Anna Korhonen[4]

[1]*Language Technology Lab, University of Cambridge, Cambridge, UK*
[2]*Department of Mathematics and Science Studies, Aarhus University, Aarhus Centrum, Denmark*
[3]*Faculty of Law, Centre for Law, Medicine and Life Sciences, University of Cambridge, Cambridge, UK*
[4]*Language Technology Lab, Centre for Human-Inspired AI, University of Cambridge, Cambridge, UK*

10.1 Introduction

In recent years, research into applying artificial intelligence (AI) for the early detection of Alzheimer's disease (AD) through speech has advanced considerably. This progress has been driven by greater availability of speech and language data, along with new technologies and methodologies that enable large-scale data collection, storage and analysis of such data. A growing body of evidence indicates that alterations in language use can serve as valuable biomarkers for cognitive decline, and that identifying these changes can support both early diagnosis and ongoing monitoring of disease progression. Language-based AI systems, therefore, hold promise as non-invasive, accessible, rapid and cost-effective tools for the early detection of AD. Although no approved cure for AD currently exists, identifying the condition early remains highly valuable. Early detection can:

- Enable timely interventions that may delay the progression of symptoms
- Provide individuals and their families with earlier access to services that ease both emotional and financial strain
- Support the introduction of lifestyle adjustments that could reduce risk or mitigate impact
- Give individuals the opportunity to plan ahead and make informed choices about their future while they are still able to do so (Calzà et al., 2015; Schicktanz et al., 2014; Brock, 1993; Porteri et al., 2017; Ursin et al., 2021).

At the community level, large-scale cognitive screening studies provide valuable evidence that can inform healthcare systems and guide policy decisions (Calzà et al., 2015). Such screening can also detect individuals with non-neurological conditions that are treatable, offering clear benefits for those participants (Mattsson et al., 2010). As new therapies and medications to slow AD-related decline emerge, the demand for more effective tools to detect early signs of AD at scale will continue to grow.

Natural language processing (NLP) methods powered by machine learning (ML) have advanced to a stage where they can effectively support clinical applications (Velupillai et al., 2018; Wu et al., 2020; Hasan & Farri, 2019). These techniques transform written language into computationally usable data (Wu et al., 2020; Harrison & Sidey-Gibbons, 2021), typically drawing on lexical features (such as part-of-speech tagging), syntactic structures (like dependency parsing) and semantic representations (including word embeddings, sentiment analysis and dictionary-based features). Unlike earlier approaches that require extensive manual feature design, ML-driven NLP automates much of this process, improving the efficiency of analysis and the accuracy of outcomes (Wu et al., 2020). In the clinical context, NLP is frequently applied to tasks such as text classification, named entity recognition and relation extraction, with data sources ranging from electronic health records to social media posts, drug reviews and speech transcripts (Wu et al., 2020).

Although NLP offers many advantages in clinical context, and research indicates that it may help identify early signs of AD from speech, its use also raises important ethical challenges that need to be addressed before real-world implementation (Calzà et al., 2015; Wu et al., 2020). Yet, the specific ethical issues linked to applying NLP for early AD detection remain underexplored.

This chapter seeks to fill this gap by outlining the key ethical challenges and offering recommendations that could inform ethical guidelines and best practices for researchers and clinicians in this field.

We organise the discussion around five main categories of concern. Section 10.2 examines autonomy, including challenges around informed consent in AD, risks of depersonalisation and the disclosure of research findings. Section 10.3 addresses privacy and data protection, with a focus on the sensitivities of handling personal speech data. Section 10.4 considers welfare, highlighting potential harms such as distress, discrimination, stigmatisation and questions of reliability. Section 10.5 explores transparency, looking at the interpretability of language features and diagnostic decisions and emphasising the need for clarity for both clinicians and developers. Section 10.6 turns to fairness, asking who stands to benefit from this research and tackling issues of bias and the inclusion of underrepresented populations.

10.2 Autonomy

Autonomy has long been a core principle in bioethics (O'Neill, 2002). Although its definition is debated, it is generally understood as respecting a person's ability to make their own choices, guided by their values and judgement, without pressure or manipulation from others (Christman, 2003). In the context of AD, questions of autonomy arise in everyday decision-making, legal representation, medical consent and participation in research (Gauthier et al., 2013).

In this section, we discuss three key areas:

- Consent and decision-making
- The risk of depersonalisation in language-based testing
- Challenges linked to sharing research results

10.2.1 Consent and Decision-making

Informed consent is a cornerstone of research involving humans. According to the Declaration of Helsinki, it must be voluntary, competent, informed and comprehensive. In AD, however, obtaining valid consent can be difficult as cognitive abilities decline, leaving participants vulnerable (Rogers & Lange, 2013; Nickel, 2006).

The ethical challenges vary with disease stage. In pre-symptomatic stage, participants must clearly understand the difference between biomarker studies and clinical trials, as well as how their data will be stored and shared, to avoid distress, stigma or discrimination (Schicktanz et al., 2014). In the early stage, decision-making capacity may already be reduced, especially due to difficulties in processing complex information (Gauthier et al., 2013). Consent forms should therefore use simple language, repetition, explanations and visual aids when necessary, and cultural factors should also be considered (Schicktanz et al., 2014). In the advanced stage, severe cognitive impairments require stronger safeguards and often proxy decision-making, which comes with its own ethical challenges (Gauthier et al., 2013; Schicktanz et al., 2014). However, it must also be remembered that AD diagnosis does not always mean incapacity (Marson, 2001) – a significant number of individuals with AD are still able to provide consent, and researchers should prioritise enrolling those participants before involving individuals who lack this capacity (Illes et al., 2007).

When designing a study and preparing consent forms, researchers must take into account both the severity of AD and the individual's decision-making capacity (Marson, 2001). Consent should also be revisited as the condition progresses (Schicktanz et al., 2014). In longitudinal research, a 'gradual informed consent transfer' is often recommended, where responsibility for providing consent shifts from the participant to a caregiver or family member over time (Schicktanz et al., 2014).

These issues are particularly important in biomarker studies. Current classification models already distinguish fairly accurately between people with a diagnosis of AD and healthy individuals (Petti et al., 2020). As a result, research has increasingly shifted towards identifying the earliest signs of decline and developing preventative approaches (Meltzer, 2020). Because this work relies heavily on long-term data to track changes over time, population-based longitudinal studies of cognitive decline are considered especially valuable for creating more effective screening tools (Calzà et al., 2015).

At every stage of AD, participants must be clearly informed about the risks and benefits of the research, what they are consenting to, and how their data will be used (Illes et al., 2007).

Further ethical concerns arise with the use of secondary speech and language data available in the public domain. Because there are few longitudinal speech datasets for AD, many studies rely on materials from well-known individuals, such as published writers (Le et al., 2011), press conferences (Berisha et al., 2015) or interviews (Petti et al., 2023). Although this content was made public, the individuals did not consent to its use for research on early signs of AD. This creates risks of their data being used in ways they might not have approved (Meyer, 2018). In many cases, participants are deceased, and obtaining consent from families is difficult. For these reasons, secondary data must be handled with caution, prioritising ethical considerations and the best interests of the individuals, supported by clear guidelines and best practices.

10.2.2 Depersonalisation and Language Testing

Speech analysis in AD is often carried out through various language and cognitive tests. However, these tests can be stressful and may draw attention to a person's communication difficulties, leading to feelings of humiliation, loss of identity and reduced autonomy.

Communication is closely tied to personhood, and when someone communicates in nonstandard ways, they risk being treated as less of a person, which can damage their self-esteem and sense of agency (Sabat & Harre, 1994; Kitwood, 1990). Some early views assumed that a lack of communication meant a lack of thought (Foley, 1992), but others argue that people with AD still act meaningfully and retain their sense of self, even if their communication follows different rules or conventions (Sabat & Harre, 1994).

Using rigid or insensitive language tests can reinforce negative stereotypes, portraying individuals with AD as confused or incapable, and blur the distinction between the person and the disease (Sabat & Harre, 1994; Lipowski, 1969). Beyond being inherently disrespectful, this depersonalisation undermines autonomy and can lead to paternalistic behaviour where decisions are made for individuals rather than with them. Moreover, diminishing a person's confidence in their decision-making capacity may increase their susceptibility to paternalistic treatment.

To prevent this, researchers recommend viewing people with AD as capable of meaningful communication, interpreting their actions as purposeful and studying not only what communication is lost but also what remains intact (Sabat & Harre, 1994; Lipowski, 1969).

10.2.3 Sharing Research Results

The question of whether to disclose signs of AD has been debated for decades. Arguments against disclosure include uncertainty in diagnosis, lack of effective treatment, potentially limited understanding of the diagnosis in advanced stages of the disease and psychological distress, while arguments in favour focus mainly on respecting patient autonomy (Ursin et al., 2021; Gauthier et al., 2013).

Gauthier and colleagues (2013) note that the key challenge now lies in whether to share biomarker results with people who show no symptoms. Although researchers are not yet required to disclose these results because their clinical value is uncertain, there is a growing consensus that participants should be informed, as also observed by Kaduszkiewicz and colleagues (2008). Information sheets and consent forms should clearly outline how results will be communicated and respect each participant's right to know or not to know (Schicktanz et al., 2014; Porteri et al., 2017).

The use of AI in AD detection introduces further complexity. Questions about model accuracy and predictive reliability affect how results are shared (Ursin et al., 2021). In line with existing research on AI-based prediction of psychiatric risk (McKernan & Clayton, 2018), participants should be told in advance that AI predictions may be wrong and informed of the model's accuracy and related concerns. The model's accuracy can influence whether participants choose to be informed of their results. For instance, their choice could differ if the model is 60% accurate rather than 95%.

10.3 Privacy and Data Protection

Everyday speech contains valuable indicators of cognitive decline in AD, especially when analysed using ML-based NLP. However, because speech often includes personal and identifiable details, automatic analysis also introduces new challenges for safely handling and analysing medical language data.

NLP relies on large amounts of data, and collecting such data has become easier with the rise of smart devices, digital text sources, voice assistants, high-quality microphones and greater data storage capacities (Robin et al., 2021; Yamada et al., 2021; Cychosz et al., 2020; Le Franc et al., 2018; Mehl, 2017). These advances have supported the development of language-based diagnostic tools.

Language data can be collected in several ways: through remote recordings or writings in everyday settings, through speech samples taken in clinical environments or by using publicly available materials such as recordings, transcripts or writings. However, technologies for collecting and analysing such data raise significant privacy and data protection concerns. Although many of these issues are shared with other areas of speech and language research, studying individuals with cognitive decline introduces further risks. These include the potential disclosure of private speech content and the possibility of identifying individuals through their voice, particularly when assessing cognitive function or handling sensitive health information.

The following sections discuss these challenges in handling personal and medical language data, along with possible strategies to address them.

10.3.1 Privacy Concerns

Natural speech often contains personal or identifying information. When data is collected remotely in everyday settings, researchers have limited control over who is being recorded or what is said (Kumar et al., 2015). Long recordings can be especially risky, particularly for people with cognitive decline, as they may unintentionally share information they would normally keep private (Cychosz et al., 2020). The importance of protecting participants' data is illustrated by a study showing that only 8% of respondents would agree to wear a device that records sound continuously (Klasnja et al., 2009).

Issues of confidentiality and data protection pose major obstacles to developing and sharing large, open-access speech datasets on cognitive decline in AD. To work around this limitation, researchers have created alternative datasets based on publicly available sources (such as online posts, interviews, books, speeches) from individuals who later developed AD. However, even though this material is publicly accessible, compiling it into research datasets can expose additional personal information and make those details permanently traceable (Meyer, 2018).

This issue becomes particularly important when research involves sensitive health topics, such as cognitive decline or AD. For instance, individuals who write public posts, publish books, or appear in press conferences and interviews are aware that their words are being recorded and shared publicly. However, they are unlikely to anticipate that their speech or writing might later be used to study, or even reveal, early signs of AD and changes in their language abilities.

Storing and distributing voice recordings also carries a risk of identifying the speaker, particularly if the listener is familiar with their voice. Sharing such data with additional research teams further increases the likelihood of participants being recognised (Cychosz et al., 2020). The potential harm resulting from identification and the association of speech with sensitive health information is difficult to predict but should be carefully considered and acknowledged.

10.3.2 Protecting Speaker Privacy

Protecting privacy in digital environments is far more difficult than in the analogue world (Müller, 2020). Strengthening dialogue between science and law across different legal systems and cultural contexts is essential, and researchers must uphold high ethical standards, especially in regions with weaker data protection and consent frameworks (Schicktanz et al., 2014).

A number of strategies have been proposed to safeguard participant confidentiality when working with language data. One approach is to give participants control over the recording process, allowing them to choose when to record, pause or delete parts they do not wish to share (Cychosz et al., 2020). However, this method carries the risk that participants may forget to restart the device.

Technical strategies have also been explored. Automatic feature extraction allows researchers to analyse derived features rather than raw audio, reducing privacy risks (Wyatt et al., 2007; Lane et al., 2012). This can be done after data collection or directly on the device, ensuring that the original recordings are never stored (Cychosz et al., 2020). Yet, this approach depends on current scientific understanding of AD, and future insights might be lost if the raw data no longer exists and new features cannot be discovered. Researchers have highlighted the need to focus efforts on manual data annotation and training models of automatic feature extraction (Cychosz et al., 2020).

Another method involves reducing speech intelligibility, protecting both participants and bystanders (Wyatt et al., 2007). Examples include:

- Analysing broad acoustic features such as speech rate, energy and pitch (Wyatt et al., 2007)
- Applying principal component analysis (PCA) to detect non-speech sounds (e.g., laughter, pauses, breaths, hums, coughs) while eliminating the possibility of recreating intelligible speech (Kumar et al., 2015; Larson et al., 2011)
- Using linear prediction to replace vowel sounds (Chen et al., 2008)
- Using an accelerometer to measure bone-conducted acoustic signals from the vocal tract instead of recording speech directly (Garrard et al., 2017). According to the authors, this technique makes it possible to identify repetitive speech behaviours in people with AD at home without compromising privacy

It is important to recognise that while Mel-frequency cepstral coefficients (MFCC) features are widely used in AD speech studies, these features can still expose private information, such as prosody, speaker identity and speech content, making them unsuitable for strong privacy protection (Kumar et al., 2015; Larson et al., 2011). When using methods that make recorded speech unintelligible, researchers should also be aware that audio modification can limit future analyses and that mobile devices may lack sufficient processing power for advanced audio transformations (Cychosz et al., 2020).

Another technique for protecting privacy is subsampling, where only short segments of high-quality audio are retained (Mehl, 2017; Mehl & Pennebaker, 2003). Shorter samples reduce privacy risks and lower ethical training requirements for researchers accessing the data (Cychosz et al., 2020). However, this approach has limitations:

- Key linguistic or behavioural features may be missed.
- Reduced context can lead to misinterpretation.
- Sensitive information may still appear in short clips (Cychosz et al., 2020).

Future research may address these challenges by developing algorithms that first identify relevant speech events and then record short segments (Cychosz et al., 2020). To further protect participant identity, pseudonymisation or anonymisation should be applied (including when working with public secondary data) (Meyer, 2018; Müller, 2020). Removing names, places and other identifiable details from transcripts helps reduce the risk of recognition. Nevertheless, many scholars argue that participants should still have the right to request exclusion from datasets, even after anonymisation (Mohammad, 2022). While transcripts can be anonymised effectively, complete anonymity of voice data cannot be guaranteed (Cychosz et al., 2020). Audio encryption can provide additional protection by concealing speech content while retaining prosodic and conversational information (Klasnja et al., 2009; Müller, 2020). Earlier studies noted that encryption was difficult to apply on mobile devices due to limited computing power (Kumar et al., 2015; Klasnja et al., 2009), but more recent research demonstrates significant advances in mobile audio encryption (Farsana et al., 2023; Eltengy, 2021).

Researchers must also carefully determine the extent and conditions of data sharing, thinking about the speakers' best interests and potential risks of harm. Decisions should consider the type of data (e.g., transcripts, quantified metrics, audio or video), the level of access (public, research-only or restricted), the consent provided and the ethical training of data users (Meyer, 2018; Cychosz et al., 2020). Participants should be clearly informed, before giving consent, about who will have access to their data and how it will be used, ideally with concrete examples (Low et al., 2020; Meyer, 2018).

It is generally discouraged to promise that data will never be shared; instead, researchers should carefully plan how and with whom it will be shared (Meyer, 2018). As Low and colleagues (2020) note, even when consent is given, participants may not fully understand what data sharing entails – an issue especially relevant in AD research. Participants should retain the right to control data access. Meyer (2018) suggests offering participants choices about how much of their data is shared, while ensuring that reproducibility of results remains possible. Low et al. (2020) further propose including an option for participants to withdraw or deactivate their consent.

Finally, while destroying data is generally discouraged, researchers should consider secure storage solutions and the use of trusted repositories (Meyer, 2018; Cychosz et al., 2020). Kumar and colleagues (2015) recommend avoiding the storage of full real-world audio recordings to protect privacy. Although complete recordings best capture spontaneous, natural speech, which is a valuable feature for research, they also pose the highest risk of revealing personal or third-party content. To reduce these risks, we recommend applying the techniques described earlier.

10.3.3 Advantages of Data Sharing

Although data sharing involves certain risks that must be carefully managed, greater data availability can significantly enhance research reproducibility and support the development of more accurate tools for detecting AD. Both the potential benefits and risks to individual participants and to the scientific community need to be weighed.

From a research standpoint, open data sharing is often seen as ideal, but researchers must also honour their commitments to participants, which may include restricting access or providing data only upon request (Meyer, 2018). Participants should always retain control over identifiable information (Cychosz et al., 2020). Importantly, researchers should not assume that all participants are opposed to sharing their data. While some prioritise privacy, others may be motivated by the prospect of their data helping to advance scientific understanding and support future studies (Meyer, 2018).

For researchers, sharing data also brings several advantages. It fosters collaboration, builds a sense of shared purpose and allows others to reanalyse or combine datasets, ultimately improving the robustness, accuracy and efficiency of research outcomes (Cychosz et al., 2020). Nonetheless, because collecting high-quality data is both time-consuming and expensive, disagreements often arise over when and how datasets should be shared or integrated (Meyer, 2018).

10.4 Welfare

Beyond the issues of autonomy and privacy, using AI-based speech technologies to detect early signs of AD also raises several concerns about participant well-being. This section examines three areas where potential harm may occur: psychological distress, discrimination and stigmatisation, and the risk of inaccurate diagnoses.

10.4.1 Distress

Undergoing cognitive testing and receiving research results may lead to distress, affecting not only participants themselves but also their families, clinicians and the researchers involved.

10.4.1.1 Cognitive Testing

Language testing can be particularly difficult for individuals with AD due to difficulties in speech production and cognitive limitations (Chien et al., 2019; Lopez-de-Ipina et al., 2018). While researchers may view such testing or recording simply as a means of collecting data on cognitive function, participants often experience it differently. Many people with AD report that being placed in situations that expose their deficits feels humiliating and meaningless (Mattsson et al., 2010; Sabat & Harre, 1994). Recordings from verbal fluency tasks, such as those in the Pitt Corpus (Becker et al., 1994), often capture moments of visible frustration and confusion when participants struggle to recall words. This frustration, arising from being asked to perform tasks they once managed with ease, can provoke strong emotional reactions, exacerbate processing difficulties and negatively affect test performance (Sabat & Harre, 1994).

Minimising participant burden is essential to encourage engagement and protect wellbeing. Efforts to reduce distress may include collecting spontaneous speech samples in relaxed, conversational settings rather than through formal testing, allowing participants to decide when recordings begin and end, and respecting their right to pause or withdraw from the process at any time (Cychosz et al., 2020; Chien et al., 2019; Lopez-de-Ipina et al., 2018).

10.4.1.2 Receiving the Research Results

It is important to recognise that being informed about signs of a progressive, incurable neurological condition can be deeply distressing, sometimes leading to negative self-perception or even suicidal thoughts (Schicktanz et al., 2014; Ursin et al., 2021; Mattsson et al., 2010; Gauthier et al., 2013). Reactions to an early diagnosis are often shaped by the availability of treatment options. When treatment exists, identifying early indicators of the disease can improve outcomes and may make the diagnosis easier to accept (Mattsson et al., 2010). Even in the absence of treatment, however, many individuals value early information, as it enables them to make informed decisions about their future and develop coping strategies for cognitive decline (Porteri et al., 2017; Ursin et al., 2021; Mattsson et al., 2010).

Conversely, a negative test result can also be upsetting for those who hoped for an explanation for their memory problems (Smith & Beattie, 2001). Moreover, the risk of false positives and false negatives (Mattsson et al., 2010; Illes et al., 2007) adds further complexity and potential emotional strain in disclosing early findings (Ursin et al., 2021).

To mitigate disclosure-related distress, it is essential to develop highly accurate diagnostic models to minimise misclassification and to establish transparent, sensitive procedures for communicating results. Clear information about the disclosure process, available follow-up support and the right to decline knowing one's results can help participants make truly informed decisions about participation or tool use. Porteri and colleagues (2010) recommend that researchers conduct personal ethical evaluations and involve participants in decisions about disclosure, while Schicktanz et al. (2014) and Mattsson et al. (2010) emphasise the need to consider cultural context and individual coping capacities when sharing prognostic information.

10.4.1.3 Emotional Impact on Others Involved

Beyond the research participants themselves, studies on the early markers of AD can also have emotional consequences for family members, clinicians and researchers. Families may feel relieved to finally understand the cause of their loved one's symptoms (Smith & Beattie, 2001), yet this knowledge can also heighten anxiety about the relative's future and their own potential genetic vulnerability, sometimes creating tension within family relationships (Cammen et al., 2004; Schicktanz et al., 2014).

Clinicians may experience discomfort or ethical conflict when faced with uncertain findings, particularly in early biomarker research, as they must decide how much information to disclose and how best to communicate it (Mattson et al., 2010; Smith & Beattie, 2001). Researchers, too, may be affected emotionally when analysing speech data that reflects cognitive decline. They should be supported with the option to step back from such work when necessary (Cychosz et al., 2020). When dealing with especially sensitive or distressing material, annotators should be informed in advance, and the volume of content assigned to each individual should be limited to prevent emotional fatigue (Mohammad, 2022).

10.4.2 Technology's Role in Discrimination and Stigma

One of the most pressing ethical risks in studying early speech markers of AD is the possibility of stigmatisation and discrimination, which may arise even before clinical symptoms appear (Calzà et al., 2015; Ursin et al., 2021; Gauthier et al., 2013; Meyer, 2018; Cychosz et al., 2020).

AD-related stigmatisation can occur in both personal and professional contexts, often driven by persistent myths and misconceptions (Alzheimer's Association, 2008) that portray individuals with AD as somehow less than persons (Gauthier et al., 2013). While public campaigns have sought to reduce fear surrounding dementia, Foley (1992) argues that such efforts may have inadvertently reinforced stigma, making life more difficult for those affected and their families, and hindering access to social support and resources. To counteract this, Schicktanz and colleagues (2014) suggest avoiding portrayals of dementia solely as a burden to healthcare and society. More research into the lived experiences of people in pre-symptomatic stages could foster understanding, promote discussion and help reduce stigmatisation.

Discrimination against individuals with early or pre-symptomatic AD can manifest in multiple areas, including employment, legal rights, health insurance, driving eligibility and access to medical care, such as organ transplantation (Schicktanz et al., 2014; Ursin et al., 2021; Illes et al., 2007). Emerging predictive technologies may amplify these risks (Cychosz et al., 2020). For example, Low et al. (2020) highlight concerns in psychiatric prediction models, warning that even unvalidated tools could influence decisions by insurers or employers – an issue equally relevant for neurodegenerative conditions such as AD. Similarly, Illes et al. (2007) note that imaging-based detection of early AD biomarkers could lead insurers to require diagnostic testing or disclosure of results. Speech-based tools, which are easier to deploy and can analyse everyday language, present even greater potential for misuse.

Nevertheless, when developed responsibly and used within clinical frameworks, such technologies could have a profoundly positive impact. This underscores the urgent need for regulation to prevent commercial exploitation and to safeguard against unethical use by third parties such as employers and insurers. The issue connects closely to the concept of dual use – a term first used for technologies developed for both civilian and military use, and now referring to technologies developed for beneficial purposes that can also be applied in harmful or unintended ways. The potential for unethical or dual use represents a key concern in technology development, underscoring the need to carefully consider the ethical implications of research (Low et al., 2020).

The growing development of tablet- or app-based tools for home screening raises additional ethical challenges. Allowing open access to applications capable of detecting AD from speech could enable unauthorised testing of individuals within microphone range, compromising privacy and exposing people to information about their cognitive health without consent (Low et al., 2020). To prevent this, such tools should only be used under clinical supervision, with informed consent from individuals who actively seek assessment (Ursin et al., 2021).

Beyond regulation, researchers themselves carry responsibility for upholding the highest professional and ethical standards. Protecting participant privacy and confidentiality, considering potential misuse and evaluating the societal consequences of new technologies must be integral to the development process (Schicktanz et al., 2014; Véliz, 2019; Karlawish, 2011).

10.4.3 Validity and Reliability of Results

Deficiencies in data quality or model design may compromise the reliability of research outcomes and pose risks to participants. This section addresses the adequacy of linguistic information captured during testing and evaluates the robustness of model predictions and reported results in speech-based AD detection.

10.4.3.1 Adequacy of Linguistic Data

Although language and psychological tests can help identify certain AD-related changes, they often demonstrate limited sensitivity and fail to capture the communication challenges that individuals with AD experience in everyday conversation (Calzà et al., 2021; Boyé et al., 2014). The repetitive nature of many testing tasks can lead to habituation, potentially distorting the assessment of cognitive ability and reducing screening accuracy (Ujiro et al., 2018). In addition, tests that rely on mean scores may oversimplify variability across participants, implying uniformity within diagnostic groups. Psychological tests also risk built-in biases, such as lack of validation across education levels and disparities in the quantity and quality of available research (Calzà et al., 2015).

Cultural background further complicates interpretation. Cognitive processes and language use differ across cultures, which can lead to misinterpretation of communicative performance. For example, patterns of confusion between individuals in the same social category (Fiske, 1993) or object recall based on background (Masuda & Nisbett, 2001) vary culturally. As both confusion and recall are often used to identify early signs of AD, cultural differences must be accounted for when recruiting participants and analysing linguistic data. Promising examples of adapting the tests and analysis methods to a certain culture can be found in the research on Thai. For example, Munthuli and colleagues (2021) used the Thammasat-NECTEC-Chula's Thai Language and Cognition Assessment (TLCA), in which many parts of the tests were made more appropriate for Thai culture, and reported good classification results between healthy, MCI and AD groups. Similarly, Sangchocanonta and colleagues (2021) employed culturally relevant picture-description tasks and a Thai-specific POS tagger that aligns with the recommendations for Thai word types to analyse the data, and Metarugcheep and colleagues (2022) adapted verbal fluency tests and the extracted features for Thai language-based cognitive assessment.

To address the limitations of traditional testing, researchers have proposed using spontaneous speech rather than formal language tasks. Collecting natural speech data reduces participant stress (Chien et al., 2019; Lopez-de-Ipina et al., 2018), reflects real-life communication more accurately (Sabat & Harre, 1994; Gosztolya et al., 2016) and can be gathered easily in everyday contexts using mobile devices with minimal instruction (Robin et al., 2021). It also enables the inclusion of participants unable to tolerate structured testing (Blanken et al., 1987). Advances in NLP and AI have further improved the scalability and automation of such analyses (Robin et al., 2021; Yamada et al., 2021; Chien et al., 2019; Luz et al., 2021a).

Nonetheless, spontaneous speech data present certain challenges regarding adequacy. Semantic accuracy can be difficult to verify – speech may be coherent but factually incorrect, or may include name confusions that are undetectable without personal knowledge of the speaker. Involving close contacts in transcript verification could mitigate this issue but raises additional privacy and ethical concerns. Similarly, while emotional content is often analysed, current automatic emotion recognition systems are not equipped to handle the emotional variability associated with AD, and their use should therefore be limited or accompanied by clear communication of their constraints (Mohammad, 2022). Other potential biases stem from the complex, multivariate nature of linguistic and acoustic data (Cychosz et al., 2020) and variability in audio quality (Balagopalan et al., 2021).

Despite these challenges, spontaneous speech tasks have shown strong performance. Lopez-de-Ipina et al. (2018) reported a 95% classification accuracy in distinguishing individuals with AD from healthy controls, outperforming structured tasks. The authors attribute this success to the natural and relaxed setting of spontaneous speech, which allows subtle cognitive changes to emerge more clearly.

10.4.3.2 Adequacy of Models

The adequacy of a model can be viewed from several angles. First, it concerns the overall performance and trustworthiness of its results. Second, it involves assessing whether model performance differs across population groups (for example, between minority and majority participants). This section focuses on issues related to general model performance, such as sample size, overfitting and the choice of evaluation metrics, while questions of bias and unequal harm are discussed later in the Fairness section (Section 10.6).

Although speech-based AD classification models have achieved high test accuracy, their reliability in real-world applications remains uncertain. Unreliable models risk producing biased or misleading results, potentially harming participants through false positives or false negatives. False negatives may lead to exclusion from treatment or clinical trials and provide false reassurance, while false positives can result in overdiagnosis, unnecessary interventions or inappropriate trial inclusion (Ursin et al., 2021).

Several factors contribute to the potential unreliability of these models. One common issue is small sample size, which can lead to overfitting. Despite the importance of large datasets in ML, many AD classification models are trained on small, highly specific samples due to limited data availability (Petti et al., 2020; Voleti et al., 2019; de la Fuente Garcia et al., 2020). In a meta-analysis of existing studies, Berisha et al. (2021) found that models trained on smaller datasets often achieve higher test accuracy but perform poorly in real-world contexts. This lack of generalisability poses significant risks in clinical use.

The shortage of large, high-quality datasets exacerbates overfitting. Expanding access to speech data would improve model robustness, support better health outcomes and reduce reliance on re-used language data collected for other purposes – which can reinforce historical biases (Müller, 2020; Voleti et al., 2019). Collaboration between researchers in speech neuroscience, neuropsychology and language technology is therefore essential (Voleti et al., 2019). Additional strategies include data synthesis and revising consent models to facilitate responsible data sharing (Velupillai et al., 2018).

Publication bias also contributes to overfitting and limited reproducibility. Studies with positive results are more likely to be published (Munafò et al., 2017), encouraging selective reporting and overemphasis on accuracy. Researchers may unintentionally generalise findings from narrow samples (Henrich et al., 2010) or interpret data in line with pre-existing beliefs – a form of cognitive bias (Müller, 2020). Techniques such as researcher blinding and integrative training can help reduce this risk (Munafo et al., 2017). However, Müller (2020) also raises an interesting question: Should AI systems emulate or avoid the cognitive biases inherent to humans?

To improve robustness, methods such as cross-validation and held-out test sets are recommended. Yet, only 10% of studies reviewed by de la Fuente Garcia et al. (2020) reported using holdout sets, and just one validated the model on an independent dataset. Evaluation metrics also warrant careful selection. Accuracy alone is unsuitable for imbalanced datasets, yet 35% of studies in de la Fuente Garcia et al. (2020) reported only accuracy. Alternative metrics such as F1 scores, UAR (Unforeseen Attack Robustness) and ROC (Receiver Operating Characteristic) curves are better suited for these tasks (de la Fuente Garcia et al., 2020; Gebru et al., 2021). For longitudinal speech data, Velupillai et al. (2018) further recommend metrics that consider time sensitivity, model interpretability and personalisation.

The lack of standardisation in data collection, model design and evaluation practices limits comparability across studies and hinders contextual interpretation (Petti et al., 2020; Balagopalan et al., 2021; Voleti et al., 2019; de la Fuente Garcia et al., 2020). For instance, only 61% of studies in de la Fuente Garcia et al.'s review (2020) compared results quantitatively with prior work or provided baselines for reference. Establishing standardised methods for data collection and analysis in cognitive assessment would enable greater consistency across studies (Voleti et al., 2019), improve the reliability of findings and reduce the likelihood of harmful decisions resulting from false positives or negatives.

10.5 Transparency

In speech-based AD detection using AI and NLP, ML algorithms are deployed to infer information about an individual's cognitive health. Ensuring that clinical decision-making is transparent and interpretable is therefore essential (Ursin et al., 2021; Voleti et al., 2019; Tu et al., 2017). Although classification models can achieve high accuracy in distinguishing individuals with AD from healthy controls, they often lack interpretability, making it unclear why a decision has been made and which linguistic features drive the classification results (Robin et al., 2021).

Here, interpretability operates on two levels: for developers and for end users such as clinicians. Developer-level interpretability enables better model refinement and error correction, while user-level interpretability is vital for clinical trust and communication. Clinicians must understand why a model has made a particular prediction before deciding whether to rely on it and to communicate the reasoning behind diagnoses or treatment recommendations to patients, thereby supporting informed consent and shared decision-making (Ursin et al., 2021; Tjoa & Guan, 2020; Keeling & Nyrup, 2022).

Explainable AI (XAI) is an active field of research that aims to address these challenges (Tjoa & Guan, 2020). Techniques to enhance interpretability include verbal, visual and mathematical explanations, but developers must also consider the audience – who the explanation is for, how understandable it is, and how much trust it inspires (Mohammad, 2022; Keeling & Nyrup, 2022). In the context of AD detection, interpretability issues arise on two fronts: understanding the extracted linguistic features and understanding the model's decision-making process. These are interrelated: clarity about language features can illuminate why the system reached a particular health assessment.

As data-driven approaches grow and larger datasets become available, models increasingly prioritise performance over feature interpretability (Voleti et al., 2019). Deep neural networks have improved performance in speech and NLP tasks but at the cost of transparency. As a result, extracted features often remain difficult for both developers and clinicians to interpret. This is particularly true for complex acoustic features, such as the MFCC features, which encode valuable information about the speaker but offer limited clinical interpretability (Tu et al., 2017). To address this, Jiao et al. (2017) proposed visual tools such as spider plots, which help clinicians compare a patient's phonological signals with those of healthy speakers, supporting understanding rather than replacing professional judgement.

The growing integration of AI into healthcare underscores the need for greater technical literacy among clinicians (Ursin et al., 2021; Wu et al., 2020; Harrison & Sidey-Gibbons, 2021; McKernan & Clayton, 2018; Le Glaz et al., 2021). Nonetheless, AI and NLP systems should function as decision-support tools, not replacements for clinical expertise (Wu et al., 2020; Le Glaz et al., 2021). Clinicians must retain the ability to recognise when model outputs contradict clinical evidence and to understand potential algorithmic limitations (Mattsson et al., 2010). When the reasoning behind model predictions cannot be explained due to algorithmic opacity, the process of obtaining informed consent is also undermined (Ursin et al., 2021). Hence, the development of clinical NLP tools should involve clinicians throughout and be designed with their practical needs in mind (Velupillai et al., 2018).

Regulatory standards for algorithmic transparency differ internationally. The European Union (EU) grants individuals the right to request explanations for algorithmic decisions (Wachter et al., 2017), whereas the United States does not have equivalent provisions (Garcia, 2016). Even within the EU, however, questions remain regarding enforceability (Wachter et al., 2017) and the inconsistency of demanding full explainability from machines when human decision-making often lacks transparency (Müller, 2020). Access to algorithmic code alone rarely provides meaningful insight, meaning that many AI-driven decisions remain opaque to both experts and those affected (Garcia, 2016; Müller, 2020).

Transparency and bias are closely linked: opacity can conceal bias, while improved transparency can help uncover and mitigate it (Müller, 2020). Issues of opacity also shape public trust. Research suggests that people often view scientists, medical professionals and biotechnology companies as acting in their own interests rather than in those of the public (O'Neill, 2002). When data, methods and results lack transparency or cannot be independently replicated, confidence in science suffers (Meyer, 2018). Since science is a collective enterprise aimed at the public good, its legitimacy and credibility ultimately depend on openness, reproducibility and transparency (Munafò et al., 2017).

10.6 Fairness and Inclusion

Many studies suggest that developing language-based AI for early AD detection could enhance the quality of life for people with dementia and promote greater equity by offering accessible and affordable screening options. However, it is important to ask whose quality of life will actually improve, and in what ways. Questions of recruitment, treatment availability, participant age and structural bias must be considered, especially as AD disproportionately affects women and minority populations (Babulal et al., 2019; Lopez et al., 2019), and clinical knowledge of AD within these groups remains limited (Mayeda et al., 2016).

Since no approved treatment currently exists, individuals participating in research are unlikely to benefit directly from its outcomes. The long timeline of screening tool and treatment development offers little to no immediate benefit to the research subjects and may reduce motivation to participate. Some participants report finding language testing distressing and unhelpful, though many express willingness to take part if research has a tangible personal benefit (Sabat & Harre, 1994). While altruism motivates some, participants should be clearly informed about the potential benefits – or lack thereof – of participation, and fair compensation mechanisms should be considered.

When designing early-detection models, it is vital to consider who will ultimately benefit from them. Responsible design (Müller, 2020) requires careful thought about participant recruitment and data use, since model performance depends on the similarity between training and application populations. Therefore, today's recruitment decisions have a major impact on future outcomes and the fairness of screening tools.

Balanced datasets are central to fairness: imbalanced data can amplify gender-, racial- and age-related biases, leading to unequal accuracy across groups (Cychosz et al., 2020; Gebru et al., 2021; Maley et al., 2020). These are especially important considerations in language-based data and clinical contexts, as speech, AD manifestations and NLP performance can be affected by these characteristics. Such bias can cause allocational harm where models perform better on well-represented and often privileged populations, as their behaviour is seen as 'typical' due to existing training data (Cychosz et al., 2020), thereby reinforcing inequity and stigmatisation.

Despite these risks, reviews show that many studies still use demographically skewed datasets. De la Fuente Garcia et al. (2020) found that only 25% of studies reported balanced age, gender and education levels, and Petti et al. (2020) observed that only 40% of datasets were gender-balanced, with education levels typically higher in control groups.

Another concern is linguistic diversity. Most existing studies focus on English-speaking populations – 41% according to the review by de la Fuente Garcia et al. (2020) and 30% by Petti et al. (2020) – limiting generalisability and primarily benefitting Western high-income countries. Dialect variation also affects model performance (Jurgens et al., 2017); for instance, removing dialect-sensitive features (MFCC features, formants) improved dementia detection accuracy in Japanese speakers by 15% (Fukuda et al., 2022). Given that older individuals often use stronger regional dialects, models must account for linguistic variation (Fukuda et al., 2022). Developing multilingual datasets, such as the eight-language AZTIAHO database (Lopez-de-Ipina et al., 2018), is crucial for equitable advancement (de la Fuente Garcia et al., 2020).

Sampling bias extends beyond AD research. Behavioural sciences have long been dominated by WEIRD (Western, Educated, Industrialised, Rich, Democratic) populations, limiting global applicability and leading to calls for cultural, linguistic and geographical inclusion (Cychosz et al., 2020). Underrepresentation often stems not from unwillingness to participate but from lack of opportunity and structural exclusion (Rogers & Lange, 2013). Barriers include restrictive recruitment methods, limited resources or assumptions about disinterest – often reflecting stereotypes (Rogers & Lange, 2013). Such exclusion results in poorer health outcomes, lower access to interventions and unjust concentration of research benefits within privileged groups.

Promoting inclusivity and dataset transparency is therefore essential. Standardised dataset descriptions should accompany publications (Henrich et al., 2010), and risk-of-bias tools such as QUADAS-2 (Whiting et al., 2011) and PROBAST (Wolff et al., 2019) can help identify and mitigate sampling bias. While subsampling may help balance datasets, data preprocessing must be handled carefully to avoid inadvertently introducing new imbalances (de la Fuente Garcia et al., 2020). Broader representation in AD research would not only improve model generalisability but also ensure that the benefits of early detection reach diverse linguistic, cultural and demographic groups.

10.7 Conclusion

Although the use of AI to detect early signs of AD from speech holds great promise, it also raises a number of ethical challenges that must be carefully examined and addressed. In this chapter, we have outlined key issues relating to autonomy, privacy and data protection, welfare, transparency and fairness. We highlight the importance of developing clear ethical guidelines and best practices and of fostering dialogue across disciplines, research areas and communities.

Particular emphasis should be placed on respecting participants' choices, ensuring that data protection decisions prioritise their best interests, and designing research methods that minimise discomfort and distress. Ethical reflection should guide every stage of technology development, with attention to model explainability, inclusivity and the equitable distribution of research benefits. Our recommendations aim to support responsible and participant-centred innovation in this field, and a summary of these suggestions is provided in Table 10.1.

Table 10.1 Summary of recommendations for developing ethical AI for speech-based AD detection.

Category	Key considerations
Autonomy	• Ensure participants understand the risks, benefits and data use before consenting. • Adapt consent procedures to disease stage (pre-symptomatic, early, advanced). • Use accessible language and visual aids; consider cultural background. • Apply safeguards for advanced-stage participants and ethical proxy decision-making. • Use gradual or transferred consent in longitudinal studies; act in participants' best interests when using secondary data. • Avoid depersonalisation – treat people with AD as full individuals, not defined by their condition. • Clearly explain disclosure procedures and respect the right not to know outcomes.
Privacy and data protection	• Maintain high standards of data governance and ethical self-regulation. • Consider giving participants control over recording devices. • Use privacy-preserving methods such as feature extraction, reduced intelligibility, anonymisation, pseudonymisation and encryption. • Avoid storing raw or full audio when possible; consider recording only short, necessary segments. • Define clear data-sharing levels and communicate who has access and how data will be used. • Consider participant choice in how much and what type of data is shared.
Welfare	• Use natural, conversational speech instead of stressful cognitive tests. • Allow participants to stop at any time. • Provide clear information about result delivery, accuracy and available support. • Involve participants in ethical decisions about disclosure. • Anticipate and minimise distress for families, clinicians and researchers. • Avoid framing dementia as a societal 'burden'; raise awareness to reduce stigma. • Guard against unethical or dual use of technology.
Transparency	• Improve model and feature interpretability; explain outputs in understandable ways. • Tailor explanations to target users (e.g., clinicians, participants, developers). • Include clinician input in AI tool design. • Promote health and technical literacy among professionals and the public. • Emphasise openness and reproducibility to strengthen trust in science.
Fairness	• Ensure research benefits are equitably distributed. • Use balanced, representative datasets across age, gender, education and ethnicity. • Include multiple languages and dialects; develop resources for underrepresented groups. • Clearly describe datasets and use bias-assessment frameworks (e.g., QUADAS-2, PROBAST). • Ensure inclusive participation of underserved populations.

References

Alzheimer's Association (2008). 2008 Alzheimer's disease facts and figures. *Alzheimer's & Dementia* 4 (2): 110–133.

Babulal, G.M., Quiroz, Y.T., Albensi, B.C. et al. (2019). Perspectives on ethnic and racial disparities in Alzheimer's disease and related dementias: update and areas of immediate need. *Alzheimer's & Dementia* 15 (2): 292–312.

Balagopalan, A., Eyre, B., Robin, J. et al. (2021). Comparing pre-trained and feature-based models for prediction of Alzheimer's disease based on speech. *Frontiers in Aging Neuroscience* 13: 635945.

Becker, J.T., Boiler, F., Lopez, O.L. et al. (1994). The natural history of Alzheimer's disease: description of study cohort and accuracy of diagnosis. *Archives of Neurology* 51 (6): 585–594.

Berisha, V., Krantsevich, C., Hahn, P.R. et al. (2021). Digital medicine and the curse of dimensionality. *NPJ Digital Medicine* 4 (1): 153.

Berisha, V., Wang, S., LaCross, A. et al. (2015). Tracking discourse complexity preceding Alzheimer's disease diagnosis: a case study comparing the press conferences of Presidents Ronald Reagan and George Herbert Walker Bush. *Journal of Alzheimer's Disease* 45 (3): 959–963.

Blanken, G., Dittmann, J., Haas, J.C. et al. (1987). Spontaneous speech in senile dementia and aphasia: implications for a neurolinguistic model of language production. *Cognition* 27 (3): 247–274.

Boyé, M., Tran, T.M., and Grabar, N. (2014). NLP-oriented contrastive study of linguistic productions of Alzheimer's and control people. In: *Advances in Natural Language Processing: 9th International Conference on NLP, PolTAL 2014,* Warsaw, Poland, September 17–19, 2014. Proceedings 9, 412–424. Springer International Publishing.

Brock, D. (1993). Quality of life measures in health care and medical ethics. *The Quality of Life* 1: 95–133.

Calzà, L., Beltrami, D., Gagliardi, G. et al. (2015). Should we screen for cognitive decline and dementia? *Maturitas* 82 (1): 28–35.

Calzà, L., Gagliardi, G., Favretti, R.R. et al. (2021) Linguistic features and automatic classifiers for identifying mild cognitive impairment and dementia. *Computer Speech & Language* 65: 101–113.

Cammen, T.J.V.D., Croes, E.A., Dermaut, B., et al. (2004). Genetic testing has no place as a routine diagnostic test in sporadic and familial cases of Alzheimer's disease. *Journal of the American Geriatrics Society* 52 (12): 2110–2113.

Chen, F., Adcock, J., and Krishnagiri, S. (2008). Audio privacy: reducing speech intelligibility while preserving environmental sounds. In: *Proceedings of the 16th ACM International Conference on Multimedia*, 733–736.

Chien, Y.W., Hong, S.Y., Cheah, W.T. et al. (2019). An automatic assessment system for Alzheimer's disease based on speech using feature sequence generator and recurrent neural network. *Scientific Reports* 9 (1): 19597.

Christman, J. (2003). Autonomy in moral and political philosophy.

Cychosz, M., Romeo, R., Soderstrom, M. et al. (2020). Longform recordings of everyday life: ethics for best practices. *Behavior Research Methods* 52: 1951–1969.

de la Fuente Garcia, S., Ritchie, C.W., and Luz, S. (2020). Artificial intelligence, speech, and language processing approaches to monitoring Alzheimer's disease: a systematic review. *Journal of Alzheimer's Disease* 78 (4): 1547–1574.

Eltengy, A.H. (2021, July). Encryption of voice calls using CryptoBin algorithm. In: *2021 International Telecommunications Conference (ITC-Egypt)*, 1–5. IEEE.

Farsana, F.J., Devi, V.R., and Gopakumar, K. (2023). An audio encryption scheme based on fast Walsh Hadamard transform and mixed chaotic keystreams. *Applied Computing and Informatics* 19 (3/4): 239–264.

Fiske, A.P. (1993). Social errors in four cultures: evidence about universal forms of social relations. *Journal of Cross-Cultural Psychology* 24 (4): 463–494.

Foley, J.M. (1992). The experience of being demented. In: *Dementia and Aging: Ethics, Values, and Policy Choices*, 30–43.

Fukuda, M., Nishimura, R., Umezawa, M. et al. (2022). Elderly conversational speech corpus with cognitive impairment test and pilot dementia detection experiment using acoustic characteristics of speech in Japanese dialects. In: *Proceedings of the Thirteenth Language Resources and Evaluation Conference*, 1016–1022.

Garcia, M. (2016). Racist in the Machine. *World Policy Journal* 33 (4): 111–117.

Garrard, P., Nemes, V., Nikolic, D. et al. (2017). Motif discovery in speech: application to monitoring Alzheimer's disease. *Current Alzheimer Research* 14 (9): 951–959.

Gauthier, S., Leuzy, A., Racine, E. et al. (2013). Diagnosis and management of Alzheimer's disease: past, present and future ethical issues. *Progress in Neurobiology* 110: 102–113.

Gebru, T., Morgenstern, J., Vecchione, B. et al. (2021). Datasheets for datasets. *Communications of the ACM* 64 (12): 86–92.

Gosztolya, G., Tóth, L., Grósz, T. et al. (2016). Detecting mild cognitive impairment from spontaneous speech by correlation-based phonetic feature selection.

Harrison, C.J. and Sidey-Gibbons, C.J. (2021). Machine learning in medicine: a practical introduction to natural language processing. *BMC Medical Research Methodology* 21 (1): 158.

Hasan, S.A. and Farri, O. (2019). Clinical natural language processing with deep learning. In: *Data Science for Healthcare: Methodologies and Applications*, 147–171.

Henrich, J., Heine, S.J., and Norenzayan, A. (2010). The weirdest people in the world? *Behavioral and Brain Sciences* 33 (2–3): 61–83.

Illes, J., Rosen, A., Greicius, M. et al. (2007). Prospects for prediction: ethics analysis of neuroimaging in Alzheimer's disease. *Annals of the New York Academy of Sciences* 1097 (1): 278–295.

Jiao, Y., Berisha, V., and Liss, J. (2017). Interpretable phonological features for clinical applications. In: *2017 IEEE International Conference on Acoustics, Speech and Signal Processing (ICASSP)*, 5045–5049. IEEE.

Jurgens, D., Tsvetkov, Y., and Jurafsky, D. (2017). Incorporating dialectal variability for socially equitable language identification. In: *Proceedings of the 55th Annual Meeting of the Association for Computational Linguistics* (Volume 2: *Short Papers)*, 51–57.

Kaduszkiewicz, H., Bachmann, C., and van den Bussche, H. (2008). Telling "the truth" in dementia—Do attitude and approach of general practitioners and specialists differ? *Patient Education and Counseling* 70 (2): 220–226.

Karlawish, J. (2011). Addressing the ethical, policy, and social challenges of preclinical Alzheimer disease. *Neurology* 77 (15): 1487–1493.

Keeling, G. and Nyrup, R. (2022). Explainable machine learning, patient autonomy, and clinical reasoning. In: *Oxford Handbook of Digital Ethics*, 528–552. Oxford University Press.

Kitwood, T. (1990). The dialectics of dementia: with particular reference to Alzheimer's disease. *Ageing & Society* 10 (2): 177–196.

Klasnja, P., Consolvo, S., Choudhury, T. et al. (2009). Exploring privacy concerns about personal sensing. In: *International Conference on Pervasive Computing*, 176–183. Berlin, Heidelberg: Springer Berlin Heidelberg.

Kumar, S., Nguyen, L.T., Zeng, M. et al. (2015). Sound shredding: privacy preserved audio sensing. In: *Proceedings of the 16th International Workshop on Mobile Computing Systems and Applications*, 135–140.

Lane, N., Mohammod, M., Lin, M. et al. (2011). Bewell: a smartphone application to monitor, model and promote wellbeing. In: *5th International ICST Conference on Pervasive Computing Technologies for Healthcare*, 23–26.

Larson, E.C., Lee, T., Liu, S. et al. (2011). Accurate and privacy preserving cough sensing using a low-cost microphone. In: *Proceedings of the 13th International Conference on Ubiquitous Computing*, 375–384.

Le Franc, A., Riebling, E., Karadayi, J. et al. (2018). The ACLEW DiViMe: an easy-to-use diarization tool. In: *Interspeech*, 1383–1387.

Le Glaz, A., Haralambous, Y., Kim-Dufor, D.H. et al. (2021). Machine learning and natural language processing in mental health: systematic review. *Journal of Medical Internet Research* 23 (5): e15708.

Le, X., Lancashire, I., Hirst, G. et al. (2011). Longitudinal detection of dementia through lexical and syntactic changes in writing: a case study of three British novelists. *Literary and Linguistic Computing* 26 (4): 435–461.

Lipowski, Z.J. (1969). Psychosocial aspects of disease. *Annals of Internal Medicine* 71 (6): 1197–1206.

Lopez, J.A.S., González, H.M., and Léger, G.C. (2019). Alzheimer's disease. *Handbook of Clinical Neurology* 167: 231–255.

Lopez-de-Ipina, K., de-Lizarduy, M.U., Calvo, P.M. et al. (2018). Advances on automatic speech analysis for early detection of Alzheimer disease: a non-linear multi-task approach. *Current Alzheimer Research* 15 (2): 139–148.

Low, D.M., Bentley, K.H., and Ghosh, S.S. (2020). Automated assessment of psychiatric disorders using speech: a systematic review. *Laryngoscope Investigative Otolaryngology* 5 (1): 96–116.

Luz, S., Haider, F., de la Fuente, S. et al. (2021a). Detecting cognitive decline using speech only: the ADReSSo challenge. *arXiv preprint arXiv:2104.09356*.

Maley, J.H., Wanis, K.N., Young, J.G. et al. (2020). Mortality prediction models, causal effects, and end-of-life decision making in the intensive care unit. *BMJ Health & Care Informatics* 27 (3): e100220.

Marson, D.C. (2001). Loss of competency in Alzheimer's disease: conceptual and psychometric approaches. *International Journal of Law and Psychiatry* 24 (2–3): 267–283.

Masuda, T. and Nisbett, R.E. (2001). Attending holistically versus analytically: comparing the context sensitivity of Japanese and Americans. *Journal of Personality and Social Psychology* 81 (5): 922.

Mattsson, N., Brax, D., and Zetterberg, H. (2010). To know or not to know: ethical issues related to early diagnosis of Alzheimer's disease. *International Journal of Alzheimer's Disease* 2010: 841941.

Mayeda, E.R., Glymour, M.M., Quesenberry, C.P. et al. (2016). Inequalities in dementia incidence between six racial and ethnic groups over 14 years. *Alzheimer's & Dementia* 12 (3): 216–224.

McKernan, L.C. and Clayton, E.W. (2018). Protecting life while preserving liberty: ethical recommendations for suicide prevention with artificial intelligence. *Frontiers in Psychiatry* 9: 424191.

Mehl, M.R. (2017). The electronically activated recorder (EAR) a method for the naturalistic observation of daily social behavior. *Current Directions in Psychological Science* 26 (2): 184–190.

Mehl, M.R. and Pennebaker, J.W. (2003). The sounds of social life: a psychometric analysis of students' daily social environments and natural conversations. *Journal of Personality and Social Psychology* 84 (4): 857.

Meltzer, J.A. (2020). Towards early prediction of Alzheimer's disease through language samples. *EClinicalMedicine* 29: 100644.

Metarugcheep, S., Punyabukkana, P., Wanvarie, D. et al. (2022). Selecting the most important features for predicting mild cognitive impairment from Thai verbal fluency assessments. *Sensors* 22 (15): 5813.

Meyer, M.N. (2018). Practical tips for ethical data sharing. *Advances in Methods and Practices in Psychological Science* 1 (1): 131–144.

Mohammad, S.M. (2022). Ethics sheet for automatic emotion recognition and sentiment analysis. *Computational Linguistics* 48 (2): 239–278.

Müller, V.C. (2020). Ethics of artificial intelligence and robotics.

Munafò, M.R., Nosek, B.A., Bishop, D.V. et al. (2017). A manifesto for reproducible science. *Nature Human Behaviour* 1 (1): 1–9.

Munthuli, A., Vongsurakrai, S., Anansiripinyo, T. et al. (2021). Thammasat-NECTEC-Chula's Thai language and cognition assessment (TLCA): the Thai Alzheimer's and mild cognitive impairment screening test. In: *2021 43rd Annual International Conference of the IEEE Engineering in Medicine & Biology Society (EMBC)*, 690–694. IEEE.

Nickel, P.J. (2006). Vulnerable populations in research: the case of the seriously ill. *Theoretical Medicine and Bioethics* 27: 245–264.

O'Neill, O. (2002). *Autonomy and Trust in Bioethics*. Cambridge University Press.

Petti, U., Baker, S., and Korhonen, A. (2020). A systematic literature review of automatic Alzheimer's disease detection from speech and language. *Journal of the American Medical Informatics Association* 27 (11): 1784–1797.

Petti, U., Baker, S., Korhonen, A. et al. (2023). The generalizability of longitudinal changes in speech before Alzheimer's disease diagnosis. *Journal of Alzheimer's Disease* 92 (2): 547–564.

Porteri, C., Albanese, E., Scerri, C. et al. (2017). The biomarker-based diagnosis of Alzheimer's disease. 1—ethical and societal issues. *Neurobiology of Aging* 52: 132–140.

Porteri, C., Galluzzi, S., Geroldi, C. et al. (2010). Diagnosis disclosure of prodromal Alzheimer disease-ethical analysis of two cases. *Canadian Journal of Neurological Sciences* 37 (1): 67–75.

Robin, J., Xu, M., Kaufman, L.D. et al. (2021). Using digital speech assessments to detect early signs of cognitive impairment. *Frontiers in Digital Health* 3: 749758.

Rogers, W. and Lange, M.M. (2013). Rethinking the vulnerability of minority populations in research. *American Journal of Public Health* 103 (12): 2141–2146.

Sabat, S.R. and Harré, R. (1994). The Alzheimer's disease sufferer as a semiotic subject. *Philosophy, Psychiatry, & Psychology* 1 (3): 145–160.

Sangchocanonta, S., Vongsurakrai, S., Sroykhumpa, K. et al. (2021). Development of Thai picture description task for Alzheimer's screening using part-of-speech tagging. In: *2021 43rd Annual International Conference of the IEEE Engineering in Medicine & Biology Society (EMBC)*, 2104–2109. IEEE.

Schicktanz, S., Schweda, M., Ballenger, J.F. et al. (2014). Before it is too late: professional responsibilities in late-onset Alzheimer's research and pre-symptomatic prediction. *Frontiers in Human Neuroscience* 8: 921.

Smith, A.P. and Beattie, B.L. (2001). Disclosing a diagnosis of Alzheimer's disease: patient and family experiences. *Canadian Journal of Neurological Sciences* 28 (S1): S67–S71.

Tjoa, E. and Guan, C. (2020). A survey on explainable artificial intelligence (xai): Toward medical xai. *IEEE Transactions on Neural Networks and Learning Systems* 32 (11): 4793–4813.

Tu, M., Berisha, V., and Liss, J. (2017). Interpretable objective assessment of dysarthric speech based on deep neural networks. In: *Interspeech*, 1849–1853.

Ujiro, T., Tanaka, H., Adachi, H. et al. (2018). Detection of dementia from responses to atypical questions asked by embodied conversational agents. In: *Interspeech* 1691–1695.

Ursin, F., Timmermann, C., and Steger, F. (2021). Ethical implications of Alzheimer's disease prediction in asymptomatic individuals through artificial intelligence. *Diagnostics* 11 (3): 440.

Véliz, C. (2019). Three things digital ethics can learn from medical ethics. *Nature Electronics* 2 (8); 316–318.

Velupillai, S., Suominen, H., Liakata, M. et al. (2018). Using clinical natural language processing for health outcomes research: overview and actionable suggestions for future advances. *Journal of Biomedical Informatics* 88: 11–19.

Voleti, R., Liss, J.M., and Berisha, V. (2019). A review of automated speech and language features for assessment of cognitive and thought disorders. *IEEE Journal of Selected Topics in Signal Processing* 14 (2): 282–298.

Wachter, S., Mittelstadt, B., and Floridi, L. (2017). Why a right to explanation of automated decision-making does not exist in the general data protection regulation. *International Data Privacy Law* 7 (2): 76–99.

Whiting, P.F., Rutjes, A.W., Westwood, M.E. et al. (2011). QUADAS-2: a revised tool for the quality assessment of diagnostic accuracy studies. *Annals of Internal Medicine* 155 (8): 529–536.

Wolff, R.F., Moons, K.G., Riley, R.D. et al. (2019). PROBAST: a tool to assess the risk of bias and applicability of prediction model studies. *Annals of Internal Medicine* 170 (1): 51–58.

Wu, S., Roberts, K., Datta, S. et al. (2020). Deep learning in clinical natural language processing: a methodical review. *Journal of the American Medical Informatics Association* 27 (3): 457–470.

Wyatt, D., Choudhury, T., and Bilmes, J.A. (2007). Conversation detection and speaker segmentation in privacy-sensitive situated speech data. In: *Interspeech*, 586–589.

Yamada, Y., Shinkawa, K., Kobayashi, M. et al. (2021). Tablet-based automatic assessment for early detection of Alzheimer's disease using speech responses to daily life questions. *Frontiers in Digital Health* 3: 653904.

11

Privacy-preserving AI Methods to Protect Patient Privacy

Lewis Hotchkiss

Medical School, Swansea University, Swansea, UK

11.1 Introduction

Artificial intelligence (AI) models often require vast amounts of training data to learn to make predictions. In neuroscience applications, this training is driven by patient data, which can contain sensitive and confidential information relating to individuals. Not only that, these models are often developed on restricted datasets that cannot be shared publicly. Therefore, any leakage of data could result in significant damage to the patients whose data has been used to train those AI models.

Recent advances of AI in neuroscience have predominantly been due to the explosion of big data collections, which have enabled such research to take place. This includes the use of wearable data to monitor and detect early signs of neurodegenerative diseases (Schalkamp et al., 2023), neuroimaging collections to model disease trajectories and aid in diagnosis (Burkhart et al., 2024), genomics to investigate the risk of developing diseases (Zhou et al., 2023) and AI-based home monitoring systems for people living with dementia (Tegou et al., 2019). However, all of these AI systems use highly sensitive data collected about individuals, which can be highly revealing. This is why these data collections are often stored in secure environments with strict access control restrictions. To enable the development of these AI solutions, this data needs to be shared securely to allow the research to take place. This has led to the rise of Trusted Research Environments (TREs) (Hubbard et al., 2020) and the NHS Secure Data Environments (SDEs) (Department of Health and Social Care, 2022), which allow access to pseudonymised data collections in a secure, controlled environment.

Although these environments have enabled the sharing of data, they have strict governance controls, which create challenges with the sharing and deployment of AI models that are trained on these datasets within secure environments. This is due to privacy risks and vulnerabilities that may allow adversaries to infer or reconstruct sensitive information about the individuals used to train those models. This presents unique disclosure control risks and challenges around data protection and anonymisation. Therefore, AI models that need to be shared or deployed from secure environments must sufficiently demonstrate that privacy is protected.

This chapter will explore the complexities around pseudonymised data and re-identification risk, how this creates privacy risks and vulnerabilities in AI models, and how they can be appropriately mitigated through privacy-preserving techniques and be deployed securely.

11.2 Anonymisation

Data owners typically de-identify data by stripping out personally identifiable information such as names, addresses and dates of birth so that individuals are no longer directly identifiable. However, the concept of 'identifiable' is challenging to define properly, as there are multiple ways that an individual can be identified, either through direct identification (where they can be identified from a single piece of information) or indirect identification (where they can be identified by combining multiple pieces of information) (Information Commissioner's Office, 2012).

Health data collected about individuals often includes three categories of variables: identifiers, quasi-identifiers and sensitive attributes. Identifiers are information that explicitly identifies a person, such as their name, phone number or biometric data, whereas quasi-identifiers refer to information that, when combined, can identify a person, for example, a combination of their date of birth, gender, employment status and city. Both direct and indirect identifying information are considered personal data under the Data Protection Act (GOV.UK, 2018). Sensitive attributes refer to information that individuals may not want to be disclosed, such as a medical condition. When these datasets are shared for research purposes, they typically undergo pseudonymisation processes that aim to remove direct identifiers; however, quasi-identifiers and sensitive attributes often remain.

Rocher et al. (2019) demonstrated the re-identification risk of individuals through the combination of attributes in de-identified and sampled data. Their study introduced a model to estimate the probability of correctly re-identifying an individual within a de-identified dataset given 15 demographic attributes. An example is demonstrated in Table 11.1, indicating that the likelihood of re-identification increases substantially as more attributes are included. When all of these attributes are included, there is a 98% probability of re-identification, but with the removal of certain attributes, such as postcode district or day of birth, this noticeably reduces that risk. However, Table 11.2 demonstrates what happens when the marital status attribute changes from 'never married' to 'same-sex partnership'. Just changing this single attribute significantly increases the risk of re-identification, even with the removal of certain attributes. This highlights that while individuals with common demographic profiles may face relatively low risk, members of minority or underrepresented groups are disproportionately vulnerable to re-identification.

The risk becomes even more prominent in large-scale health and neuroscience research, where cohort datasets often contain extensive information about individuals, including lifestyle, diet, medical and family history, and sociodemographic data. Culnane et al. (2017) further demonstrated this vulnerability by successfully re-identifying patients

Table 11.1 Risk of identification – never married.

Gender	DoB Year	DoB Month	DoB Day	Marital status	Postcode district	Risk
Male	1962	May	28	Never married	SA1	
✓	✓	✓	✓	✓	✓	98%
✓	✓	✓	✗	✓	✓	53%
✗	✓	✓	✗	✓	✓	33%
✓	✓	✗	✗	✓	✓	6%
✗	✓	✗	✗	✓	✓	3%
✓	✓	✓	✓	✓	✗	2%
✗	✗	✗	✗	✓	✗	1%

Table 11.2 Risk of identification – same-sex partnership.

Gender	DoB Year	DoB Month	DoB Day	Marital status	Postcode district	Risk
Male	1962	May	28	Same-sex partnership	SA1	
✓	✓	✓	✓	✓	✓	100%
✓	✓	✓	✗	✓	✓	99%
✗	✓	✓	✗	✓	✓	99%
✓	✓	✗	✗	✓	✓	93%
✗	✓	✗	✗	✓	✓	87%
✓	✓	✓	✓	✓	✗	82%
✗	✗	✗	✗	✓	✗	1%

from an Australian de-identified health dataset using only publicly available information. Collectively, these studies highlight that conventional anonymisation methods are frequently insufficient to guarantee privacy protection.

There are typically three main indicators for determining whether data is identifiable or not: singling out, linkability and inference.

11.2.1 Singling Out

Singling out occurs when an individual's record can be distinguished from all others within a dataset. In these cases, a person's data can be isolated as unique through distinctive attributes, enabling that individual to potentially be re-identified. For example, if only one individual has a weight of 90 kg and a height of 161 cm. Although there may be other individuals who share either of those attributes, if the combination is unique, then that individual can be distinguished from the rest. As the amount of information available about an individual increases, so too does the likelihood of unique combinations that could be used to single participants out.

However, if data comes from a subset of the population, for example cohort data, it is often argued that these types of incomplete datasets provide plausible deniability – that even if a record appears to match a known individual, you cannot be certain that it belongs to them. Since these datasets do not contain information on everyone, it is possible that a supposed match could refer to someone outside of the cohort. Many organisations and agencies release sampled versions of data, such as only releasing 1% of census data, assuming that incompleteness offers stronger privacy protection. However, Rocher et al. (2019) demonstrated that individuals could still be singled out within these subsets. While plausible deniability introduces uncertainty, the ability to isolate a unique record and associate it with external information still constitutes a potential privacy breach. As the number of attributes available about an individual increases, so too does the likelihood of generating unique combinations that can single participants out.

11.2.2 Linkability

Linkability refers to combining multiple records about an individual, either from the same dataset or multiple different datasets, to identify an individual. This means that one dataset may not be identifiable on its own, but if it is combined with an external dataset or with other information, then the individuals within that data could be identified. For example, if you have data such as age ranges, gender and medical condition (e.g., 'male, aged 45–50, admitted on 25th of October with a leg fracture'), on its own it does not identify anyone. However, external information such as news articles could be used to infer the patient's identity. People like politicians and celebrities may be at higher risk of linkability due to the greater amount of information existing about them in the public domain. However, it's not just information in the public domain that can be used to link and infer. If we combine datasets such as health data with census data or administrative data, then the two datasets on their own might not be identifiable, but once you combine them, you increase the risk of singling out or inference. This has been demonstrated several times where non-identifying information can be combined with data in the public domain to re-identify individuals from seemingly anonymised datasets, such as in the study by Culnane et al. (2017). This also means that data containing any records that could be linked to other information that enables identification are pseudonymised, rather than anonymised.

11.2.3 Inference

Inference refers to the ability to infer, predict or deduce details about individuals. For example, family, friends, doctors or colleagues could use information that they know about an individual to deduce the identity of them in a dataset. Publicly available data or external datasets, such as census data, could additionally be used to infer the identities of individuals by putting together pieces of information that are not directly linked. Inference-based attacks are particularly relevant in the context of AI models trained on sensitive data. Machine learning models can inadvertently encode patterns that reveal private information, and adversaries may exploit these patterns to infer details about individuals included in the training dataset.

Because most health datasets are not fully anonymised and remain susceptible to singling out, linkability and inference, it is critical to adopt a 'worst-case scenario' approach when assessing privacy risk. This entails assuming that an adversary may have access to auxiliary information and may attempt to exploit AI models to re-identify individuals. This approach is essential to ensure compliance with data protection regulations, including GDPR and other privacy frameworks, and to maintain ethical standards in research involving sensitive human data.

11.2.4 Motivated Intruder Test

One way to assess the risk of re-identification is to determine whether an adversary (intruder) would be able to re-identify individuals if they were motivated to do so. In this scenario, a motivated intruder is someone who has no prior knowledge but aims to identify an individual from a dataset and has the means (competency, resources and use of investigative techniques) to succeed. The intruder is therefore someone who is not an ordinary person but a determined person with a particular reason to identify an individual. The test should cover the type and volume of data, the possibility of someone attempting to identify an individual, the capabilities they may have and the information they may already know or have access to.

There are different types of potential attackers, and depending on the type of data being used, there will be different motivations behind why they would attempt to attack. For example, an attacker may want to gain health conditions or reveal information about politicians or public figures for the news. This identifiability risk is often highly contextual and depends on what information the attacker may know to be able to infer an anonymised individual. Minority groups or people with rare conditions are at a much higher risk of identifiability due to having less common attributes that can be exploited to reveal them. It is also important to note that the risk of re-identification can change as more information about individuals enters the public domain, or as the resources/techniques for attacks improve. Therefore, data that is considered safe now may not be in a few years.

11.2.5 Challenges with Complex Data Modalities

Complex data modalities such as genomics and neuroimaging pose distinct challenges for anonymisation and disclosure control, particularly in AI-based neuroscience research. These data types are inherently high-dimensional, deeply personal and often highly distinctive to each individual. Their richness makes them valuable for research but simultaneously increases their susceptibility to singling out, linkability and inference risks.

Genomic data has become central to biomedical and dementia research, supported by advances in sequencing technologies and reduced costs that enable large-scale data collection. However, genomic sequences are intrinsically identifiable; they are unique to each person, remain unchanged throughout life and are shared across all tissues. Even aggregate genomic information can be sufficient to identify an individual within a dataset, and this risk further increases when genomics data is linked with phenotypic or demographic data.

Genomics data is mainly susceptible to two main types of attacks: identification and phenotype inference attacks (Bonomi et al., 2020). For instance, phenotype inference attacks can use genomic information to predict visible or biological traits such as eye colour, height or disease risk. If an adversary already knows an individual's identity, they can use techniques like genotype imputation or linkage disequilibrium analysis to infer masked variants or familial predispositions. Because genomic information is shared within families, such inference not only compromises the individual's privacy but also that of their relatives.

The linkability of genomic data also presents a major privacy concern. Genomic datasets can be cross-referenced with external sources such as genealogical databases, biobanks or public DNA repositories, allowing individuals or their relatives to be traced across datasets. Well-known cases have shown that individuals have been re-identified through familial matching in public genealogy data.

Neuroimaging represents another complex modality frequently used in AI applications in neuroscience. Data sharing in this domain commonly involves defacing brain scans to remove identifiable facial features. However, depending on the defacing method, some features, such as the eyes or parts of the facial contour, may remain, enabling potential facial reconstruction (Schwarz et al., 2021). Moreover, researchers often require access to non-defaced scans to preserve analytical precision, which further elevates the risk of re-identification.

Even beyond facial features, neuroimaging data carries significant privacy challenges because of its individual-specific nature. The structural and functional organisation of the brain has been shown to be as distinctive as a fingerprint, meaning that a person's scan can be uniquely identified even within large cohorts. This means that neuroimaging data also enables linkability across datasets. If an adversary has access to one scan of an individual, they may be able to match it to that same individual in another dataset based on overlapping brain patterns. Studies have demonstrated that both structural MRI features and functional connectivity matrices are sufficiently stable and unique to identify individuals across sessions or datasets (Puglisi et al., 2023; Venkatesh et al., 2020).

In both genomics and neuroimaging, the very features that make these data modalities valuable for scientific and AI-driven discovery also heighten privacy risks. Their ability to enable singling out of individuals, linking across datasets and inferring sensitive information fundamentally challenges traditional anonymisation approaches. Consequently, most datasets of this kind are more accurately described as pseudonymised rather than truly anonymised. To ensure privacy protection while maintaining scientific utility, advanced privacy-preserving methods are increasingly essential in neuroscience and health data research.

11.3 Privacy Concerns in AI Models

Because of the risks of re-identification in health datasets, there are privacy concerns in AI models, which may reveal the data used to train them either through attacks or vulnerabilities, meaning that data could potentially be exposed. Additionally, these risks are highly dependent on the scenarios in which they are shared or deployed due to factors such as accessibility.

11.3.1 Attack Models

The main privacy concerns in these AI models are privacy attacks that can infer or reconstruct the original data. These attack models can be categorised into three types: membership inference, attribute inference and reconstruction attacks, each having different goals (Table 11.3).

11.3.1.1 Attribute Inference Attack

An attribute inference attack (AIA) aims to infer unknown attributes of an individual given some prior knowledge about some of their other attributes (Fredrikson et al., 2015). This relies on an adversary already knowing some information about an individual to make inferences about the unknown data. For example, an AI model is developed to classify healthy vs. dementia patients, where it provides a confidence score of its classification. You might already know some information about an individual, such as their age, gender, whether they are a smoker, if they have diabetes, etc., which are variables in the training dataset. You can use this information to query the AI model with the variables you know, along with random values for the variables that you don't know, and observe the confidence score. It is a common feature of AI models that they perform better on samples they have seen before in their training set, and it is this feature that is exploited in these attacks. By testing different values for the variables you don't know, an adversary aims to get the highest possible confidence score. If they manage to get a high confidence score, then that likely means that those are the values of the individual used to train the model (Table 11.4).

Table 11.3 AI privacy attacks.

Privacy attack	Knowledge of data	Attack goal
Attribute inference	Partial	Retrieve unknown attributes about an individual, given the adversary already knows some attributes
Membership inference	Partial/entire	Determine if an individual is present in the training data
Reconstruction	None	Recover the entire data used to train the AI model

Table 11.4 Attribute inference attack example.

Age (years)	Gender	MMSE	Smoker	Diabetes	Depression	Confidence score
87	Female	25	No	Yes	No	57.4%
87	Female	22	No	Yes	Yes	62.3%
87	Female	19	No	Yes	Yes	98.7%

The success of an AIA is dependent on the type of data and the volume of data used to train the AI model. For example, in a training dataset of derived brain volumes, an adversary is unlikely to know those values, and being able to infer them is difficult due to the values being continuous. These attacks are much more likely in smaller datasets that contain discrete or categorical data about individuals, which could be known.

11.3.1.2 Membership Inference Attack

An adversary who may already know a full record of an individual could potentially conduct a membership inference attack (MIA) to determine whether the individual was included in the dataset used to train an AI model (Shokri et al., n.d.). This attack allows the adversary to infer sensitive information about the individual, but only if the dataset is composed of a specific group of people. For instance, if an AI model is trained to predict treatment responses of participants with depression, then you know that everyone in the training dataset has depression; therefore, by inferring that an individual was part of that dataset, you can infer that they have depression. However, if the model is trained on a diverse dataset representing the general population, inferring an individual's inclusion in the dataset does not provide any additional information about their personal characteristics or health conditions.

MIA works by training an attack model to infer whether a particular individual was part of the training set of the target model. First, a series of shadow models (either exact copies if the attacker has access to the model, or models that are comparable to the target model in terms of hyperparameters and architecture, etc.) are trained on a shadow dataset that is similar to the data used in the target model. These shadow models are trained to generate an attack dataset composed of the confidence scores given by the model and whether they were part of the training set or not. This attack dataset is then used to train an attack model to determine whether that confidence score was gained from a sample in the training set, given the confidence score for a given class (Figure 11.1).

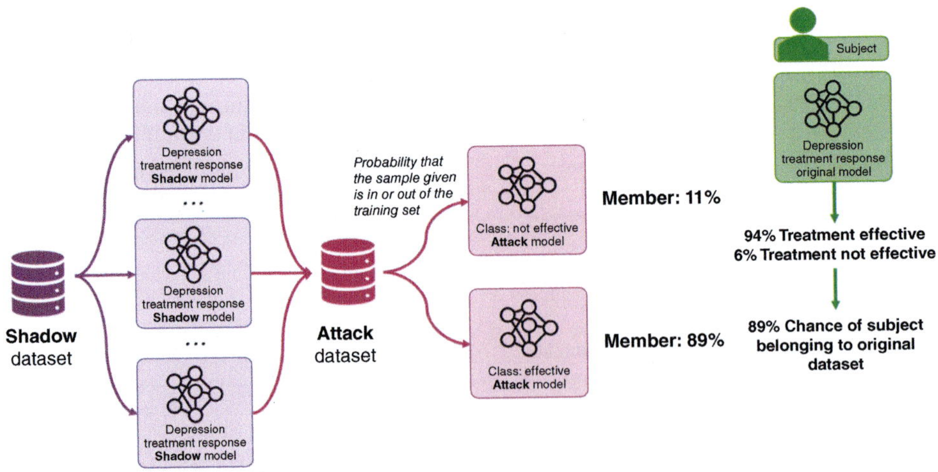

Figure 11.1 Membership inference attack.

11.3.1.3 Reconstruction/Inversion Attack

A reconstruction attack poses a significant threat to the privacy of individuals whose data is used to train a machine learning model. Unlike inference attacks, where the adversary may possess some information about the individuals to infer specific information about them, reconstruction attacks seek to retrieve entire records from the training data without requiring any prior information about them. This makes reconstruction attacks a particular concern, as they can potentially expose entire records of sensitive information.

Image data is particularly prone to reconstruction attacks and has been demonstrated several times (Fredrikson et al., 2015; Yoshimura et al., n.d.; Han et al., n.d.), even in federated learning contexts where model gradients are shared. A common example of this type of attack is demonstrated by exploiting a facial recognition system that provides a class confidence score along with its prediction. An adversary is able to input features into the model and observe the prediction with its confidence value. By exploiting this confidence score, like in inference attacks, an adversary is able to adjust their input to increase the confidence score until the input resembles an image used in the original training data. There are two approaches to this attack depending on whether the adversary has direct access to the model or not. If the adversary does not have direct access to the model but can query it and receive a prediction, then the attack performed on this model is known as a black-box attack. If the adversary does have direct access to the model, then they are able to perform white-box attacks, which are more efficient, as the adversary has access to more information about the model and can exploit it in other ways. A black-box attack is fairly easy to mitigate by simply rounding the confidence score, as demonstrated in Figure 11.2. However, if the adversary has direct access, then this cannot be achieved.

Additional methods for performing reconstruction attacks have been developed that don't rely on confidence scores. Zhu et al. (n.d.) developed a method called Deep Leakage from Gradients (DLG), which uses an optimisation algorithm to obtain training inputs (Figure 11.3). Typical reconstruction attacks often rely on generative models and confidence scores to generate inputs and attempt to obtain samples of the training set (Hitaj et al., 2017); however, since DLG is based on optimisation, it doesn't depend on these and can obtain more accurate reconstructions. DLG first randomly initialises random pixels with Gaussian noise for the image (Iters = 0 in Figure 11.3), then tries to match the gradients produced by the random pixel data to the real ones. This is achieved by minimising the distance between the fake gradients and the real gradients through optimisation until the images are recovered. Black-and-white images with clean backgrounds were easier to retrieve compared to more complex images such as faces, which took more iterations to

Figure 11.2 Effect of rounding the confidence score (r) on a black-box reconstruction attack. *Source*: Fredrikson et al. (2015)/with permission of Association for Computing Machinery (ACM).

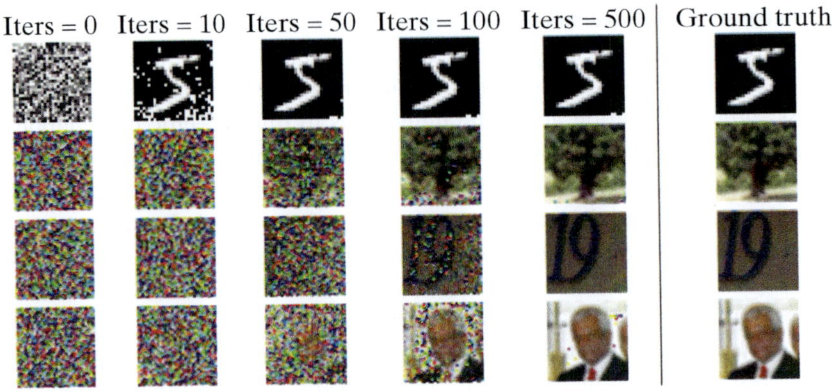

Figure 11.3 Reconstruction of images from model gradients. *Source*: Zhu et al. (n.d.)/arXiv/ CC BY 4.0.

recover. They were also able to apply the same technique to a language model task where discrete words are processed into embeddings, completely different from vision models where RGB inputs are continuous values. DLG was applied to the embedding space to minimise the gradients between the fake embeddings and the real ones, and they were able to successfully derive the original words. Although these attacks were very successful, they also acknowledged limitations and mitigations of these attacks by adjusting factors such as increasing the batch size, meaning more variables need to be solved during the optimisation, making it more difficult or upscaling the input images, as their technique currently only works on image resolutions of up to 64 × 64.

Similar attacks have been demonstrated in 3D neuroimaging data to reconstruct brain scans from encoder–decoder-based segmentation models by reverse-engineering the model's internal representations (Subbanna et al., 2021). In this case, an inversion decoder is trained to reconstruct scans from their latent space representations based on a separate dataset of scans and aims to minimise the difference between the original image and the reconstructed image. This essentially means that the inversion decoder inverts the encoder to reconstruct images from the training set that the adversary does not have access to. From this, they were able to successfully reconstruct scans, but contained some blurriness (Figure 11.4).

These attacks highlight the significant risks of disclosing sensitive data from AI models by reconstructing the original training data with no/minimal prior knowledge needed.

11.3.2 Vulnerabilities

In addition to the privacy attacks discussed earlier, AI models may have certain vulnerabilities that may lead to data disclosure without the need to attack or facilitate easier attacks for the adversary. These vulnerabilities may be due to the architecture of the model, how it was trained or additional measures added by the model developer to make predictions more interpretable. When developing AI models, these vulnerabilities have to be carefully considered in the context of privacy to ensure models aren't vulnerable to attacks or data disclosure.

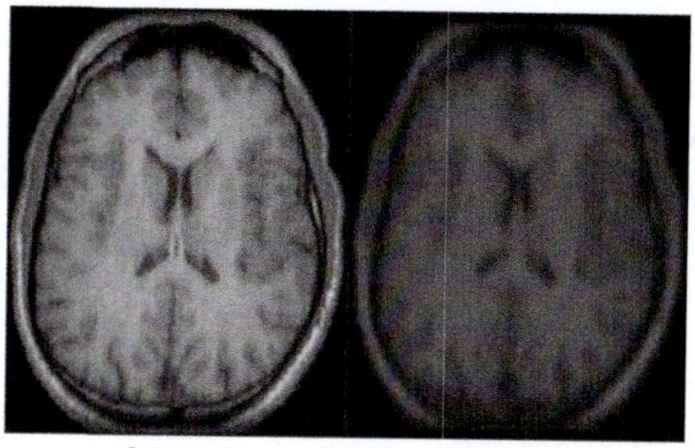

Figure 11.4 Reconstruction of an MRI scan. *Source*: Subbanna et al. (2021)/MDPI/CC BY 4.0.

11.3.2.1 Instance-based Models

When it comes to AI model types, instance-based models are most likely to disclose training data in a white-box setting without the need for attacks. This is because instance-based learning actually stores instances of the training data as the architecture of the model, so when a new data point is seen, it makes a prediction based on the proximity of other instances. This is contrary to model-based learning, which instead learns the relationships between input features and output labels. An example of an instance-based model is a k-nearest neighbours (KNN) algorithm, which makes predictions based on the distance of a given data point to other instances. Because of how these types of models represent the training data in their architecture, exact rows of the training data can be extracted from these models. In a support vector machine (SVM), this can be achieved simply by calling the method `svm.support_vectors_` to retrieve the stored data points. These models present a high risk to privacy, and because they store data, a transfer of this type of model would have to be treated as a transfer of data.

11.3.2.2 Overfitting

The reason many AI models are vulnerable to attacks is their susceptibility to memorise specific details about the training data rather than generalising patterns. This essentially means that the model may remember particular data points in the training set, making AI models more prone to attacks as they are more likely to reveal information about the original training data. This typically happens when there are many more features and too few samples to learn from, which leads to the model not being able to generalise well on new, unseen data points. This is a common problem in AI models, especially in the health domain, where there is a lack of access to large enough datasets to train AI models on. Additionally, there is a lack of real-world data for training AI models on health data, which means they are often developed on specific groups of people collected in cohort studies.

11.3.2.3 Hyperparameters

The architecture and learning process of an AI model can be controlled by adjusting hyperparameters, such as learning rate, batch size, number of layers, etc., to ensure optimal performance. These hyperparameter values have to be carefully chosen to avoid overfitting the training data. For example, in a decision tree model classifying patients as non-demented or demented, there are several hyperparameters that can be chosen to adjust the configuration of the model. If you set the maximum depth of the tree too high, then the tree will become too deep and start capturing more details of the training data, leading to overfitting. Similarly, increasing the maximum features allows the model to consider more features during splitting, which can also lead to overfitting. If some hyperparameters are set too low, then this can lead to small sample sizes at the leaves, causing the model to capture details about specific data points. Figure 11.5 shows an example of setting the `min_samples_leaf` hyperparameter to 1, which causes leaves to have very small sample sizes, some of which are below 5.

By increasing this hyperparameter to 6, you can make sure that there are no leaves with small sample sizes that could potentially disclose information about those individuals. This highlights the importance of choosing hyperparameters when configuring AI models to ensure that the model does not overfit.

11.3.2.4 Explainability

Explainability is an important part of making an AI model and its decisions interpretable, which is especially important in a clinical setting where the user needs to understand how the AI model arrived at its prediction. However, in black-box settings, explainability can make it easier to conduct attacks by revealing information about how the model works and by giving extra results to be exploited in attacks (Shokri et al., 2021), as well as potentially revealing information about the training set. There are two different types of explainability in AI models: (1) backpropagation-based and (2) perturbation-based. Backpropagation-based methods utilise the model's gradients to trace the model's decisions and identify which features contribute most to a prediction, such as saliency maps. Perturbation-based methods involve modifying the input and observing the effect on the model's predictions, such as Shapley values. Shokri et al. (2021) demonstrated that backpropagation-based methods of explainability can leak a significant amount of data about samples in the training dataset and improve the effectiveness of MIA; however, perturbation-based methods such as LIME were not considered vulnerable.

11.3.3 Risk Scenarios

AI models trained on sensitive healthcare data are often developed in secure environments where that data cannot leave. For instance, they might be developed on secure servers in a hospital or a TRE, where the data used to train the model cannot be shared outside those servers. This means the disclosure risks of AI models have to be carefully considered when sharing and releasing them so that data is not revealed when brought outside of those secure environments.

There are several reasons why an AI model may need to be shared outside of its original environment – that might be to implement it in real-world settings, to publish it alongside a paper or to share it as part of something like federated learning. These different scenarios pose different risks due to aspects such as accessibility, so depending on the scenario, different mitigations may need to be employed.

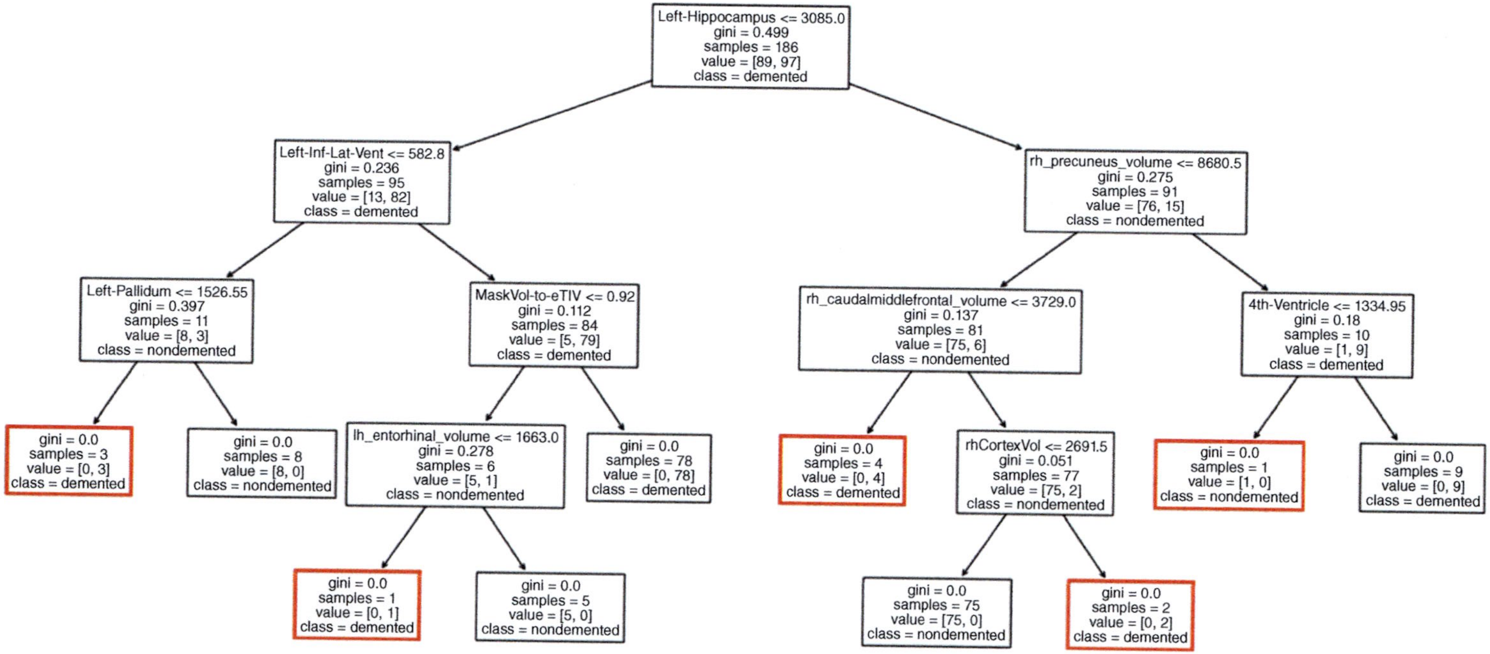

Figure 11.5 Decision tree with minimum samples leaf set to 1, red shows leaves with sample sizes under 5, which would be considered disclosive.

11.3.3.1 Publishing into Public Domain

One of the highest risk scenarios is releasing an AI model into the public domain on a platform like GitHub (GitHub, 2024a), where it is easily accessible for anyone to use and potentially attack. GitHub is a widely used platform for publishing code and AI models to enable open and reproducible science. Some journals or funders may even require researchers to make their results accessible, and it is encouraged in the community. However, this poses significant concerns for data privacy in AI models published in this way where there are no access controls or privacy protections. Once the model becomes publicly available, it becomes vulnerable to unauthorised usage, as anyone can access that model and potentially exploit it to gain sensitive patient information. This lack of control over the model and how it is used presents difficult challenges, and once it is released, it becomes virtually impossible to monitor or regulate its use. Additionally, it is encouraged that the AI model is well documented with details on how it was trained, what sort of data was used, how external users can query the model, etc., which provides adversaries with a wealth of information to use to run attacks. This poses the question of how much we let privacy affect open and reproducible science. This is already a big problem in AI research, so we should not be discouraging models from being published; we just need to do it in a safer way with appropriate safeguards.

11.3.3.2 Deploying into Clinical Practice

If an AI model is ready to be deployed into a real-world setting, such as in a memory clinic, then the privacy risks are reduced if there is controlled access and secure storage of the AI model. This is one of the lowest risk scenarios, as there are typically restrictions on who can access and query the model. However, the owners of the original data server where the model was developed need to be able to trust wherever that model is being sent to be deployed, as it becomes out of their control once released. Additionally, depending on who has access to query the model, there still exists the vulnerability to black-box attacks if a malicious user were able to have access. Another privacy consideration in this scenario is the security of the input data being sent to a deployed model on a separate server. If a user is sending data off to query a model, then there need to be assurances that the data being sent is protected or deleted after inference.

11.3.3.3 Environment Transfer and Federated Learning

In certain instances, AI models may need to be transferred from their initial training environment to a different one. This might occur for various reasons, such as training on additional datasets or enabling researchers to maintain control over their models for future applications. However, once the AI model exists outside its original environment, the data owners relinquish control over its usage and further distribution. This raises the possibility of the model later being publicly released or clinically implemented further down the line.

But there is a real need to train models on multiple data sources. AI models require large amounts of data to be able to effectively learn and make predictions; however, health data collections are often small and siloed, meaning that researchers have to combine datasets from different sources to be able to train their models. This need has led to solutions such as federated learning (FL) where an AI model can be developed across

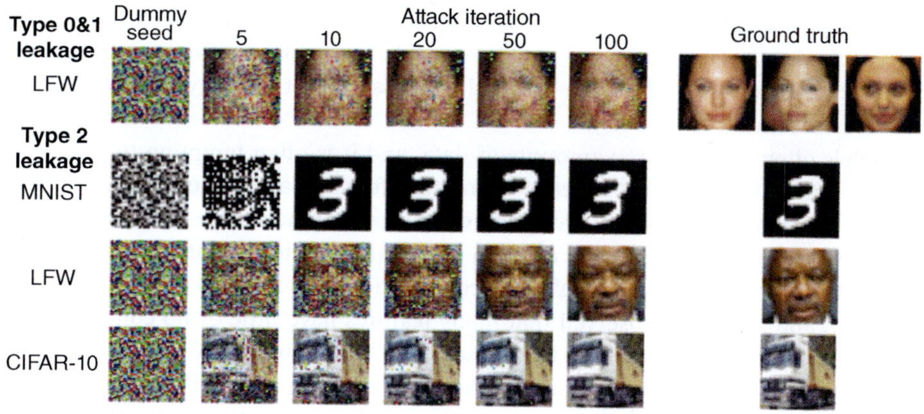

Figure 11.6 FL gradient-based attacks. *Source*: Wei et al. (2020)/with permission of Springer Nature.

multiple data sources without having direct access to them. There are typically two ways of doing this: centralised FL, where model updates from each server are sent to a central server where they are aggregated and sent back, and decentralised FL, where model updates are sent between all servers involved to individually aggregate, which means no central server is involved. In a centralised approach, you only have to trust the central server aggregating the model updates, whereas in a decentralised approach, you have to trust every participating server that is involved. The privacy risk of AI models in such a scenario is very much dependent on whether or not you trust the server(s). In FL, you are still sharing the model gradients and updates, so are still prone to the same attacks mentioned previously.

Wei et al. (2021) demonstrated gradient-based attacks in FL in three different leakage scenarios: type 0, where the attack is performed on intercepted gradient updates for each client; type 1, where the attack is performed on gradient updates resulting from the local training at a client; and type 2, where the attack is performed during the local training at a client on each training example. Figure 11.6 demonstrates example attacks across four different datasets.

11.4 Privacy-preserving Techniques

To mitigate these attacks, privacy-preserving techniques can be implemented to reduce the risk of re-identification and reconstruction of the training data. These mitigations can be implemented either at the training stage or the inference stage, depending on how the model is going to be released and used.

11.4.1 Training Stage

11.4.1.1 Data Minimisation

Data minimisation is a crucial aspect of the Data Protection Act (GOV.UK, 2018) practices and plays an important role in using personal data to train AI models. Data minimisation refers to the principle that requires you to only process information that is relevant, limited and necessary for the intended purpose. Therefore, before implementing any mitigations, it is important to consider what data is actually necessary for training your model to gain the desired outcome.

AI models often require large amounts of data, and it is often tempting to give an AI model the entire dataset to try to learn from, but often not all of the data is necessary to make predictions. Additionally, using more features than necessary not only increases disclosure risks from leaked data but also can potentially lead to overfitting. Plotting a feature importance graph simply highlights the contributions that features make to predicting an outcome, and it is often observed that many features have little or no contribution to the models' decisions. This is where feature selection techniques come in – to reduce the amount of features needed for making a prediction, which can help with the data minimisation principle. However, it is also important to consider, at the start of the project, what features you truly need and how they relate to the purpose.

11.4.1.2 Synthetic Data

The core principle behind synthetic data generation is to produce a new dataset that statistically resembles the original data while ensuring that no record corresponds directly to a real individual. By doing so, synthetic data seeks to maintain the utility of the data, the ability to support valid research and AI development, while guaranteeing privacy for individuals whose information was used in the model training process.

Synthetic datasets are typically produced using generative AI models, such as generative adversarial networks (GANs) or variational autoencoders (VAEs), which learn from the original data to generate new samples that mimic its patterns and distributions (Xu et al., 2019). In a GAN architecture, there are two neural networks: a generator and a discriminator. The generator's objective is to produce synthetic records that are indistinguishable from the real data, while the discriminator attempts to detect whether a given record is genuine or synthetic. Through this iterative process, both networks improve their performance, resulting in increasingly realistic synthetic data. Beyond machine learning approaches, other methods for generating synthetic data include statistical techniques, such as sampling from fitted distributions, and Bayesian networks, which model probabilistic dependencies between variables to produce new records that preserve relationships in the data. Regardless of the method, as the generated data becomes more faithful to the original, the risk of reproducing or approximating individual records rises (Figure 11.7).

Figure 11.7 Privacy vs. utility in synthetic data.

To mitigate this, additional privacy-preserving mechanisms are often applied during the generation process to carefully balance fidelity with the required privacy guarantees.

The creation of synthetic data inherently involves managing the trade-off between privacy and utility. Fidelity refers to the degree to which synthetic data preserves the structure and relationships within the original dataset. Low-fidelity data might replicate only the overall format and broad statistical distribution of the source data, failing to retain meaningful intervariable correlations and therefore limiting its analytical value. Conversely, high-fidelity data captures these relationships more accurately, providing greater utility for downstream analysis or AI model development. However, this increase in fidelity also heightens the potential risk of disclosure, as the synthetic data may inadvertently reflect characteristics of specific individuals from the source dataset.

It is important to note that this trade-off between privacy and utility is not always linear. Improvements in privacy protection do not necessarily lead to proportional losses in data utility, and vice versa. Depending on the complexity of the data and the model used, certain frameworks can maintain high analytical validity while still offering substantial privacy guarantees. In some cases, small reductions in fidelity can yield significant privacy benefits with minimal impact on model performance. Thus, finding the optimal balance between privacy and utility requires an empirical, data-specific approach rather than assuming a simple trade-off.

Evaluation of synthetic data quality and privacy typically considers three interrelated dimensions: fidelity, utility and privacy. Fidelity measures how closely the synthetic dataset resembles the real data, often through statistical comparisons such as distributional similarity, correlation preservation and boundary consistency. These metrics assess whether the generation process successfully captures key structures without overfitting to individual-level data. Utility assesses whether the synthetic data is sufficiently representative to support meaningful analytical outcomes. This can be evaluated by testing the performance of AI or statistical models trained on synthetic data compared to those trained on the original dataset. If performance remains comparable, the synthetic dataset is considered to have high utility. Privacy, however, is the most critical and complex dimension. It requires not only testing for direct duplication, or similarity, of records but also for broader re-identification risks of singling out, linkability and inference, which determine whether a dataset can be considered fully anonymised under frameworks such as the GDPR.

To evaluate these risks, privacy evaluations often employ tests such as nearest-neighbour analysis to detect overfitting and potential duplication, membership inference attacks to assess whether an individual's data influenced the generation process, and AIAs to evaluate if sensitive attributes can be predicted from the synthetic data.

11.4.1.3 Differential Privacy

Differential privacy (DP) is a framework designed to provide quantifiable privacy guarantees by ensuring that the inclusion or exclusion of any single individual's record in a dataset does not significantly affect the outcome of any analysis, thereby limiting the amount of information that can be inferred about any individual (Dwork & Roth, 2014). In other words, a differentially private algorithm ensures that the results do not depend too heavily on any single record, thereby protecting individuals' contributions within the dataset. This is achieved by introducing carefully calibrated random noise into queries or computations on the dataset, such that statistical results remain accurate at the population level while the privacy of individuals is protected. This means DP allows researchers to perform analysis, train machine learning models or generate synthetic datasets without exposing sensitive information about participants.

There are several ways to implement DP, and it can be applied at different stages of model training. The two main types of DP are local and global. In local DP, random noise is added directly to the raw data before any processing takes place. In global DP, the noise is introduced later in the process, typically during model training by adding it to the gradient updates or at the prediction or output stage to mask sensitive contributions.

The level of privacy protection is controlled by a parameter known as ε (epsilon), which quantifies the privacy loss: smaller values of ε correspond to stronger privacy guarantees, but may reduce the utility or accuracy of the analysis, while larger ε values provide less stringent privacy protection but higher-fidelity results. Determining an appropriate ε value is therefore critical but also highly challenging, as the optimal setting depends on the characteristics of the dataset, the task and the acceptable level of privacy risk.

While there is no universally accepted standard for setting ε, general guidance suggests that:

- ε values between 0 and 1 often provide strong privacy protection.
- ε values between 1 and 10 can offer protection in certain contexts.
- ε values greater than 10 are unlikely to provide substantial privacy guarantees.

DP can be implemented at three distinct stages: the data stage, the training stage or the prediction stage. Among these, the training stage is the most commonly targeted for implementation, as it directly governs how sensitive information influences model parameters. At this stage, two main approaches are widely used: Differentially Private Stochastic Gradient Descent (DP-SGD) and the Private Aggregation of Teacher Ensembles (PATE) framework.

DP-SGD is one of the more established approaches for training differentially private deep learning models (Ziller et al., 2021). It modifies the conventional stochastic gradient descent algorithm by introducing noise to the gradient updates during training. The process typically involves:

1. Randomly selecting a batch of samples from the training data
2. Computing and clipping the gradient for each sample to limit the influence of any individual record
3. Adding random noise to each clipped gradient
4. Updating the model weights using the average of the noisy gradients across the batch

This approach ensures that individual samples have a limited impact on model updates, thereby preserving privacy while still enabling effective learning (Figure 11.8). However, the introduction of noise inevitably affects model performance, requiring careful calibration of the privacy budget (ε) to balance accuracy and privacy guarantees.

An alternative method, known as PATE, uses a different approach (Papernot et al., 2018). Instead of perturbing gradients, PATE trains multiple non-private teacher models on disjoint subsets of the sensitive data. These teachers then 'vote' on the correct label for each query, and noise is added to the aggregated vote counts to ensure DP. The resulting noisy,

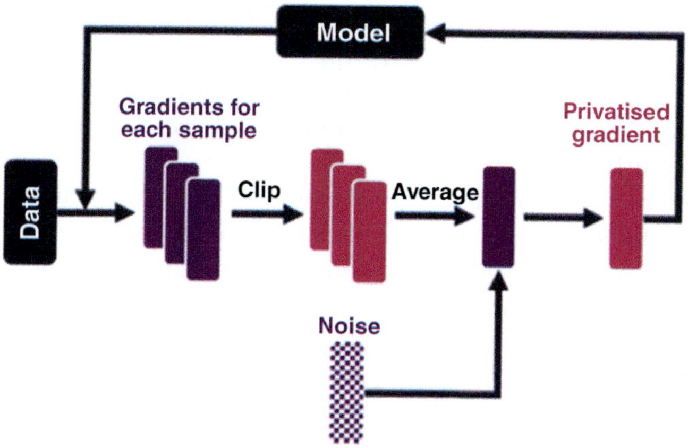

Figure 11.8 Differentially Private Stochastic Gradient Descent.

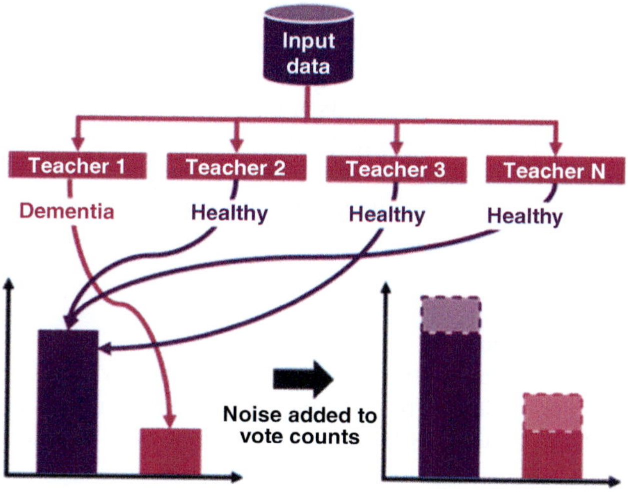

Figure 11.9 Private Aggregation of Teacher Ensembles.

aggregated labels are used to train a student model, which is the only model shared externally (Figure 11.9). This structure prevents the exposure of any individual teacher's predictions, which limits information leakage from the sensitive data. PATE generally maintains higher model accuracy compared to DP-SGD but relies on the availability of additional public data for training the student model. It primarily protects the privacy of the teachers' training data but does not extend this protection to the student model's training data.

Beyond deep learning, DP can also be applied to other machine learning methods. For example, in decision tree-based models, privacy can be introduced by adding randomisation to the decision thresholds or by generating split nodes with added noise (Li et al. 2019). These techniques ensure that the model structure and outputs remain differentially private, with varying impacts on performance.

Over recent years, DP methods have matured, with large-scale deployments by companies such as Google and Apple (Google, 2021; Apple Machine Learning Research, 2017). This has led to the availability of open-source software libraries supporting DP implementation in widely used AI frameworks such as TensorFlow Privacy (TensorFlow, 2019) and Opacus for PyTorch (GitHub, 2023b).

Most demonstrations and real-world applications of DP have focused on tabular data, where noise addition is relatively straightforward. However, applying DP to high-dimensional and complex data types, such as medical imaging or genomics, presents additional challenges. These data types are typically large and computationally intensive, making it more difficult to achieve meaningful privacy guarantees without severely impacting model performance. For such data, global DP is generally preferred, as local DP becomes less feasible due to the excessive noise required to protect individual records. In the genomics domain, although traditional local DP is inadequate, an adapted version known as (ε, T)-dependent local DP has been proposed to account for dependencies between genomic variants and repeated measurements (Yilmaz et al., 2022).

In the imaging domain, several studies have demonstrated the feasibility of DP-SGD for protecting patient privacy. For instance, one implementation using a PyTorch DP framework achieved competitive results in chest radiography classification and computed tomography (CT) segmentation tasks (www.ons.gov.uk. n.d.). In segmentation, the model's performance under DP was nearly identical to its non-private counterpart, whereas in classification, the area under the ROC curve (ROC-AUC) declined from 0.96 to 0.85. When the privacy guarantee was relaxed (higher ε), the score improved to 0.88. This example illustrates the privacy–accuracy trade-off inherent to DP and highlights the need for careful tuning of privacy parameters, as the impact of DP varies across datasets, model architectures and analytical objectives.

11.4.1.4 Homomorphic Encryption

Homomorphic encryption (HE) provides one of the strongest privacy protections currently available, as it enables computations to be performed directly on encrypted data. In theory, HE offers the ideal balance between privacy and utility; however, in practice, its application within AI remains limited due to computational complexity and implementation challenges.

HE uses a public–private key framework, in which a public key is used to encrypt data and a private key is required to decrypt the computed results. Unlike conventional encryption algorithms such as AES or RSA, which make data unusable for computation while encrypted, HE uses specialised cryptographic schemes that allow mathematical operations to be executed on ciphertexts. These operations produce encrypted outputs that, once decrypted, correspond to the same result as if the operations had been performed on the plaintext data. Depending on the complexity of computations supported, HE can be applied at different stages of the AI pipeline, including data preprocessing, model training or even for encrypting the model parameters themselves.

While fully HE (FHE) represents the most powerful and theoretically complete form of HE, it is also the most computationally expensive (Table 11.5). Current FHE implementations are still relatively slow, requiring extensive memory and processing power, which limits their scalability to large datasets or complex AI models. In addition, HE schemes can

Table 11.5 Types of HE.

Type of HE	Description
Fully HE	Allows an unlimited number of computations on encrypted data but is computationally intensive and resource demanding.
Somewhat HE	Supports a limited number of additions and multiplications.
Partially HE	Allows only one type of operation (either addition or multiplication). For example, the Paillier encryption scheme supports additive operations.

be challenging to implement, often requiring specialist cryptographic expertise and careful optimisation to manage numerical precision and computational overhead. Practical applications of HE in AI so far have primarily focused on relatively simple models, such as logistic regression (Chen et al., 2018) and small-scale neural networks (Vizitiu et al., 2020). These examples demonstrate that while HE offers strong privacy guarantees, the method is not yet mature enough for widespread deployment in complex deep learning tasks.

11.4.1.5 Secure Multi-party Computation

As discussed, methods for shared learning, such as FL, are still prone to attacks, as local gradients are transmitted between servers. Because of this, privacy-preserving techniques are recommended to be implemented in combination with FL solutions to protect the privacy of local data used to train a global model. For example, TensorFlow's federated framework (TensorFlow n.d.) integrates the Secure Aggregation protocol, which is a form of secure multi-party computation (SMPC).

SMPC enables multiple parties to jointly train an AI model using their respective private data, without ever exposing that data to one another or to a central coordinator. Each party contributes to the overall computation while keeping its input confidential. A common implementation technique is secret sharing, where each party divides its data into multiple encrypted 'shares' that are distributed among the other participants. The computation is then performed collaboratively on these shares rather than on the raw data itself. At the end of the process, only the final aggregated output, such as an updated model or statistical result, is revealed, ensuring that no participant learns the others' private inputs. However, both SMPC and HE approaches face similar challenges: they require significant computational resources, specialised expertise and careful protocol design to ensure efficiency and scalability. Another consideration is that if the number of shares or participating parties falls below a defined reconstruction threshold, or if a sufficient number of shares are compromised, it may become possible to partially or fully reconstruct the original input data.

11.4.2 Inference/Sharing Stage

11.4.2.1 Homomorphic Encryption

Given the computational and implementation challenges of applying HE during model training, this technique is more commonly employed at the inference or sharing stage of AI workflows. In the context of predictions, HE enables encrypted inference, where

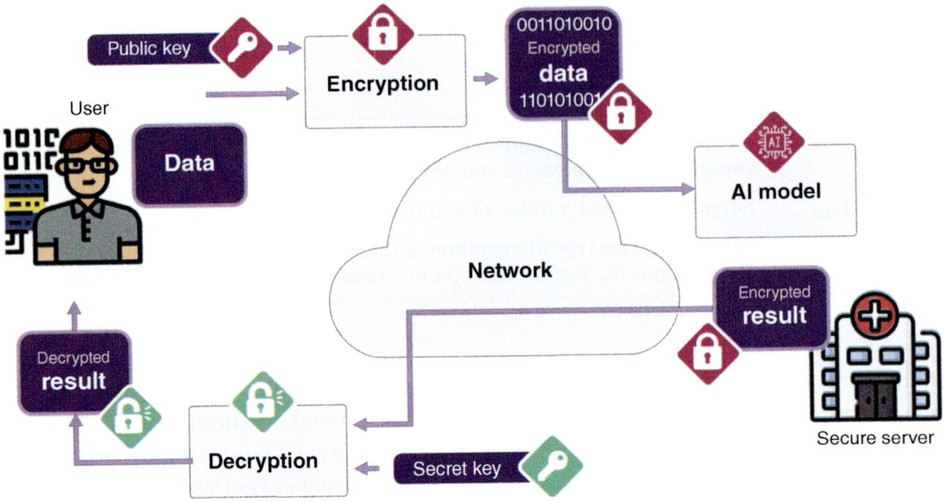

Figure 11.10 HE at inference.

encrypted input data can be sent to a model to obtain encrypted predictions, without ever revealing the input values themselves. The model performs all computations directly on the encrypted data, and only the data owner, who possesses the private decryption key, can access the final output in a readable form. This approach offers a significant privacy advantage: it allows individuals or institutions to query an AI model securely without disclosing their sensitive data (e.g., medical scans) to the model provider (Figure 11.10).

In this case, an AI model would be held in a secure environment but would be able to be queried externally to provide a prediction. A user sending data to the model in a secure environment may want to protect the data they are sending and can therefore encrypt it using HE. The AI model is able to take the encrypted data and provide an encrypted result, which is then sent back to the user, who can then decrypt the response. This provides a way to keep input data and inferences secure in a deployed model.

11.4.2.2 Secure Hosting/Access Controls

Once an AI model is ready for deployment, one approach to protecting both the model and sensitive data is to host the model within a secure environment or secure query server, with restricted access and controlled queries. In this setup, the model remains within a trusted server or platform and can only be queried through controlled channels, such as a web interface or an API. By imposing access restrictions and query limitations, the system ensures that only approved users can interact with the model, thereby reducing the likelihood of malicious attacks. Even if an adversary were able to issue queries, they would be limited to black-box attacks, as they do not have direct access to the model's parameters or internal architecture. This restriction significantly constrains the adversary's ability to perform model inversion, membership inference or other attacks that rely on full model visibility.

AI models can be hosted using serving frameworks such as TensorFlow Serving (GitHub, 2022), TorchServe (GitHub, 2024b) or MLflow (GitHub, 2023a). These frameworks allow models to be deployed for inference via an API, handling both prediction requests and the necessary pre-processing of input data (Figure 11.11). When deployed within a TRE

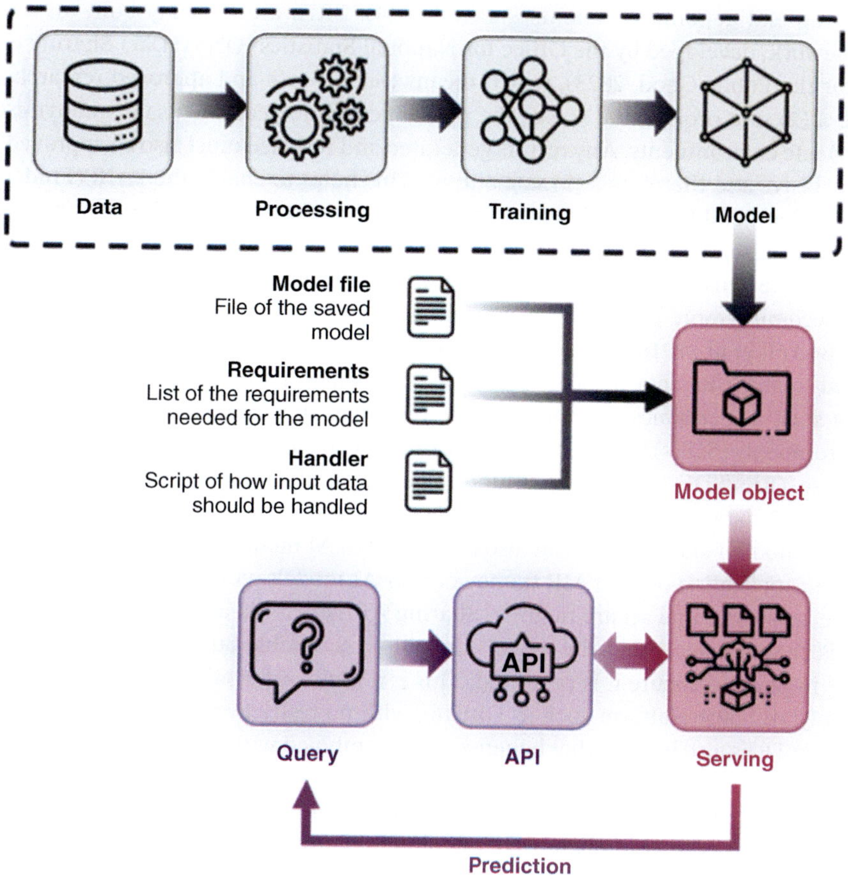

Figure 11.11 Model serving.

or other secure infrastructure, the server can include the saved model, supporting scripts and any dependencies required for execution. API-based access allows clients to send queries securely to the server, receive predictions, and ensures that only authorised users can interact with the model. Additional controls, such as limiting the number or frequency of queries, can be implemented to reduce the risk of inference or adversarial attacks.

For situations where external data cannot be directly shared with the server, secure hosting can be combined with encrypted inference using HE. In this configuration, users submit encrypted data to the hosted model, which performs computations on the encrypted inputs and returns an encrypted prediction. The user can then decrypt the result locally, ensuring that sensitive input data never leaves their environment.

11.4.2.3 Trusted Research Environments

TREs are a common way for data owners to share data for researchers to access via secure environments. The Dementias Platform UK Data Portal (Bauermeister et al., 2020) is one such TRE that provides access to over 100 different datasets related to ageing and neurodegenerative diseases and facilitates access to complex data modalities such as neuroimaging and genomics (Hotchkiss et al., 2024). TREs are typically built on the 5 Safes framework, developed by the Office for National Statistics (ONS) (Data Sharing and Linkage for the Public Good, 2023), which means that (1) safe and approved researchers undertake a (2) safe project that serves the public benefit to access (3) safe, anonymised data via (4) safe environments. Any results generated and released must also be approved to be non-disclosive and therefore a (5) safe output. This helps to enable the FAIR (Findable Accessible Interoperable Reusable) principles to ensure that data can be accessed and reused by other researchers (Wilkinson et al., 2016).

Because of the sensitivity and identifiability of healthcare datasets such as genomics, TREs are becoming more popular for accessing data used in training AI models. The UK BioBank (Bycroft et al., 2018), a popular health data resource in the United Kingdom, has also started to share its data through a TRE, as it provides assurances that complex sensitive data such as genomics is protected. The NHS has also recently established Secure Data Environments (SDEs) (Department of Health and Social Care, 2022) to share data for access. These TREs/SDEs are mainly behind the concern of sharing AI models developed on secure health data due to the strict governance and restrictions around them. However, this same method for sharing data can also be applied to AI models. Hotchkiss et al. (2024) proposed the application of the FAIR framework for AI models within TREs to ensure that models can be shared in a secure manner. Sharing AI models via a TRE, in the same way that data is shared, means that the model can still be accessible but with the appropriate safeguards in place to ensure it is protected. This is a much safer solution than publishing on an openly available platform such as GitHub, with no restrictions and security. Storing AI models via TREs also brings other benefits when combined with federated data solutions such as the DARE UK TELEPORT (Orton et al., 2023) or TRE-FX (Giles et al., 2023) which allow researchers to access data from multiple data environments in one place to fine-tune or validate AI models on. This means that researchers can access AI models within a secure environment, allowing them to test for reproducibility and fine-tune models on different data sources while protecting patient privacy. Even when AI models need to be deployed into real-world settings, this can be done by hosting the model via a TRE where it can be safely queried and inferences sent back. By doing it in this way, access controls and query restrictions can be imposed, therefore limiting the chances of attacks. Therefore, TREs can be used to provide suitable, secure alternatives to the risk scenarios identified in Section 3.3.

11.4.2.4 Agreements and Licences

Agreements and licences play an important role in the responsible release and governance of AI models, providing legal and procedural mechanisms to control their use. Similar to how software licences regulate distribution and usage, AI model licences can impose restrictions on utilisation, redistribution and access, specifying permitted purposes and prohibiting actions that could compromise the model or its underlying training data. These measures help safeguard privacy by discouraging attempts to reverse-engineer the model, extract sensitive information or perform adversarial attacks.

As an AI model moves from a controlled research environment to broader accessibility, adherence to licence terms becomes increasingly important. Clear licence agreements establish enforceable expectations for users, ensuring that the model is used only for approved applications and in accordance with privacy-preserving practices. In collaborative contexts, such as federated learning or environment transfer, agreements can define protocols for participation, outline responsibilities and explicitly prohibit attempts to access or compromise the data or other participants' models. By formalising these rules, agreements provide an additional layer of governance to complement technical privacy-preserving measures.

Despite their importance, enforcing licence agreements presents significant challenges, particularly in the context of international deployment. AI models are often shared across jurisdictions with varying legal frameworks, making compliance monitoring and enforcement complex. Moreover, the rapidly evolving landscape of AI privacy attacks means that licence terms must be continually updated to address new threats. Consequently, while agreements and licences are essential tools for governing AI model usage, they are most effective when integrated with technical safeguards to provide a comprehensive approach to privacy protection.

11.5 Conclusion

AI has the potential to revolutionise multiple aspects in diagnosis, prognosis and treatment of neurodegenerative diseases, but this cannot be achieved without effectively mitigating privacy risks to enable them to be used in real-world settings. Ensuring privacy in AI models for neuroscience is therefore crucial to enable open and reproducible science, as well as the responsible deployment of models into clinical practice. Privacy-preserving methods such as DP, secure federated learning and HE offer promising solutions for training AI models on sensitive data without compromising individual privacy. Additionally, secure deployment strategies such as trusted research environments and licence agreements play a crucial role in governing the use and distribution of AI models. By embracing these privacy-preserving techniques and mitigations, the full potential of AI in neuroscience can be unlocked while safeguarding the privacy of individuals.

References

Apple Machine Learning Research (2017). Learning with privacy at scale. [online] https://machinelearning.apple.com/research/learning-with-privacy-at-scale.

Bauermeister, S., Orton, C., Thompson, S. et al. (2020). The Dementias Platform UK (DPUK) data portal. *European Journal of Epidemiology* 35 (6): 601–611. https://doi.org/10.1007/s10654-020-00633-4.

Bonomi, L., Huang, Y., and Ohno-Machado, L. (2020). Privacy challenges and research opportunities for genomic data sharing. *Nature Genetics* 52 (7): 646–654. https://doi.org/10.1038/s41588-020-0651-0.

Burkhart, M.C., Lee, L.Y., Vaghari, D. et al. (2024). Unsupervised multimodal modeling of cognitive and brain health trajectories for early dementia prediction. *Scientific Reports* 14 (1): 10755. https://doi.org/10.1038/s41598-024-60914-w.

Bycroft, C., Freeman, C., Petkova, D. et al. (2018). The UK Biobank resource with deep phenotyping and genomic data. *Nature* 562 (7726): 203–209. https://doi.org/10.1038/s41586-018-0579-z.

Chen, H., Gilad-Bachrach, R., Han, K. et al. (2018). Logistic regression over encrypted data from fully homomorphic encryption. *BMC Medical Genomics* 11 (S4). https://doi.org/10.1186/s12920-018-0397-z.

Culnane, C., Rubinstein, B.I.P., and Teague, V. (2017). Health data in an open world. arXiv (Cornell University). https://doi.org/10.48550/arxiv.1712.05627.

Data Sharing and Linkage for the Public Good (2023). https://osr.statisticsauthority.gov.uk/wp-content/uploads/2023/07/202307_office_statistics_regulation_data_sharing_linkage_report.pdf.

Department of Health and Social Care (2022). Data saves lives: reshaping health and social care with data. [online] GOV.UK. https://www.gov.uk/government/publications/data-saves-lives-reshaping-health-and-social-care-with-data/data-saves-lives-reshaping-health-and-social-care-with-data.

Dwork, C. and Roth, A. (2014). The algorithmic foundations of differential privacy. *Foundations and Trends® in Theoretical Computer Science* 9 (3–4): 211–407. https://doi.org/10.1561/0400000042.

Fredrikson, M., Jha, S., and Ristenpart, T. (2015). Model inversion attacks that exploit confidence information and basic countermeasures. Proceedings of the 22nd ACM SIGSAC Conference on Computer and Communications Security - CCS '15 (1322–1333). [online] https://doi.org/10.1145/2810103.2813677.

Giles, D.T., Soiland-Reyes, S., Couldridge, J. et al. (2023). TRE-FX: delivering a federated network of trusted research environments to enable safe data analytics. [online] Zenodo. https://zenodo.org/records/10055354 (accessed 2 July 2024).

GitHub (2022). TensorFlow Serving. [online] https://github.com/tensorflow/serving.

GitHub (2023a). MLflow: a machine learning lifecycle platform. [online] https://github.com/mlflow/mlflow.

GitHub (2023b). pytorch/opacus. [online] https://github.com/pytorch/opacus.

GitHub (2024a). GitHub. [online] GitHub. https://github.com/.

GitHub (2024b). pytorch/serve. [online] https://github.com/pytorch/serve.

Google (2021). How we're helping developers with differential privacy. [online] https://developers.googleblog.com/2021/01/how-were-helping-developers-with-differential-privacy.html.

GOV.UK (2018). Data Protection Act 2018. [online] Legislation.gov.uk. https://www.legislation.gov.uk/ukpga/2018/12/contents/enacted.

Han, G., Choi, J., Lee, H. et al. (n.d.). Reinforcement Learning-Based Black-Box Model Inversion Attacks. [online] https://arxiv.org/pdf/2304.04625.

Hitaj, B., Ateniese, G., and Perez-Cruz, F. (2017). Deep models under the GAN: information leakage from collaborative deep learning. arXiv (Cornell University). https://doi.org/10.48550/arxiv.1702.07464.

Hotchkiss, L., Squires, E., Adeoye, K. et al. (2024). Perspectives & recommendations on the development of safe ai in sensitive healthcare data. [online] Zenodo. https://zenodo.org/records/11045640.

Hotchkiss, L., Squires, E., Gallacher, J.E. et al. (2024). Enabling advanced multi-modal neuroimaging analysis within a trusted research environment. medRxiv (Cold Spring Harbor Laboratory). https://doi.org/10.1101/2024.02.13.24302751.

Hubbard, T., Reilly, G., Varma, S. et al. (2020). Trusted Research Environments (TRE) green paper. [online] Zenodo. https://zenodo.org/records/4594704#.Yd8RRdHP1zq.

Information Commissioner's Office (2012). Anonymisation: managing data protection risk code of practice. [online] https://ico.org.uk/media/1061/anonymisation-code.pdf.

Li, Q., Wu, Z., Wen, Z. et al. (2019). Privacy-preserving gradient boosting decision trees. [online] arXiv.org. https://arxiv.org/abs/1911.04209 (accessed 18 September 2025).

Orton, C., Thompson, P.S., Lee, A. et al. (2023). TELEPORT: connecting researchers to big data at light speed. [online] Zenodo. https://zenodo.org/records/10055358 (accessed 2 July 2024).

Papernot, N., Song, S., Mironov, I. et al. (2018). Scalable private learning with PATE. [online] arXiv.org. https://doi.org/10.48550/arXiv.1802.08908.

Puglisi, L., Barkhof, F., Alexander, D.C. et al. (2023). DeepBrainPrint: a novel contrastive framework for brain MRI Re-identification. arXiv (Cornell University). https://doi.org/10.48550/arxiv.2302.13057.

Rocher, L., Hendrickx, J.M., and de Montjoye, Y.-A. (2019). Estimating the success of re-identifications in incomplete datasets using generative models. *Nature Communications* 10 (1): 3069. https://doi.org/10.1038/s41467-019-10933-3.

Schalkamp, A.-K., Peall, K.J., Harrison, N.A. et al. (2023). Wearable movement-tracking data identify Parkinson's disease years before clinical diagnosis. *Nature Medicine* 29 (8): 2048–2056. https://doi.org/10.1038/s41591-023-02440-2.

Schwarz, C.G., Kremers, W.K., Wiste, H.J. et al. (2021). Changing the face of neuroimaging research: comparing a new MRI de-facing technique with popular alternatives. *NeuroImage* 231: 117845. https://doi.org/10.1016/j.neuroimage.2021.117845.

Shokri, R., Strobel, M., and Zick, Y. (2021). On the privacy risks of model explanations. [online] arXiv.org. https://doi.org/10.48550/arXiv.1907.00164.

Shokri, R., Tech, C., Stronati, M. et al. (n.d.). Membership inference attacks against machine learning models. [online] https://www.cs.cornell.edu/~shmat/shmat_oak17.pdf (accessed 31 March 2023).

Subbanna, N., Wilms, M., Tuladhar, A. et al. (2021). An analysis of the vulnerability of two common deep learning-based medical image segmentation techniques to model inversion attacks. *Sensors* 21 (11): 3874. https://doi.org/10.3390/s21113874.

Tegou, T., Kalamaras, I., Tsipouras, M. et al. (2019). A low-cost indoor activity monitoring system for detecting frailty in older adults. *Sensors* 19 (3): 452. https://doi.org/10.3390/s19030452.

TensorFlow (2019). TensorFlow/privacy. [online] GitHub. https://github.com/tensorflow/privacy.

TensorFlow (n.d.). TensorFlow Federated. [online] https://www.tensorflow.org/federated.

Venkatesh, M., Jaja, J., and Pessoa, L. (2020). Comparing functional connectivity matrices: a geometry-aware approach applied to participant identification. *NeuroImage* 207: 116398. https://doi.org/10.1016/j.neuroimage.2019.116398.

Vizitiu, A., Niţă, C.I., Puiu, A. et al. (2020). Applying deep neural networks over homomorphic encrypted medical data. *Computational and Mathematical Methods in Medicine* 2020: 1–26. https://doi.org/10.1155/2020/3910250.

Wei, W., Liu, L., Wu, Y. et al. (2021). Gradient-leakage resilient federated learning. [online] arXiv.org. https://doi.org/10.48550/arXiv.2107.01154.

Wei, W., Liu, L., Margaret, L. et al. (2020). A framework for evaluating gradient leakage attacks in federated learning. [online] arXiv.org. https://doi.org/10.48550/arXiv.2004.10397.

Wilkinson, M.D., Dumontier, M., Aalbersberg, I.J.J. et al. (2016). The FAIR guiding principles for scientific data management and stewardship. *Scientific Data* 15 (3): 160018. https://www.nature.com/articles/sdata201618.

www.ons.gov.uk. (n.d.). ONS methodology working paper series number 16 - Synthetic data pilot - Office for National Statistics. [online] https://www.ons.gov.uk/methodology/methodologicalpublications/generalmethodology/onsworkingpaperseries/onsmethodologyworkingpaperseriesnumber16syntheticdatapilot.

Xu, L., Skoularidou, M., Cuesta-Infante, A. et al. (2019). Modeling tabular data using conditional GAN. arXiv:1907.00503 [cs, stat]. [online] Available at: https://arxiv.org/abs/1907.00503.

Yilmaz, E., Ji, T., Ayday, E. et al. (2022). Genomic data sharing under dependent local differential privacy. *CODASPY* 2022: 77–88. https://doi.org/10.1145/3508398.3511519.

Yoshimura, S., Nakamura, K., Nitta, N. et al. (n.d.). Model Inversion Attack against a Face Recognition System in a Black-Box Setting. [online] https://www.apsipa.org/proceedings/2021/pdfs/0001800.pdf.

Zhou, X., Chen, Y., Ip, F.C.F. et al. (2023). Deep learning-based polygenic risk analysis for Alzheimer's disease prediction. *Communications Medicine* 3 (1): 1–20. https://doi.org/10.1038/s43856-023-00269-x.

Zhu, L., Liu, Z., and Han, S. (n.d.). Deep Leakage from Gradients. [online] Available at: https://arxiv.org/pdf/1906.08935.

Ziller, A., Usynin, D., Braren, R. et al. (2021). Medical imaging deep learning with differential privacy. *Scientific Reports* 11 (1): 13524. https://doi.org/10.1038/s41598-021-93030-0.

Index

Note: Page numbers in *italics* indicate a figure and page numbers in **bold** indicate a table on the corresponding page.